Modeling and Application of Flexible Electronics Packaging

YongAn Huang · Zhouping Yin ·
Xiaodong Wan

Modeling and Application of Flexible Electronics Packaging

Science Press
Beijing

Springer

YongAn Huang
State Key Laboratory of Digital Manufacturing Equipment and Technology
Huazhong University of Science and Technology (HUST)
Wuhan, People's Republic of China

Zhouping Yin
State Key Laboratory of Digital Manufacturing Equipment and Technology
Huazhong University of Science and Technology (HUST)
Wuhan, People's Republic of China

Xiaodong Wan
State Key Laboratory of Digital Manufacturing Equipment and Technology
Huazhong University of Science and Technology (HUST)
Wuhan, People's Republic of China

ISBN 978-981-13-3626-3 ISBN 978-981-13-3627-0 (eBook)
https://doi.org/10.1007/978-981-13-3627-0

Jointly published with Science Press, Beijing, China
The print edition is not for sale in China Mainland. Customers from China Mainland please order the print book from: Science Press, Beijing, China.

Library of Congress Control Number: 2018968102

This Springer imprint is published by the registered company Springer Nature Singapore Pte Ltd.
The registered company address is: 152 Beach Road, #21-01/04 Gateway East, Singapore 189721, Singapore

Foreword

Flexible electronics represent the next-generation microelectronics that offer the electrical functions of conventional, rigid IC technologies but with the ability to be stretched, compressed, twisted, bent, and deformed into arbitrary shapes. They overcome the fundamental mismatch in mechanics and form and have enabled applications that are impossible to achieve with hard, planar integrated circuits. Their representative applications include epidermis electronics that are conformal to human skin, wearable microfluidic device for health monitoring systems, RFID tags made up of rigid chips, and flexible antennas. Some of these devices (e.g., RFID tags) are obtained by packaging high-modulus, rigid, state-of-the-art chip-scale elements (e.g., IC chips, MEMS, sensors, and power sources) that are fabricated on the donor wafer onto flexible/stretchable substrates. Therefore, the chip-scale elements have to be peeled off from the donor wafer, picked up, and placed on the receptor flexible/stretchable substrates, where the interfacial mechanics plays a critical role in the process.

The transfer manipulation (peeling-picking-placing) of chips, such as resistor, capacitor, sensor, and interconnector is a key technology for advanced packaging. Major advances in transfer manipulation of electronic components and devices have been made over the past decade. For flexible electronics, the chips tend to be much smaller, thinner, or more flexible to meet the deformability and conformability. It becomes challenging for transfer manipulation, especially in the large-scale roll-to-roll production.

This book is the first one about the interfacial mechanics of transfer manipulation for flexible electronics packaging. It covers the state-of-the-art and comprehensive work on modeling, simulations, and experiments of the authors, from the theoretical and experimental studies on the interfacial mechanics, to the design and fabrication of stamp for transfer manipulation, and the applications in roll-to-roll assembly of flexible electronics. Multiply processes are developed for the transfer manipulation of the chips with wide ranges of the mechanical and interfacial properties for flexible electronics, including single/multiple-needle peeling, conformal peeling, laser-induced peeling, vacuum-based picking-up, and low residual stress placing-on. The theoretical model and experimental verification are conducted collaboratively to reveal the underlying mechanisms and optimize the peeling process. A competing index is proposed to characterize the competing fracture mechanism, namely fracture of the chip and interfacial delamination of the adhesive. These comprehensive analytical, computational, and experimental studies are very useful for advanced packaging of flexible electronics, as have been demonstrated in the flip-chip in RFID packaging, membrane electrode assembly in fuel cell stack, and laser lift-off in flexible devices. This is an excellent reference book for both academic research and industrial design of flexible electronics packaging.

Yonggang Huang
Member, US National Academy of Engineering
Foreign Member, Academia Europaea
Foreign Member, Chinese Academy of Sciences
Northwestern University
Evanston, IL, USA

Preface

Over the past decade, technological developments have enabled the flip-chip packaging for applications in computing, communications, and consumer electronics. With the developing trends of electronic devices towards large-area and ultra-thin direction, the demand for the flip-chip packaging continues to grow to meet the requirements for performance, size, and flexibility, e.g., the flexible display, radio frequency identification (RFID), wearable electronics, and bio-integrated electronics.

The transfer manipulation (peeling-picking-placing) of microscale and thin chip is one of the key technologies of advanced packaging to accurately transfer a single chip from the wafer to a target circuit. Especially as the chips tend to be much thinner than before, a reliable peeling-picking-placing process is essential to the electrical performance, cost-effectiveness, service reliability, and life span of electronic devices. This book addresses important issues concerning the nondestructive peeling-picking-placing technique for ultra-thin chip, including single/multi-needle peeling, conformal peeling, laser lift-off, and vacuum-based picking-up and placing-on. The theoretical model, physical mechanism, and experimental verification are conducted. The review on the developments of the transfer manipulation is discussed in Chap. 1.

In Chap. 2, the interfacial peeling mechanism of the chip-on-substrate structure subjected to a transversely concentrated load resulting from the single ejector needle is first investigated, where the adhesive layer is neglected. A finite element simulation technique is also performed to obtain the energy release rate (ERR) of interfacial peeling. Then, an analytical model for the chip-adhesive-substrate structure is established. The corresponding analytical solutions for internal forces, displacements, and stresses are obtained. The theoretically and numerically estimated ERR of interfacial peeling is used to indicate the peeling behavior at the adhesive layer.

In Chap. 3, the fracture strength of the ultra-thin silicon chip is first evaluated combining the geometrically nonlinear theory. The classical linear theory on the three-point bending test is found failed to match the geometrically nonlinear characteristics of ultra-thin silicon chip. The peeling behavior of the adhesive tape from the adherend is then investigated experimentally. A novel homemade angle/speed controlled peeling platform is developed to assist in estimating the adhesive fracture energy, where the stretchability of the adhesive tape is considered in the evaluation model.

In Chap. 4, an analytical solution is first presented based on the chip-adhesive-substrate structure model to determine the adhesive stresses in balanced and unbalanced stiffened plate, single-strap, and single-lap joints under mixed force loading and/or displacement boundary conditions. The adhesive stresses are expressed in terms of geometric dimensions and material properties, combined with integration constants obtained numerically. Then, a mechanical chip-adhesive-substrate model is proposed to predict the peeling behavior of a periodic array of IC chips adhesively bonded to a stretched substrate via the multi-segment analysis. The process parameters (substrate tension), the geometric parameters (chip thickness, and chip distance), and the material parameters (Young's moduli of the chip and substrate) are considered to predict the peeling status at the adhesive layer.

In Chap. 5, the interfacial peeling mechanism of chip-on-substrate and chip-adhesive-substrate structures subjected to a transversely concentrated load resulting from single ejector needle is first investigated from the fracture mechanics point of view. Effects of key factors including the chip size, initial crack length, and substrate material are uncovered. Analytical expressions for adhesive stresses and ERR of interfacial peeling are derived by using structural parameters and integration constants. Process limitation of the single-needle ejecting technique is discussed. The competing relationship between the chip peeling and chip cracking in the single-needle ejecting process is then investigated combining the chip-on-substrate structure. The mechanism of competing fracture is revealed fundamentally using the proposed competing index. Finally, the finite element analysis is adopted to investigate the contact-impact effect during the single-needle ejecting process considering the impact speed, the distance from the contact center, and the substrate penetration.

In Chap. 6, an analytical model is first presented to estimate the bending normal stress of the chip layer and ERR of interfacial peeling in the multi-needle ejecting process. A competing fracture model and failure criterion of the chip-adhesive-substrate structure based on chip peeling and chip cracking indexes are put forward to solve the conflict between the chip peeling and chip cracking for ultra-thin chips. Improvement on the ultra-thin chip peeling-off through the multi-needle adoption is discussed and compared with the single-needle ejecting process. The process window is preliminarily explored considering the chip geometry, the strength of the adhesive layer, and the layout of ejector needles. A modified multi-needle ejecting process using independently controlled ejector needles is finally proposed to improve the peelability of ultra-thin chips.

In Chap. 7, a mechanical model is first established to address the roll-to-roll (R2R) conformal peeling of the device-adhesive-substrate structure. Analytical expressions of the peel stress and shear stress at the intermediate adhesive layer are derived under the conformal condition of the curved surface. The allowable ranges of the thickness of the device and substrate layers are determined for a safe device peeling-off. A competition model is established combining the conformal criterion to distinguish the device breakage and successful peeling-off. The peelability criterion for the conformal peeling process is proposed through the theoretical analysis and experimental verification. The design of the peel blade is studied combining the peel radius and validated through the R2R peeling experiment of RFID tags.

In Chap. 8, the laser lift-off (LLO) mechanism of ultra-thin polyimide (PI) film from the rigid glass carrier is revealed experimentally. The ultra-thin PI film is separated completely with low internal stress. Process parameters of the laser fluence and irradiation time are considered to determine the process window. The process mechanism is interpreted combining the gas products generated by the laser irradiation, interface adhesion strength, and interface microstructure.

In Chap. 9, a theoretical model of the chip-adhesive-substrate structure is first developed to reveal the process mechanism of the ultra-thin chip picking-up under the condition of the vacuum adsorption with fixed adhesive tape length. An improved model considering the effective tape length is further proposed using the bisection algorithm. The theoretical process window for the ultra-thin chip picking-up is determined. The mechanism of the R2R transferring with low interfacial residual stress is then investigated. The collaborative optimization of material properties and structural parameters is performed. The feasibility of reducing the interfacial residual stress is validated by using the R2R experimental platform. Additionally, with respect to the chip placing-on process, a layerwise thermal-mechanical model of the ultra-thin chip-on-flex structure considering the adhesive curing and cooling is developed to predict the ultra-thin chip warpage.

We deeply acknowledge contributors of theoretical and experimental studies, including Jiankui Chen, Bo Peng, Zunxu Liu, Jing Bian, Zhoulong Xu, Huimin Liu, and Pengpeng Tang. In addition, we want to express our gratitude to the support from the national natural science foundation of China (No. 51635007, 51705180).

Wuhan, China
August 2018

YongAn Huang
Zhouping Yin
Xiaodong Wan

Contents

Abbreviations

ERR	Energy release rate
IC	Integrated circuit
ITO	Indium tin oxide
LIIR	Laser-induced interfacial reaction
LLO	Laser lift-off
LSCM	Laser scanning confocal microscope
LWT	Layerwise theory
OLED	Organic light-emitting diode
OTFT	Oxide thin film transistor
PI	Polyimide
PMAC	Programmable multi-axis controller
R2R	Roll-to-roll
RFID	Radio frequency identification
RP	Reference point
SEM	Scanning electron microscopy
TFT	Thin film transistor
UTCOF	Ultra-thin chip-on-flex
UV	Ultraviolet
VAP	Vacuum adsorption pressure
VCCT	Virtual crack closure technique

Chapter 1
Advanced Electronic Packaging

1.1 Introduction

Flexible electronics is an emerging technology that makes organic/inorganic thin film electronic devices on flexible substrates available to achieve the unique flexibility/stretchability. A schematic description of typical applications of flexible electronics based on the flexible thin film transistor (TFT) array technology is shown in Fig. 1.1. However, conventional lithography, electron beam etching, and other processes cannot meet the requirements for large-area, high-efficiency, and low-temperature manufacturing due to high manufacturing costs, harsh environmental requirements, difficulties in compatibility with organic materials, etc. The developments of high-efficient and low-cost manufacturing processes are the key to pushing flexible electronics from the laboratory to practical application.

Flexible electronics is generally composed of a substrate, backplane electronics, frontplane, and encapsulation. To make the structure flexible/stretchable, all components must comply with bending/stretching to some degree without losing their functions. Two basic approaches have been employed for the flexible electronics manufacturing: direct fabrication of the devices on the flexible substrate and transfer printing of completed devices to the flexible substrate [2]. The former seems a more direct and innovative approach to the manufacturing of flexible electronics. However, the commonly used flexible substrates are not fully compatible with existing planar silicon micro-fabrication processes. So this technology may require new processes and materials. In the transfer printing approach, the whole structure is fabricated by standard methods on a carrier substrate such as the silicon wafer or glass plate. Then, it is retrieved from the carrier substrate via an elastomeric stamp and printed onto a flexible substrate [3–17]. The technique is valuable for the construction of unusual test structures for a variety of basic scientific studies, and also for advanced engineering development of systems such as flexible/stretchable semiconductor devices including the flexible display [18–20], deformable circuits [21, 22], and flexible inorganic solar cells [23], etc. The transfer printing enables the integration of high-performance inorganic semiconductor materials in ultra-thin geometries

Y. Huang et al., *Modeling and Application of Flexible Electronics Packaging*,
https://doi.org/10.1007/978-981-13-3627-0_1

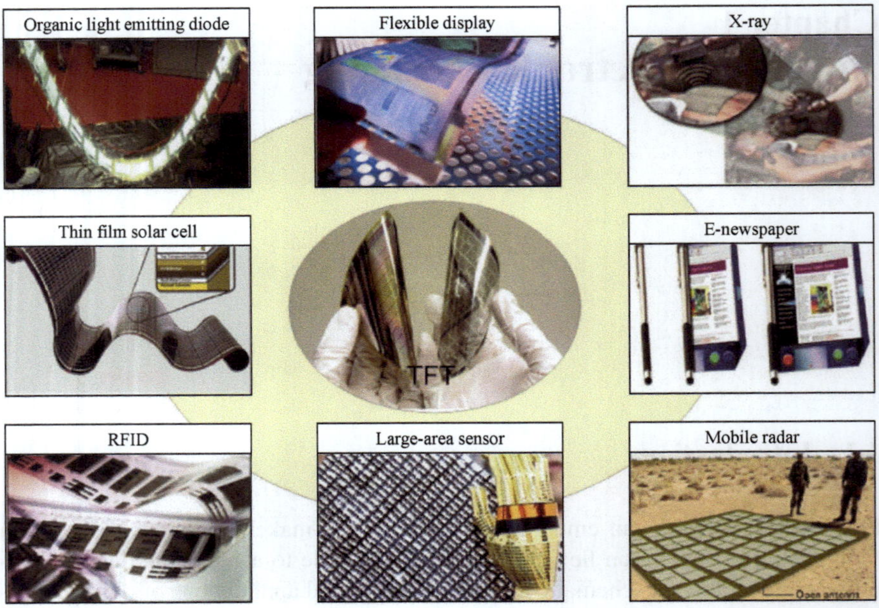

Fig. 1.1 A series applications of flexible electronics based on the flexible TFT array technology [1]

with substrates of interest such as the plastic sheet. However, the small surface area coverage, high cost, and short life span of the transfer stamp are the main obstacles to the mass production application of the transfer printing process. To address the flexible electronics packaging, the chip/device transferring issue of the electronic packaging in the integrated circuit (IC) industry cannot be neglected.

The electronic packaging is responsible for interconnecting the chip and the circuit board, acting as a bridge connecting the design and manufacturing of chips to industrial applications, and directly determines the final electronic performance, production cost, service life, and reliability of electronic products [24]. The chip/device transferring is one of the key steps in the electronic packaging process. It needs to complete the transfer of a single bare die or flexible device from the wafer to the circuit board [25]. The chip transferring in the IC packaging typically includes the chip peeling-off, picking-up, and placing-on, as illustrated in Fig. 1.2. In the procedure of chip pick-up, which includes the chip peeling-off and chip picking-up, the diced wafer is first adsorbed and fixed on the ejector cap. Afterwards, a pick-up head moves to the top surface of the chip. Then, an ejector needle rises to push the chip and adhesive tape upwards. The overall adhesively bonded structure deforms as a whole under the condition of small ejector needle displacement. The interfacial peeling initiates and propagates with further increase of the ejector needle displacement. When about 70–80% of the adhesively bonded region is peeled-off. The pick-up head draws the chip up, and the ejector needle returns back to the original position.

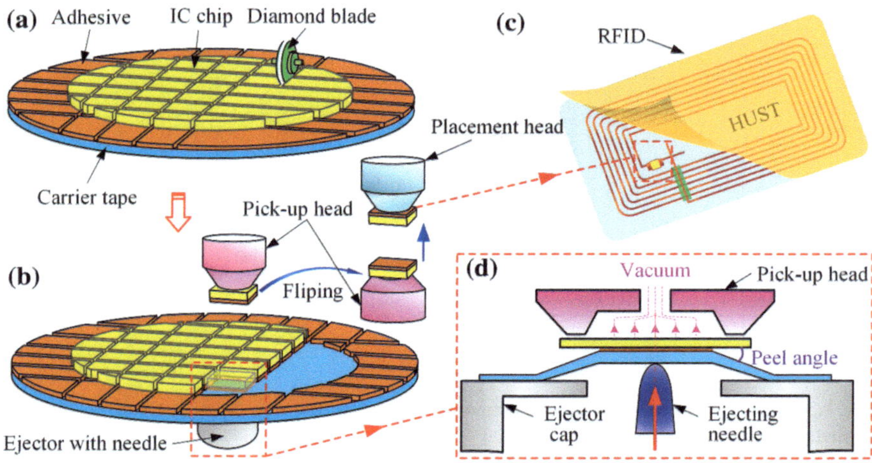

Fig. 1.2 Typical electronic packaging: **a** wafer dicing, **b** chip peeling-off and picking-up, **c** radio frequency identification (RFID) packaging (chip placing-on), **d** chip peeling-off from the adhesive tape with a single ejector needle. © 2014 IEEE. Reprinted, with permission, from Ref. [26]

After the chip pick-up process, the flipped chip will be grasped by the mounting head at the backside to make it front down and transferred to the circuit to be placed on, which is termed as the chip placing-on process.

Current electronic products are developing along the thin and lightweight trend [27–29] to improve the flexibility [23, 30–32], stretchability [33–35], and foldability [21, 36]. Integrated ultra-thin silicon-based chips (<50 μm) are increasingly applied to allow for new applications in flexible/stretchable electronics [21, 28, 35]. The advanced electronic systems are attractive because of not only the powerful electronic performance [27] but also the excellent mechanical performance [23, 31]. In the past decade, the thickness of IC chips widely used in the industry has been reduced from 120 to 40 μm [28, 37, 38], and may further reduce to 10–20 μm level [39, 40], as shown in Fig. 1.3. A very small thickness can greatly reduce the thermal resistance of chips, thereby improving the energy consumption of high-performance silicon-based chips. However, the existing chip manufacturing and packaging processes are challenged by the ultra-thin silicon-based chips despite the advantages [25, 29, 37, 41, 42]. Ultra-thin silicon-based chips require extremely high surface quality and edge quality to guarantee the excellent mechanical performance [38, 43]. Due to the intrinsic properties of easy bending and fracturing, the ultra-thin silicon-based chips are more difficult to be securely transferred compared with the traditional thick ones [41]. It has been stated that the assembly of ultra-thin chips is still one of the crucial challenges in the current electronic packaging industry [44]. The chip cracking, chip warping, and structure delamination during the ultra-thin chip transferring should pay special attention.

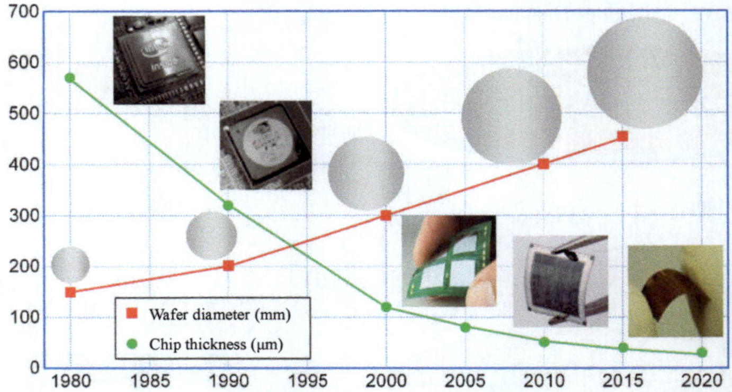

Fig. 1.3 International semiconductor technology roadmap: the development trend of the wafer diameter and chip thickness [44]

With the rapid development of ultra-thin consumer electronics and flexible electronic devices [23, 27–32], the fast, nondestructive, and reliable transfer of the ultra-thin silicon-based chip from the donor adhesive tape to the acceptor substrate plays an important role in improving the electronic packaging productivity. Considering the large-area, ultra-thinned, and miniaturized electrical structure of flexible devices, the multilayer hetero-structures are challenged by the fragile functional layer, low peeling success rate, crucial process parameters, etc. The corresponding chip/device peeling processes are also continuously developing from the single-needle ejecting, multi-needle ejecting, roll-to-roll (R2R) peeling to laser-induced peeling. Specially, the needle ejecting process may be no longer applicable due to the process limitation. The R2R process, which employs the mismatched deformation between the adhesive tape layer and the device layer when passing the peel blade, is supposed as an appropriate alternative approach to achieve the mass production of flexible devices with low costs. It can employ the deformation characteristics of flexible devices to achieve the purpose of continuous film patterning, peeling-off, and transfer [45, 46]. The specific process discussion of the R2R will be presented in Chap. 7. As to the laser-induced peeling process, the laser acts on the adhesive interface, leading to the interface melting or ablation, which reduces or eliminates the interfacial adhesion strength to achieve the final interface peeling-off. The process is highly efficient and available for large-area processing. Related process discussion of the laser lift-off (LLO) will be presented in Chap. 8.

1.2 Adhesively Bonded Multilayer Structure

Adhesively bonded multilayer structures have found extensive applications for load transfer and element connection in aerospace, mechanical and civil engineering structures [47], structural repairing [48], and surface-bonded piezoelectric smart structures [49], etc. Recently, they play increasingly important roles in flexible electronics design [50, 51] and microelectronics packaging [52, 53]. Due to the discontinuity of the layers at the edges, complicated stress fields and high-stress concentrations usually occur in the vicinity of the corners of adhesive layers, which may finally lead to the overall failure of the joint structure. The accurate estimation of the stress state in adhesively bonded multilayer joints is crucial for the joint design and structural safety evaluation as well as for a better understanding of the underlying failure mechanisms. There are many researches on the analysis of adhesively bonded multilayer joint structure typically including stiffened plate joints [52–55], single-strap joints [55–58], and single-lap joints [54, 55, 57, 59–67].

Approaches have been proposed to evaluate the stresses and energy release rate (ERR) at the adhesively bonded interface [53, 59, 64, 68–75]. In early analytical studies, all layers were modeled as elastic beams or plates based on the elastic theory [76]. It is generally difficult to obtain a precise closed-form solution for the problem [76]. One strategy is to approximately obtain the closed-form solutions for relatively simple and extremely simplified bonded structures [52–55, 57, 59]. A classical solution of the stress in adhesively bonded joints was given by Goland and Reissner [59] and improved by Tsai et al. [64] considering the shear deformation of adherends. Jiang et al. [52] examined the interfacial peel stress and shear stress due to the thermal expansion mismatch of layered electronics with thin adhesives and proposed the analytical expressions. Wang et al. [53] modeled all layers as Euler–Bernoulli beams and obtained an approximate closed-form solution for the adhesive peel stress and shear stress in tri-layer electronic assemblies based on the method of singular perturbation. Luo and Tong [69] calculated the contributions of mode I and mode II ERR to a straight interlaminar crack in a composite laminate combining the Timoshenko beam theory. Bennati et al. [70] developed a mechanical model where two Timoshenko beams were connected by a two-parameter elastic interface, which can be used to analyze the mixed-mode fracture of adhesive joints, composite laminates, and layered structures. Besides, the contributions of mode I and mode II ERR to the steady-state debonding and convergent debonding can be calculated using the complex variable method or stress-function variational method [77–82]. Yang and Chadegani et al. [72, 74, 75, 83] presented an analytical model to determine the ERR of a crack in an adhesively bonded composite joint with thin bondlines, where the governing equations were derived and solved using a Fourier series.

The adhesively bonded multilayer joints can also be modeled as an assembly of sublaminates, connected by a deformable interlaminar interface [84, 85]. In such cases, the interface does not necessarily represent a physical adhesive layer but is introduced into the model to take into account the transverse deformability of the laminates, which is neglected in models based on the Euler–Bernoulli simple beam theory. This enhanced beam theory model has already been proposed for the asymmetric double cantilever beam test [70] and the mixed-mode bending test [86, 87]. The model has been extended to generalized layered structures, showing how the same analytical solution can be applied to adhesively bonded joints [88].

Finite element simulation is commonly adopted to calculate the stress and ERR at the adhesive layer [89–95]. The ERR can be evaluated by using the finite element approach combining the virtual crack closure technique (VCCT) [96], which can be employed to solve fracture problems such as the interfacial cracking, delamination of sandwich structures, peeling-off in electronic packaging, etc. [97–99]. The VCCT has been considered as an effective way to calculate the ERR of interfacial peeling [72, 74, 75, 83, 89, 90]. Xie and Biggers [100, 101] developed a kind of interface element called the fracture element with dummy nodes based on the VCCT, through which the ERR can be calculated simultaneously as the finite element analysis was performed. Besides, particular attention should be paid when the VCCT is applied in problems with bi-material interface cracks [97] or highly asymmetric cracks [102].

1.3 Interfacial Peeling-off

1.3.1 Needle Ejecting for Thin Chips

A description of the typical chip pick-up through needle ejecting process is shown in Fig. 1.4. The practical peeling-off module is also depicted. As shown in Fig. 1.4a, the pick-up process can be conveniently divided into three stages according to the operation flow. The three stages of contact-impact, peeling-off, and picking-up are featured by transient local impact, long quasi-static peeling-off, and convergent debonding, respectively.

The chip peeling-off, which completes the transfer of the chip from the adhesive tape to the pick-up head, has to adapt to the flexibility and thinning of IC chips. The peeling-off stage occupies the longest time among these three stages and gets the own characteristic of quasi-static behavior. In order to realize the high reliability of the product packaging, it must satisfy the high requirements of the packaging technology [23, 27, 28]. Thinner chips are more difficult to be peeled off, and the corresponding process window is relatively small [103]. Additionally, properties of the adhesive and substrate significantly affect the success rate of pick-up process [104]. Improper process parameters and material selection easily lead to the failure of chips. Therefore, the success rate and efficiency of the chip peeling-off have always been the focus of academic and industrial attention [37, 41, 42, 89, 103, 105–109].

Arrays of IC chips adhesively bonded to stretched substrates have found extensive applications in the field of flexible electronics, where large-scale TFT arrays on flexible substrates are widely employed [110]. Such multilayer structures are generally put into tension during typical manufacturing processes, where the substrate is normally

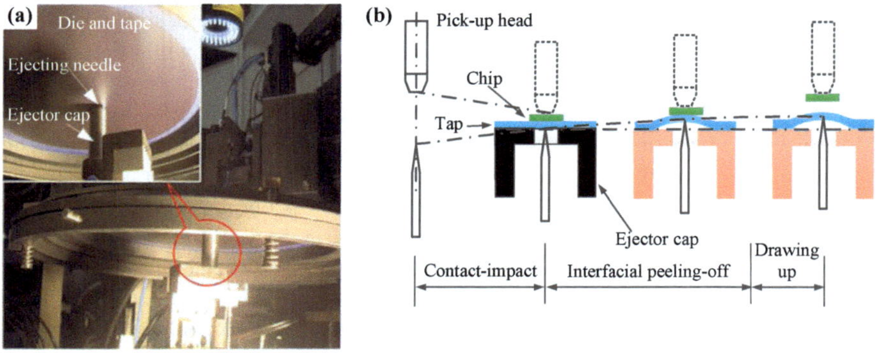

Fig. 1.4 Chip pick-up process through needle ejecting process: **a** the pick-up module with an amplified inset of the ejector needle, **b** three stages of the chip pick-up process. Reprinted from Ref. [89], with the permission of AIP Publishing

subjected to a fixed prestrain [89, 111–113]. However, an excess of prestrain may lead to high stress concentration at free edges of the bonded interfaces between chips and substrate, which in turn may lead to the premature cracking of the adhesive and debonding of the IC chips [114, 115]. In addition, local breakage or scratch marks on the backside of the semiconductor dice during the electronic packaging are potential risks leading to the failure in subsequent packaging processes. The contact-impact effect is one of the main reasons leading to the defects. Generally, it is inadequate to analyze the peeling-off process based on classical strength theory [52, 103–106, 115, 116] because of the stress singularity at both the edge and interfacial crack tip of a layered bonded structure [117–119]. Accurate modeling and efficient estimation of the adhesive stresses and ERR of interfacial peeling in the needle ejecting process are of utmost importance for the design and manufacturing of flexible/stretchable electronics.

Processes have been investigated combining the fracture mechanics theory [89], finite element simulation [90, 103, 105, 107–109, 120], experimental measurement [42], semi-experimental methods [103–106, 108, 121, 122], etc. To reduce the occurrence of the chip breakage, Lin and Hwang [105] investigated effects of key parameters including the needle ejecting speed, vacuum pressure, ejector needle force, and needle radius by the static finite element approach. Cheng et al. [103] measured the mechanical properties and analyzed numerically the adhesive peeling between the chip layer and substrate layer to enhance the success rate of the chip peeling-off process, which was found significantly affected by the chip length and chip thickness. Cheng et al. [106] examined analytically distributions of the peel stress and the shear stress in the adhesive layer and discussed parameter effects using the strength theory. Peng and Liu et al. [89, 90, 92] investigated analytically the peeling mechanism of the single-needle ejecting process based on the theory of the fracture mechanics. Results indicated that the ERR of interfacial peeling decreased as the chips became thinner and larger, namely harder to be peeled off. Further, there was a competing relationship between the chip peeling and chip cracking, and the design rules of the process window were to enhance the ERR of interfacial peeling while reducing the maximum tensile stress of the chip layer [90].

Previous studies were mainly focused on the technology with a single needle. Multi-needle adoption can be used as a practical and efficient alternative to peel the thin/ultra-thin chip off the adhesive tape. Liu et al. [93] established a model for the chip-adhesive-substrate structure to analyze the chip peeling-off with two ejector needles and revealed that multiple ejector needles can be used to reduce the occurrence of the chip cracking compared with the single-needle ejecting process [92]. Liu et al. [107] simulated the chip peeling behavior in double-needle ejecting. Saiki et al. [108] studied numerically and experimentally the chip peeling-off behavior with four ejector needles. By now, researches are still lacked with respect to the difference between the normal single-needle ejecting and the multi-needle ejecting process.

1.3.2 Conformal Peeling for Large-Area Devices

Single-needle peeling for chip with submillimeter [89], multi-needle peeling and vacuum-based picking-up for chip with several millimeters [123], and kinetically controlled transfer printing for device with dozens of millimeters [3] have been used to peel chip with size from dozens of micrometers to several millimeters. Besides, the transfer printing technique can be used for devices with thickness smaller than 20 μm. However, all the above methods have limited processing capabilities for peeling large-area devices with thickness larger than 100 μm and size larger than 100 mm. Moreover, the above processes are generally carried out individually, which is difficult to achieve the purpose of high-efficiency and low-cost manufacturing [124–126].

The successful peeling-off of large-area devices can be achieved by the conformal peeling process, through which the device layer can be peeled off from the substrate layer continuously by adopting the peel blade [127, 128]. The principle of the conformal peeling process is shown in Fig. 1.5. When the receiving roller drives the left substrate through the peel blade, the left substrate is forced to deform, and its curvature radius is the same as the radius of the peel blade. The device layer bonded onto the left substrate will also be forced to deform. When the adhesion strength of the adhesive layer is not enough to maintain the device layer conformal to the left substrate, the device layer will be peeled from the left substrate.

In the R2R manufacturing process, the flexible devices pass through several rollers with different radii, and the substrate is subjected to different tensions. Considering that flexible devices are made up of several layers with different material properties and different geometric dimensions, the load on the carrier layer will lead to interfacial stress which can be used to drive the interface delamination. The structure thickness, substrate tension, and substrate speed play key roles for a successful peeling-off [129]. Wang [130] found that the peel strength increased with Young's modulus of

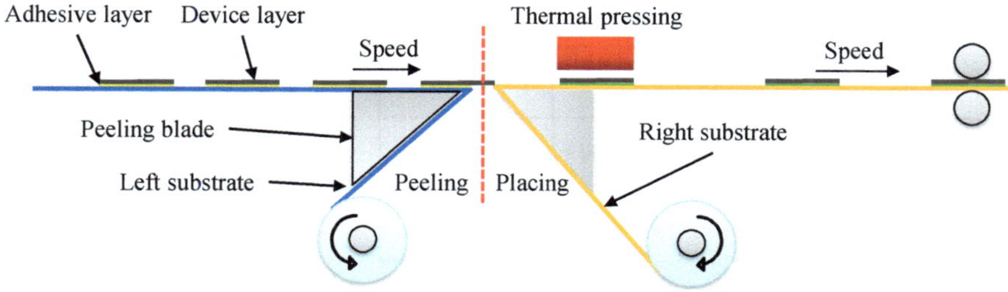

Fig. 1.5 Schematic diagram of the conformal peeling process through the R2R manufacturing system

the samples to be peeled off. Wacker et al. [131] investigated the bending-induced uniaxial stress at the top of ultra-thin single-crystal silicon chips adhesively bonded to the soft polymeric substrate. Cortet et al. [132] studied the peeling dynamics of an adhesive tape under a constant load with a special focus on the stick-slip regime of the peeling. Wang et al. [116] studied the effect of the interfacial peel stress and the shear stress on delamination. Many other theoretical and experimental researches have been carried out on the interface fracture considering the peel force [133], the peel angle [134], the peel speed [130, 135], and the adhesive layer thickness [136]. The substrate speed affects not only the production efficiency but also the peeling success rate. Researches showed that the ERR of interfacial peeling changed with the peel speed [22, 131, 137, 138], since the adhesive layer can be treated as a kind of viscoelastic material. Cheng [139] proposed the viscoelastic model for the transfer printing, and the results showed that the interfacial adhesion strength increased with the pulling speed. Eremeyev [133] stated that the nonlinearity of the stress–stain relationship must be taken into account in computing the effective work of adhesion. Peng et al. [140] found that the ERR of interfacial peeling was increased with the peel speed or peel angle. Xiong and Tay [141] studied the effect of viscoelasticity on the interfacial delamination in the IC packaging.

1.3.3 LLO for Large-Area Flexible Electronics

The LLO technique may provide a much simple and reliable way for the fabrication of large-area flexible electronics [142–145]. The flexible devices can be gently separated after the laser irradiation at the adhesion interface. Control of the interface can be achieved by the laser pulse that irradiates at the interface with high scalability [145–148]. The flexibility can be achieved by separating the device fabricated on the rigid carrier through the LLO [149–152].

Since the first successful separation of a GaN layer from a sapphire substrate using ultraviolet (UV) laser scanning [153], the LLO process has been used extensively to transfer GaN-based LED structures from sapphire substrates to silicon or metal

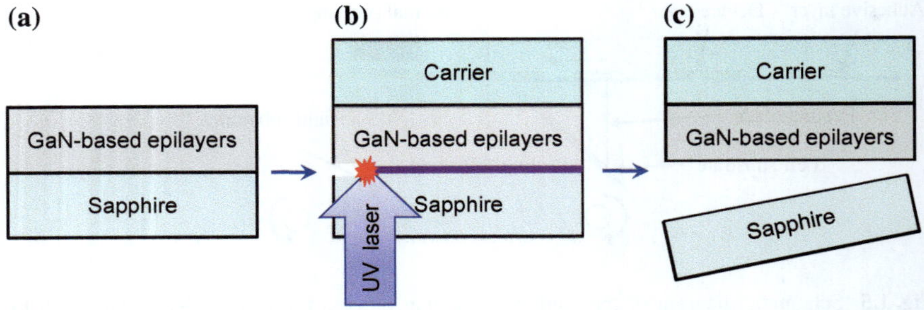

Fig. 1.6 Steps of thin film LED fabrication: **a** device growth, **b** flipping the chip, eutectic bonding of the bottom to a carrier wafer, and exposure to UV laser, **c** removal of the sapphire substrate

substrates for enhanced high current operation capabilities [154–156]. A schematic illustration of the application of the LLO in thin film LED fabrication is shown in Fig. 1.6. Approaches have presented GaN-based flexible LEDs through the LLO processing [156–165]. In consideration of the application in large-area microelectronics processing, the LLO can be preliminarily classified as the inorganic-based and organic-based processes.

In aspect of the inorganic-based LLO, the laser-induced heating and cooling enable melting/vaporization, morphological change, and structural and compositional transformations of the interface, leading to the separation of thin film from supported substrates [147, 166–169]. The melting at high temperature during the laser irradiation will weaken the interface. It is necessary to remove the substrate before solidification. Tsakalakos [170] stated theoretically that the thermal-mechanically induced deformation and stress due to the large temperature gradient promoted the film separation. Experimental observations of the microstructure and measurements of the dielectric and chemical compositions confirmed the quenching of molten phase near the interface, which provided strong evidence that the film separation by the laser irradiation resulted from the thermal stress at the interface during quenching [145, 166–168]. The functional thin film layer may be damaged due to thermal and physical shocking during the laser irradiation [145, 169], which cannot be neglected in case the film is relatively thin. A sacrificial layer between the functional layer and the substrate can be introduced to protect the functional thin film layer. The lead-based piezoelectric materials [145, 147, 171], indium tin oxide (ITO) [145], ZnO [145], GaON [172], amorphous gallium oxide (α-GaO_x) [173], and amorphous silicon (a-Si) [143] can be generally adopted as the sacrificial layer.

The stable and large-area transfer of piezoelectric ultra-thin films annealed at high temperature onto flexible substrates can be achieved through the abovementioned LLO process, which are difficult to be handled by other processes considering the brittleness of the free-standing ceramic thin film [144]. High-efficient, flexible, and large-area piezoelectric nanogenerators [144, 169, 174, 175], self-powered wireless temperature sensor [176], self-powered fully flexible light-emitting system [177], and self-powered flexible touch sensor [171] can be fabricated using such piezoelec-

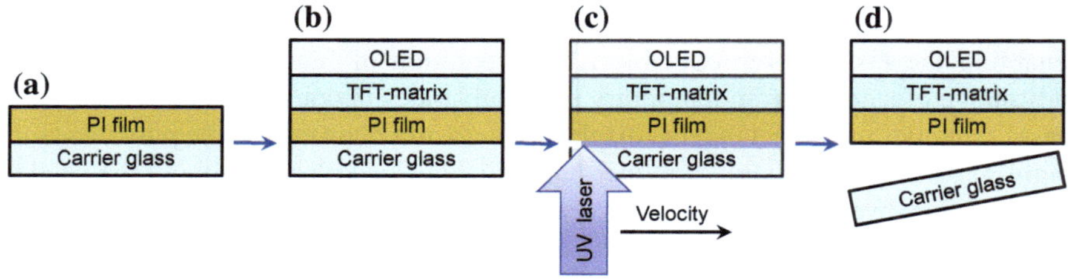

Fig. 1.7 An illustration of the OLED fabrication through the LLO: **a** PI-coated display glass, **b** buildup of functional layers, **c** LLO processing, **d** flexible display on the PI film

tric thin films, which make the concept of self-powered flexible electronics system easy to achieve. Potential applications in the biomedical field can also be found with these piezoelectric devices obtained through the LLO separation. Lee et al. [178] proposed a strategy to mimic the function of natural hair cells using a flexible inorganic piezoelectric acoustic sensor, which was able to convert sound energy into piezoelectric signals in the audible frequency range of living noise effectively. A self-powered piezoelectric pulse sensor conformally contacted with the complex texture of the rugged skin was recently demonstrated by Park et al. [179], tiny change of the pulse signal on the surface of epidermis can be detected. Lee et al. [143, 146, 180] deposited an oxide layer as a thermal buffer to block heat flow generated during the laser-induced exfoliation as well as compensating internal stress, the weak adhesion enabled entire device layers to be easily and safely separated from the substrate without damage. Flexible crossbar-structured resistive memory arrays [143] and high-performance, skin-like, transparent oxide thin film transistor (OTFT) arrays can also be fabricated in this way [180]. Kim et al. [146] proposed the process of laser multi-scanning lift-off to overcome the limitation of dehydration of the a-Si:H exfoliation layer. A high-performance flexible thermoelectric power generator was fabricated in combination with the screen printing technique.

In aspect of the organic-based LLO, the laser ablation of the polymer plays the dominant role and mainly applied in the field of the flexible display. In the production industry, the polymer-based flexible display has to be detached from the rigid temporal carrier, which is best done via the LLO to ensure both high throughput and display quality [151, 181]. The steps in fabrication of organic light-emitting diode (OLED) are schematically illustrated in Fig. 1.7. A temporary carrier glass substrate is first spin-coated with thin polyimide (PI) precursor, which is cured thereafter. The TFT circuit backplane will be built on top of the polymer layer. Followed by is the display frontplane, which contains the layers of electrodes and microcapsules. Finally, the transition from rigid substrate to flexible substrate is achieved by the LLO. The laser ablation is supposed accompanied by chemical and/or physical modifications of the polymer.

Technically, the excimer laser line beam transmits through the glass carrier substrate onto the polymer. The LLO separation occurs in a highly selective manner entirely unnoticed by the adjacent performance-determining functional layers. Nowadays, large line beam length has been made possible to the display production floor, enabling the fast and reproducible LLO separation of functional layers. A huge advantage of starting from temporary glass substrates in combination with the LLO processing lies in the fact that the display manufacturer can extend the portfolio from rigid-glass-based to flexible-polymer-based display backplanes, without the need to make major investments in entirely new production equipment [182]. The ablated surface morphologies examined by the optical microscopy, peel strength before/after the laser irradiation, and simulation of thermal pulse evolution during the laser irradiation have been investigated to get suitable process parameters and conditions [183–185]. Some novel laser-sensitive polymer release layer [186] and laser equipment [187] were also developed to survive in more rigorous process requirement.

1.4 Vacuum-Based Chip Picking-up

Chip picking-up with vacuum adsorption is a critical process in flip-chip packaging to separate the chip intactly from the donor adhesive tape. A schematic illustration of the chip picking-up process is shown in Fig. 1.8. As previously mentioned, the mechanical flexibility of thin/ultra-thin chip has been exploited to improve the performance of flexible electronics products [28, 35, 188–190]. Such thin silicon chips are a challenge to the current flip-chip packaging because they are particularly prone to fracture before being picked up [29, 37]. The failure of the chip picking-up due to the incomplete separation between the chip and substrate seriously affects the efficiency of the chip assembly [37, 41, 42]. Actually, the chip picking-up process relies on the capability of crack propagation along the adhesive layer, which can be enhanced by the optimization of vacuum adsorption tool and material properties of the adhesively bonded multilayer structure [92, 103, 105]. It is thus essential to understand fundamentally the mechanism of the chip delamination from the adhesive tape and reveal the effects of main factors on the chip picking-up process.

In practice, an anticipated crack length in the adhesive layer should be first generated by the ejector needle through the chip peeling-off process, followed by is the picking-up process with vacuum adsorption for full separation of the chip from the remaining bonded region at the adhesive tape [42]. Even though the chip peeling-off and picking-up are two successive procedures for propagating a crack in the adhesive layer, the corresponding models are independent of each other due to the transformation of external loads and boundary conditions [105]. In picking-up process, vacuum adsorption is essential to hold the chip and tape. It was pointed out that the dimension and materials of the adhesively bonded multilayer structure had a great influence on the success rate of the chip picking-up process. Typically, when the chip became thinner and bigger, the chip picking-up failed easily due to the occurrence of chip breakage and incomplete separation [103]. It was considered that the failure might be caused by the larger remaining bonded region [103]. However, there are few reports on further investigation of the chip picking-up, especially with respect to process parameters including the vacuum conditions and picking-up displacement [105].

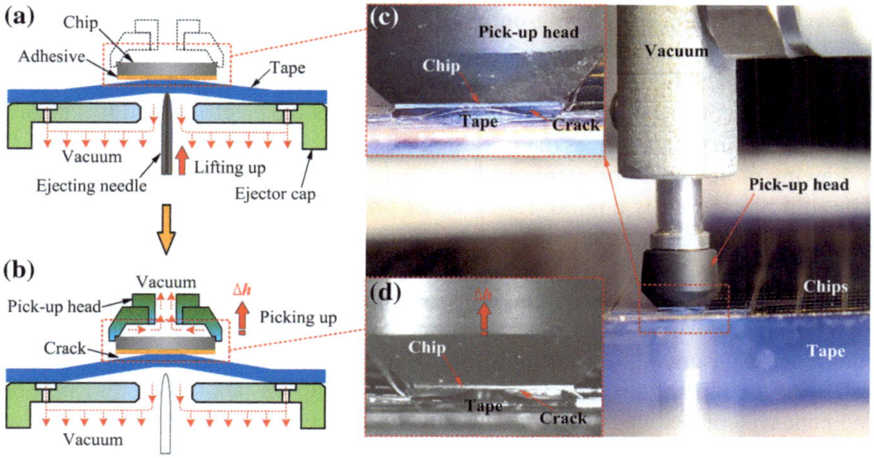

Fig. 1.8 Schematic illustration of the chip picking-up including: **a** the chip peeling-off process with single ejector needle, and **b** chip picking-up process with vacuum adsorption, practical modules of which can be observed in **c** and **d**, respectively. Reprinted from Ref. [123] by permission of Taylor & Francis Ltd.

In practical picking-up process, the picking-up head holds the target chip by vacuum adsorption and picks it away from the adhesive tape through upward picking-up displacement and picking-up force [37]. According to the fracture mechanics, the ERR of interfacial peeling must exceed the adhesive fracture energy of the adhesive layer to propagate the interfacial crack. Meanwhile, the picking-up force required for picking up a chip from the adhesive tape should be less than the vacuum adsorption force above the chip. It was found that the picking-up force changed with the increase of the picking-up displacement and propagation of the chip delamination [122]. Both the ERR of interfacial peeling and picking-up force are key parameters to evaluate the capability of chip picking-up with vacuum adsorption. Enhancing the reliability of the chip picking-up is to increase the ERR of interfacial peeling while reducing the picking-up force.

1.5 Vacuum-Based Chip Placing-on

After the vacuum-based pick-up process is completed, next is the placing-on of the ultra-thin IC chip to the flexible substrate [191]. The chip placing-on is mainly a process in which the anisotropic conductive adhesive (ACA) fluid and conductive particles are compressed and cured under the synergetic effect of pressure and temperature loads. The electrical and mechanical interconnections between the chip and the flexible substrate circuit can be achieved by using the ACA. In the pick-up process, the chip is completely peeled-off from the adhesive tape combining the vacuum adsorption. Then, in the placing-on process, the chip is bonded to a new substrate through the ACA solidification.

The ultra-thin chip placing-on is the most critical technology in the ACA-based flip-chip interconnection process, and directly affects the electrical and mechanical interconnection performances of ultra-thin IC devices. The ultra-thin chip placing-on is drawing more and more attentions in the field of the flip-chip bonding. The ultra-thin chip placing-on is a relatively complex process involving key operations like the adjustment and alignment of the chip position before placing-on, control of process parameters, and control of the ultra-thin chip warpage after the thermal pressing. With respect to the placing-on of high-density ultra-thin chips, especially with a large length-to-thickness ratio, the process requires precise placement and direct placing operations to ensure high-level alignment accuracy and assembly efficiency. An illustration of the placing-on process of the ultra-thin chip-on-flex (UTCOF) assembly and concave/convex warpage by the thermally induced deformation mismatch during the process is depicted in Fig. 1.9. The chip placing-on can be summarized in four specific steps: (a) ACA gluing, (b) ACA curing, (c) pressure releasing, and (d) natural cooling.

The thickness of the adhesive layer determined by the height of the bump and circuit is at a range of about 20–50 μm, approximating to the thickness of the ultra-thin chip which is below 50 μm. Under this condition, the inconstant stress distribution along the direction of the adhesive thickness cannot be neglected [192]. In order to obtain the interfacial stresses of the adhesively bonded structure, the displacement field predicted by the full layerwise theory (LWT) has been recently developed referring to the layerwise laminated plate theory [193–195]. Within the LWT framework, not only the solution accuracy can be improved by dividing the thickness of physical laminates into mathematical plies but also the analytical solution under mixed boundary conditions can be obtained [193].

In the thermal compressing process, the heating is used to cure the adhesive, while the compressing action is used to form the electrical interconnection between chip bumps and circuit board. The ultra-thin chip warpage yields just after the adhesive curing [196, 197]. It is assumed that the adhesive curing completes promptly once the adhesive temperature exceeds the glass transition temperature. The critical time can be obtained by transient thermal analysis. Temperature variation after the critical time due to continuous heating or cooling will lead to the mismatch of thermal stress in the bonded joint, as well as the chip warpage depicted in Fig. 1.9c, d. The serious warpage issue of the UTCOF assembly induced by mismatched thermal stresses greatly affects the mechanical stability and reliability of emerging ultra-thin chip packaging technology. Currently, the theoretical prediction is still lacked to describe effectively the thermal-mechanical behavior of the UTCOF assembly during the adhesive curing and cooling processes.

The adhesive-based packaging essentially implements an epoxy-based mechanical joint between the chip and thin film circuit by the thermal compressing process. Considering the complicated thermal-mechanical effect, both the experimental and numerical approaches were adopted to investigate the mechanism of thermal-induced warpage. Cheng et al. [196, 198, 199] established a three-dimensional tran-

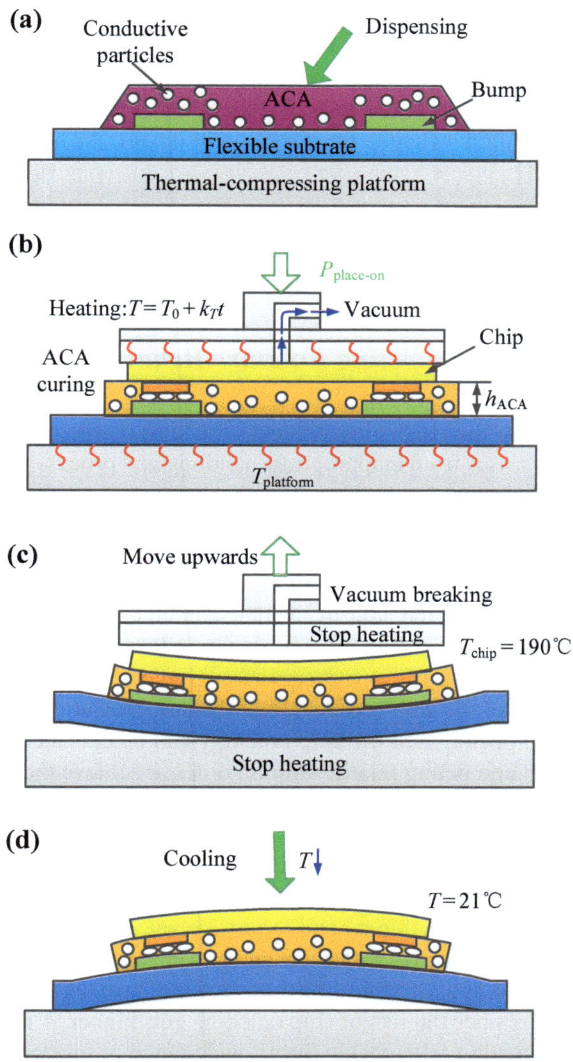

Fig. 1.9 Thermally induced warpage during the placing-on of the UTCOF assembly: **a** the ACA gluing, **b** ACA cured by thermal compressing, **c** concave warpage of the assembly after the ACA curing and pressure releasing, **d** convex warpage of the assembly after the cooling process. Reprinted by permission from Springer Nature, Ref. [191], Copyright 2016

sient model to study the thermal-mechanical behavior of thin chip-on-PI packaging with ACA/film, where the process modeling consisted of one transient thermal and four static thermal-mechanical simulations. The chip warpage after the bonding process was also verified by experiment. Lu and Chen [197] systematically analyzed the thermal-induced warpage during the ACA-based UTCOF bonding process by finite element simulation and experiment. Results indicated that the ultra-thin chip warpage was highly dependent on the bonding temperature. Tsai et al. [200] presented an analytical solution of thermal stresses of the thick chip assembly based on Suhir's theory, which required that the adhesive layer much thinner and more compliant than the chip and the substrate [201, 202]. Wang et al. [53] and Xu et al. [123, 203] introduced the Timoshenko beam theory to model the three-layered electronic structure, where an assumption of constant stress along the adhesive thickness direction resulted in the deviation of the interfacial stresses of the UTCOF assembly [131].

1.6 Competing Fracture

1.6.1 Competing Fracture Behavior

During the packaging process, cracking on the top surface center of the chip usually arises from inappropriate handling and process parameters. The high peeling-off requirement for the ultra-thin chip brings new challenges for the needle ejecting process, considering that the chip becomes fragile with the reduction of the chip thickness, and the risk of the chip cracking during the needle ejecting process increases rapidly. It is crucially important that the ultra-thin chip should be peeled off completely from the adhesive wafer without high bending tensile stress [24].

The chip peeling requires that the ERR of interfacial peeling should be larger than the adhesive fracture energy. The chip cracking occurs when the maximum tensile stress of the chip layer is larger than the fracture strength of the chip. Generally, the chip peeling and the chip cracking can be considered as two kinds of fracture, but with competing relationship. In order to achieve the successful peeling-off, the chip peeling along the adhesive layer should be promoted and the chip cracking across the chip layer should be prevented. The success rate of the chip peeling-off is significantly affected by the geometric dimensions and mechanical properties of the adhesively bonded multilayer structure. The competing fracture behavior gets more serious as the chip becomes thinner and larger. Therefore, the competing fracture behavior should be investigated to gain an insight into the underlying mechanism of the needle ejecting process and determine the process window of the chip transferring, especially for thin chips since the risk of chip cracking gets higher. A theoretical model to establish the competing relationship over a wide range of process parameters is very essential. It is also important to define a reliable criterion to explore the chip peelability, predict the failure mode, and provide guidelines for the ultra-thin chip applied in future electronic packaging.

1.6.2 Fracture Strength and Adhesive Fracture Energy

Generally, experimental approach is one of the practical methodologies [204] to evaluate the fracture strength of silicon wafers, which includes the ball-on-ring test

[39, 205], three/four-point bending test [38, 40, 206–212], and cantilever beam test [213, 214], etc. As to the flexible electronic systems [215], the fracture strength of the ultra-thin silicon chip plays a key role in determining the stability of the mechanical system. The precise evaluation of the fracture strength of the ultra-thin silicon chip is very crucial. The three-point bending (TPB) test is more frequently used due to the simplicity and high reliability. However, the classical linear theory converting the experimental load into the fracture strength fails to match the geometrically nonlinear characteristics of the ultra-thin silicon chip [212].

Actually, the geometric nonlinearity is attributed to the big deflection of the ultra-thin silicon chip. The linear solution without the square of the first derivative is no longer applicable for the ultra-thin silicon chip. Therefore, it is of significance to apply a nonlinear solution of the TPB test to get a more precise evaluation of the fracture strength. The classical nonlinear problems of the cantilever beam [216–218] and the simply supported beam [219–221] with large deformation under various boundary conditions have already been studied. Results indicated that it was not straightforward to obtain the analytical solution in closed form. Generally, the numerical method [218, 219] or approximation analysis [220] can be used to solve the nonlinear differential equations. The analytical solutions can be obtained in terms of elliptic integrals [221, 222] and nondimensional parameters [223] after appropriate assumptions. The above solution strategies for geometrically nonlinear problems of beams are also useful for the evaluation of the fracture strength of the ultra-thin silicon chip. Therefore, the large deflection theory of the beam is recommended to more deeply investigate the nonlinear characteristics of the ultra-thin silicon chip in the TPB test and provide a more reasonable evaluation of the fracture strength [212].

Besides, compliant adhesive tapes have been commonly used to tightly hold the silicon wafer during dicing for the fabrication of IC chips. After dicing, the separated chip needs to be picked up and adhered to the electronic device. The efficient peeling of the chip from the adhesive tape directly determines the qualified product ratio [89, 92], especially with the rapid development of thin/ultra-thin chip technology [28, 29, 41, 42]. Therefore, it is of significance to determine the adhesive fracture energy of the adhesive tape.

References

1. Yin Z, Huang Y, Bu N, Wang X, Xiong Y (2010) Inkjet printing for flexible electronics: materials, processes and equipments. Chin Sci Bull 55(30):3383–3407
2. Wong WS, Salleo A (2009) Flexible electronics: materials and applications, vol 11. Springer, US
3. Meitl MA, Zhu Z-T, Kumar V, Lee KJ, Feng X, Huang YY, Adesida I, Nuzzo RG, Rogers JA (2006) Transfer printing by kinetic control of adhesion to an elastomeric stamp. Nat Mater 5(1):33–38
4. Feng X, Meitl MA, Bowen AM, Huang Y, Nuzzo RG, Rogers JA (2007) Competing fracture in kinetically controlled transfer printing. Langmuir 23(25):12555–12560

5. Yu J, Bulović V (2007) Micropatterning metal electrode of organic light emitting devices using rapid polydimethylsiloxane lift-off. Appl Phys Lett 91(4):043102
6. Fan Z, Ho JC, Takahashi T, Yerushalmi R, Takei K, Ford AC, Chueh YL, Javey A (2009) Toward the development of printable nanowire electronics and sensors. Adv Mater 21(37):3730–3743
7. Kim T-H, Carlson A, Ahn J-H, Won SM, Wang S, Huang Y, Rogers JA (2009) Kinetically controlled, adhesiveless transfer printing using microstructured stamps. Appl Phys Lett 94(11):113502
8. Park S-I, Xiong Y, Kim R-H, Elvikis P, Meitl M, Kim D-H, Wu J, Yoon J, Yu C-J, Liu Z (2009) Printed assemblies of inorganic light-emitting diodes for deformable and semitransparent displays. Science 325(5943):977–981
9. Caldwell JD, Anderson TJ, Culbertson JC, Jernigan GG, Hobart KD, Kub FJ, Tadjer MJ, Tedesco JL, Hite JK, Mastro MA (2010) Technique for the dry transfer of epitaxial graphene onto arbitrary substrates. ACS Nano 4(2):1108–1114
10. Kim S, Wu J, Carlson A, Jin SH, Kovalsky A, Glass P, Liu Z, Ahmed N, Elgan SL, Chen W (2010) Microstructured elastomeric surfaces with reversible adhesion and examples of their use in deterministic assembly by transfer printing. Proc Natl Acad Sci 107(40):17095–17100
11. Ko H, Takei K, Kapadia R, Chuang S, Fang H, Leu PW, Ganapathi K, Plis E, Kim HS, Chen S-Y (2010) Ultrathin compound semiconductor on insulator layers for high-performance nanoscale transistors. Nature 468(7321):286–289
12. Liu H, Takagi D, Chiashi S, Homma Y (2010) Transfer and alignment of random single-walled carbon nanotube films by contact printing. ACS Nano 4(2):933–938
13. Packard CE, Murarka A, Lam EW, Schmidt MA, Bulović V (2010) Contact-printed microelectromechanical systems. Adv Mater 22(16):1840–1844
14. Wu CC, Liu CH, Zhong Z (2010) One-step direct transfer of pristine single-walled carbon nanotubes for functional nanoelectronics. Nano Lett 10(3):1032–1036
15. Carlson A, Kim-Lee H-J, Wu J, Elvikis P, Cheng H, Kovalsky A, Elgan S, Yu Q, Ferreira PM, Huang Y (2011) Shear-enhanced adhesiveless transfer printing for use in deterministic materials assembly. Appl Phys Lett 98(26):264104
16. Qi Y, Kim J, Nguyen TD, Lisko B, Purohit PK, McAlpine MC (2011) Enhanced piezoelectricity and stretchability in energy harvesting devices fabricated from buckled PZT ribbons. Nano Lett 11(3):1331–1336
17. Wu J, Kim S, Chen W, Carlson A, Hwang K-C, Huang Y, Rogers JA (2011) Mechanics of reversible adhesion. Soft Matter 7(18):8657–8662
18. Crawford GP (2005) Flexible flat panel display technology. Wiley Online Library
19. Forrest SR (2004) The path to ubiquitous and low-cost organic electronic appliances on plastic. Nature 428(6986):911–918
20. Gelinck GH, Huitema HEA, van Veenendaal E, Cantatore E, Schrijnemakers L, van der Putten JB, Geuns TC, Beenhakkers M, Giesbers JB, Huisman B-H (2004) Flexible active-matrix displays and shift registers based on solution-processed organic transistors. Nat Mater 3(2):106–110
21. Kim D-H, Ahn J-H, Choi WM, Kim H-S, Kim T-H, Song J, Huang YY, Liu Z, Lu C, Rogers JA (2008) Stretchable and foldable silicon integrated circuits. Science 320(5875):507–511
22. Sekitani T, Zschieschang U, Klauk H, Someya T (2010) Flexible organic transistors and circuits with extreme bending stability. Nat Mater 9(12):1015–1022
23. Yoon J, Baca AJ, Park S-I, Elvikis P, Geddes JB, Li L, Kim RH, Xiao J, Wang S, Kim T-H (2008) Ultrathin silicon solar microcells for semitransparent, mechanically flexible and microconcentrator module designs. Nat Mater 7(11):907–915
24. Greig W (2007) Integrated circuit packaging, assembly and interconnections. Springer, US
25. Tong H-M, Lai Y-S, Wong C (2013) Advanced flip chip packaging. Springer, US
26. Liu Z, Tang P, Huang Y, Yin Z (2014) Experimental estimation of adhesive fracture energy of compliant adhesive tape. In: 15th international conference on electronic packaging technology, 12–15 Aug 2014, pp 842–846

27. Kim DH, Kim YS, Wu J, Liu Z, Song J, Kim HS, Huang YY, Hwang KC, Rogers JA (2009) Ultrathin silicon circuits with strain-isolation layers and mesh layouts for high-performance electronics on fabric, vinyl, leather, and paper. Adv Mater 21(36):3703–3707
28. Burghartz JN, Appel W, Harendt C, Rempp H, Richter H, Zimmermann M (2010) Ultra-thin chip technology and applications, a new paradigm in silicon technology. Solid-State Electron 54(9):818–829
29. Burghartz JN, Harendt C, Hoang T, Kiss A, Zimmermann M (2009) Ultra-thin chip fabrication for next-generation silicon processes. In: IEEE Bipolar/BiCMOS circuits and technology meeting, 12–14 Oct 2009, pp 131–137
30. Harendt C, Kostelnik J, Kugler A, Lorenz E, Saller S, Schreivogel A, Yu ZL, Burghartz JN (2015) Hybrid Systems in Foil (HySiF) exploiting ultra-thin flexible chips. Solid-State Electron 113:101–108
31. Shahrjerdi D, Bedell SW (2013) Extremely flexible nanoscale ultrathin body silicon integrated circuits on plastic. Nano Lett 13(1):315–320
32. Kuo T-Y, Shih Y-C, Lee Y-C, Chang H-H, Hsiao Z-C, Chiang C-W, Li S-M, Hwang Y-J, Ko C-T, Chen Y-H (2009) Flexible and ultra-thin embedded chip package. In: 59th electronic components and technology conference, 26–29 May 2009, pp 1749–1753
33. Brand Jvd, Kok Md, Sridhar A, Cauwe M, Verplancke R, Bossuyt F, De Baets J, Vanfleteren J (2014) Flexible and stretchable electronics for wearable healthcare. In: 44th European solid state device research conference (ESSDERC), 22–26 Sept 2014, pp 206–209
34. Su Y, Wu J, Fan Z, Hwang K-C, Song J, Huang Y, Rogers JA (2012) Postbuckling analysis and its application to stretchable electronics. J Mech Phys Solids 60(3):487–508
35. Rogers JA, Someya T, Huang Y (2010) Materials and mechanics for stretchable electronics. Science 327(5973):1603–1607
36. Chiang C-J, Winscom C, Bull S, Monkman A (2009) Mechanical modeling of flexible OLED devices. Org Electron 10(7):1268–1274
37. Burghartz JN (2010) Ultra-thin chip technology and applications. Springer-Verlag, New York
38. Porter D, Berfield T (2013) Die separation and rupture strength for deep reactive ion etched silicon wafers. J Micromech Microeng 23(8):085020
39. Jeon E-B, Park J-D, Song JH, Lee HJ, Kim H-S (2012) Bi-axial fracture strength characteristic of an ultra-thin flash memory chip. J Micromech Microeng 22(10):105014
40. Schoenfelder S, Ebert M, Landesberger C, Bock K, Bagdahn J (2007) Investigations of the influence of dicing techniques on the strength properties of thin silicon. Microelectron Reliab 47(2):168–178
41. Feil M, Adler C, Hemmetzberger D, Konig M, Bock K (2004) The challenge of ultra thin chip assembly. In: 54th electronic components and technology conference (IEEE Cat. No. 04CH37546), 1–4 June 2004, vol 31, pp 35–40
42. Medding J, Stalder R, Niederhauser M, Stoessel P (2004) Thin die bonding techniques. In: IEEE/CPMT/SEMI 29th international electronics manufacturing technology symposium (IEEE Cat. No. 04CH37585), 14–16 July 2004, pp 68–73
43. Gambino JP (2013) Thin silicon wafer processing and strength characterization. In: Proceedings of the 20th IEEE international symposium on the physical and failure analysis of integrated circuits (IPFA), 15–19 July 2013, pp 199–207
44. Wilson L (2013) International technology roadmap for semiconductors (ITRS). Semiconductor Industry Association
45. Bock K (2005) Polymer electronics systems-polytronics. Proc IEEE 93(8):1400–1406
46. Allen KJ (2005) Reel to real: Prospects for flexible displays. Proc IEEE 93(8):1394–1399
47. Deng J, Lee MM, Moy SS (2004) Stress analysis of steel beams reinforced with a bonded CFRP plate. Compos Struct 65(2):205–215
48. Campilho RD, De Moura M, Domingues J (2008) Using a cohesive damage model to predict the tensile behaviour of CFRP single-strap repairs. Int J Solids Struct 45(5):1497–1512
49. Cheng J, Taheri F (2006) A smart single-lap adhesive joint integrated with partially distributed piezoelectric patches. Int J Solids Struct 43(5):1079–1092

50. Cheng H, Wu J, Li M, Kim D-H, Kim Y-S, Huang Y, Kang Z, Hwang K, Rogers J (2011) An analytical model of strain isolation for stretchable and flexible electronics. Appl Phys Lett 98(6):061902
51. Sun J-Y, Lu N, Yoon J, Oh K-H, Suo Z, Vlassak JJ (2012) Debonding and fracture of ceramic islands on polymer substrates. J Appl Phys 111(1):013517
52. Jiang Z, Huang Y, Chandra A (1997) Thermal stresses in layered electronic assemblies. J Electron Packag 119(2):127–132
53. Wang K, Huang Y, Chandra A, Hu KX (2000) Interfacial shear stress, peeling stress, and die cracking stress in trilayer electronic assemblies. In: ITHERM 2000. The seventh intersociety conference on thermal and thermomechanical phenomena in electronic systems (Cat. No. 00CH37069), vol 52, pp 56–64
54. Delale F, Erdogan F, Aydinoglu M (1981) Stresses in adhesively bonded joints: a closed-form solution. J Compos Mater 15(3):249–271
55. Zou G, Shahin K, Taheri F (2004) An analytical solution for the analysis of symmetric composite adhesively bonded joints. Compos Struct 65(3):499–510
56. Shahin K, Taheri F (2007) Analysis of deformations and stresses in balanced and unbalanced adhesively bonded single-strap joints. Compos Struct 81(4):511–524
57. Shahin K, Kember G, Taheri F (2008) An asymptotic solution for evaluation of stresses in balanced and unbalanced adhesively bonded joints. Mech Adv Mater Struct 15(2):88–103
58. Li G (2010) Elastic analysis of closed-form solutions for adhesive stresses in bonded single-strap butt joints. J Mech Mater Struct 5(3):409–426
59. Goland M, Reissner E (1944) The stresses in cemented joints. J Appl Mech 11(1):A17–A27
60. Hart-Smith LJ (1973) Adhesive-bonded single-lap joints. National Aeronautics and Space Administration (NASA)
61. Cheng S, Chen D, Shi Y (1991) Analysis of adhesive-bonded joints with nonidentical adherends. J Eng Mech 117(3):605–623
62. Oplinger D (1994) Effects of adherend deflections in single lap joints. Int J Solids Struct 31(18):2565–2587
63. Wu Z, Romeijn A, Wardenier J (1997) Stress expressions of single-lap adhesive joints of dissimilar adherends. Compos Struct 38(1):273–280
64. Tsai M, Oplinger D, Morton J (1998) Improved theoretical solutions for adhesive lap joints. Int J Solids Struct 35(12):1163–1185
65. Her S-C (1999) Stress analysis of adhesively-bonded lap joints. Compos Struct 47(1):673–678
66. Weißgraeber P, Becker W (2013) Finite fracture mechanics model for mixed mode fracture in adhesive joints. Int J Solids Struct 50(14):2383–2394
67. Adams RD, Peppiatt NA (1974) Stress analysis of adhesive-bonded lap joints. J Strain Anal 9(3):185–196
68. Luo Q, Tong L (2004) Linear and higher order displacement theories for adhesively bonded lap joints. Int J Solids Struct 41(22):6351–6381
69. Luo Q, Tong L (2009) Energy release rates for interlaminar delamination in laminates considering transverse shear effects. Compos Struct 89(2):235–244
70. Bennati S, Colleluori M, Corigliano D, Valvo PS (2009) An enhanced beam-theory model of the asymmetric double cantilever beam (ADCB) test for composite laminates. Compos Sci Technol 69(11):1735–1745
71. Shahin K, Taheri F (2008) The strain energy release rates in adhesively bonded balanced and unbalanced specimens and lap joints. Int J Solids Struct 45(25):6284–6300
72. Yang C, Pang S-S (1996) Stress-strain analysis of single-lap composite joints under tension. J Eng Mater Technol 118(2):247–255
73. Wang J, Qiao P (2004) On the energy release rate and mode mix of delaminated shear deformable composite plates. Int J Solids Struct 41(9):2757–2779
74. Chadegani A, Batra RC (2011) Analysis of adhesive-bonded single-lap joint with an interfacial crack and a void. Int J Adhes Adhes 31(6):455–465
75. Chadegani A, Yang C, Smeltzer SS (2012) Adhesive-bonded composite joints analysis with delaminated surface ply using strain-energy release rate. J Aircr 49(2):503–520

76. da Silva LF, das Neves PJ, Adams R, Spelt J (2009) Analytical models of adhesively bonded joints—Part I: Literature survey. Int J Adhes Adhes 29(3):319–330
77. Hutchinson JW, Suo Z (1991) Mixed mode cracking in layered materials. In: Hutchinson JW, Wu TY (eds) Advances in applied mechanics, vol 29. Elsevier, pp 63–191
78. He M, Evans A, Hutchinson J (1997) Convergent debonding of films and fibers. Acta Mater 45(8):3481–3489
79. Li S, Wang J, Thouless M (2004) The effects of shear on delamination in layered materials. J Mech Phys Solids 52(1):193–214
80. Qiao P, Wang J (2004) Mechanics and fracture of crack tip deformable bi-material interface. Int J Solids Struct 41(26):7423–7444
81. Wang J, Zhang C (2009) Energy release rate and phase angle of delamination in sandwich beams and symmetric adhesively bonded joints. Int J Solids Struct 46(25):4409–4418
82. Lu N, Yoon J, Suo Z (2007) Delamination of stiff islands patterned on stretchable substrates. Int J Mater Res 98(8):717–722
83. Yang C, Chadegani A, Tomblin JS (2008) Strain energy release rate determination of prescribed cracks in adhesively-bonded single-lap composite joints with thick bondlines. Compos B Eng 39(5):863–873
84. Allix O, Ladevèze P (1992) Interlaminar interface modelling for the prediction of delamination. Compos Struct 22(4):235–242
85. Corigliano A (1993) Formulation, identification and use of interface models in the numerical analysis of composite delamination. Int J Solids Struct 30(20):2779–2811
86. Bennati S, Fisicaro P, Valvo PS (2013) An enhanced beam-theory model of the mixed-mode bending (MMB) test—Part I: Literature review and mechanical model. Meccanica 48(2):443–462
87. Bennati S, Fisicaro P, Valvo PS (2013) An enhanced beam-theory model of the mixed-mode bending (MMB) test—Part II: Applications and results. Meccanica 48(2):465–484
88. Bennati S, Taglialegne L, Valvo PS (2010) Modelling of interfacial fracture of layered structures. In: Proceedings of the 18th European conference on fracture: fracture of materials and structures from micro to macro scale, ECF, Dresden, Germany, 30 Aug 3 Sept 2010
89. Peng B, Huang Y, Yin Z, Xiong Y (2011) Analysis of interfacial peeling in IC chip pick-up process. J Appl Phys 110(7):073508
90. Peng B, Huang Y, Yin Z, Xiong Y (2012) Competing fracture modeling of thin chip pick-up process. IEEE Trans Compon Packag Manuf Technol 2(7):1217–1225
91. Liu Z, Valvo PS, Huang Y, Yin Z (2013) Cohesive failure analysis of an array of IC chips bonded to a stretched substrate. Int J Solids Struct 50(22–23):3528–3538
92. Liu Z, Huang Y, Chen J, Yin Z (2014) Tunable peeling technique and mechanism of thin chip from compliant adhesive tapes. IEEE Trans Compon Packag Manuf Technol 4(4):560–568
93. Liu Z, Huang YA, Liu H, Chen J, Yin Z (2014) Reliable peeling of ultrathin die with multi-needle ejector. IEEE Trans Compon Packag Manuf Technol 4(9):1545–1554
94. Liu Z, Huang Y, Yin Z, Bennati S, Valvo PS (2014) A general solution for the two-dimensional stress analysis of balanced and unbalanced adhesively bonded joints. Int J Adhes Adhes 54:112–123
95. Liu Z, Wan X, Huang Y, Chen J, Yin Z (2018) Theoretical and experimental studies of competing fracture for flexible chip-adhesive-substrate composite structure. IEEE Trans Compon Packag Manuf Technol 8(1):57–64
96. Camanho PP, Dávila CG (2002) Mixed-mode decohesion finite elements for the simulation of delamination in composite materials. National Aeronautics and Space Administration (NASA)
97. Agrawal A, Karlsson AM (2006) Obtaining mode mixity for a bimaterial interface crack using the virtual crack closure technique. Int J Fract 141(1):75–98
98. Berggreen C, Simonsen BC, Borum KK (2007) Experimental and numerical study of interface crack propagation in foam-cored sandwich beams. J Compos Mater 41(4):493–520
99. Auersperg J, Kieselstein E, Schubert A, Michel B (2002) Delamination risk evaluation for plastic packages based on mixed mode fracture mechanics approaches. J Electron Packag 124(4):318–322

100. Xie D, Biggers SB (2006) Progressive crack growth analysis using interface element based on the virtual crack closure technique. Finite Elem Anal Des 42(11):977–984
101. Xie D, Biggers SB (2007) Calculation of transient strain energy release rates under impact loading based on the virtual crack closure technique. Int J Impact Eng 34(6):1047–1060
102. Valvo PS (2012) A revised virtual crack closure technique for physically consistent fracture mode partitioning. Int J Fract 173(1):1–20
103. Cheng T-H, Du C-C, Tseng C-H (2006) Study in IC chip failure during pick-up process by using experimental and finite element methods. J Mater Process Technol 172(3):407–416
104. Cheng T-H, Tseng C-H, Hung C-H (2008) Application of a genetic algorithm associated with adhesive joint analysis to the IC chip pick-up process. J Adhes Sci Technol 22(10–11):1057–1072
105. Lin Y-J, Hwang S-J (2005) Static analysis of the die picking process. IEEE Trans Electron Packag Manuf 28(2):142–149
106. Cheng T-H, Tseng C-H, Hung C-H (2006) Analysis of stresses in adhesive joints applicable to IC chips using symbolic manipulation and the numerical method. J Adhes Sci Technol 20(15):1669–1692
107. Liu Y, Irving S, Desbiens D, Timwah L, Qiuxiao Q (2006) Simulation and analysis for typical package assembly manufacture. In: EuroSime 2006—7th international conference on thermal, mechanical and multiphysics simulation and experiments in micro-electronics and micro-systems, 24–26 April 2006, pp 1–10
108. Saiki N, Inaba K, Kishimoto K, Seno H, Ebe K (2010) Study on peeling behavior in pick-up process of IC chip with adhesive tapes. J Solid Mech Mater Eng 4(7):1051–1060
109. Saiki N, Inaba K, Kishimoto K, Seno H, Ichikawa I (2010) Modeling and validation of evaluation method on IC chip pick-up performance of dicing/die bonding tape. In: 3rd electronics system integration technology conference ESTC, 13–16 Sept 2010, pp 1–6
110. Ko HC, Stoykovich MP, Song J, Malyarchuk V, Choi WM, Yu C-J, Geddes Iii JB, Xiao J, Wang S, Huang Y (2008) A hemispherical electronic eye camera based on compressible silicon optoelectronics. Nature 454(7205):748–753
111. Huang Y, Chen J, Yin Z, Xiong Y (2011) Roll-to-roll processing of flexible heterogeneous electronics with low interfacial residual stress. IEEE Trans Compon Packag Manuf Technol 1(9):1368–1377
112. Huang Y, Yin Z, Xiong Y (2010) Thermomechanical analysis of film-on-substrate system with temperature-dependent properties. J Appl Mech-T Asme 77(4):041016
113. Huang Y, Yin Z, Xiong Y (2010) Thermomechanical analysis of thin films on temperature-dependent elastomeric substrates in flexible heterogeneous electronics. Thin Solid Films 518(6):1698–1702
114. Feng Y, Wu L (2001) Analysis of interfacial thermal stresses of chip-substrate structure. Int J Solids Struct 38(9):1551–1562
115. Park SI, Ahn JH, Feng X, Wang S, Huang Y, Rogers JA (2008) Theoretical and experimental studies of bending of inorganic electronic materials on plastic substrates. Adv Func Mater 18(18):2673–2684
116. Wang KP, Huang YY, Chandra A, Hu KX (2000) Interfacial shear stress, peeling stress, and die cracking stress in trilayer electronic assemblies. IEEE Trans Compon Packag Technol 23(2):309–316
117. Zhang Z, Yoon J, Suo Z (2006) Method to analyze dislocation injection from sharp features in strained silicon structures. Appl Phys Lett 89(26):261912
118. Liu X, Suo Z, Ma Q (1998) Split singularities: stress field near the edge of a silicon die on a polymer substrate. Acta Mater 47(1):67–76
119. Feron M, Zhang Z, Suo Z, Feron M (2007) Split singularities and dislocation injection in strained silicon. J Appl Phys 102(2):023502
120. Peng B, Huang Y, Yin Z, Xiong Y (2011) On the analysis of dynamic effect in the die pick-up process. In: 12th international conference on electronic packaging technology and high density packaging, 8–11 Aug 2011. IEEE, pp 651–654

121. Chong ACM, Cheung YM (2003) Finite element stress analysis of thin die detachment process. In: Fifth international conference on electronic packaging technology proceedings, 28–30 Oct 2003, pp 44–51
122. Cheung YM, Chong ACM, Huang B (2006) Determination of the interfacial fracture toughness of laminated silicon die on adhesive dicing tape from stud pull measurement. In: International conference on electronic materials and packaging, 11–14 Dec 2006, pp 1–10
123. Xu Z, Liu Z, Huang Y, Chen J, Liu H, Yin Z (2015) Vacuum-based picking-up of thin chip from adhesive tape. J Adhes Sci Technol 29(13):1315–1329
124. Hassinen T, Ruotsalainen T, Laakso P, Penttilä R, Sandberg HGO (2014) Roll-to-roll compatible organic thin film transistor manufacturing technique by printing, lamination, and laser ablation. Thin Solid Films 571:212–217
125. Roth B, Søndergaard RR, Krebs FC (2015) Roll-to-roll printing and coating techniques for manufacturing large-area flexible organic electronics. In: Logothetidis S (ed) Handbook of flexible organic electronics. Woodhead Publishing, Oxford, pp 171–197
126. Dupont SR, Oliver M, Krebs FC, Dauskardt RH (2012) Interlayer adhesion in roll-to-roll processed flexible inverted polymer solar cells. Sol Energy Mater Sol Cells 97:171–175
127. Chen J, Liu H, Huang Y, Yin Z (2016) High-rate roll-to-roll stack and lamination of multilayer structured membrane electrode assembly. J Manuf Process 23:175–182
128. Huang Y, Liu H, Xu Z, Chen J, Yin Z (2018) Conformal peeling of device-on-substrate system in flexible electronic assembly. IEEE Trans Compon Packag Manuf Technol 8(8):1496–1506
129. Yurenka S (1962) Peel testing of adhesive bonded metal. J Appl Polym Sci 6(20):136–144
130. Wang Y (2014) The effect of peeling rate and peeling angle on the peeling strength. The University of Akron
131. Wacker N, Richter H, Hoang T, Gazdzicki P, Schulze M, Angelopoulos EA, Hassan M-U, Burghartz JN (2014) Stress analysis of ultra-thin silicon chip-on-foil electronic assembly under bending. Semicond Sci Technol 29:095007 (095012pp)
132. Cortet PP, Ciccotti M, Vanel L (2007) Imaging the stick–slip peeling of an adhesive tape under a constant load. J Stat Mech Theory Exp 2007(03):P03005
133. Eremeyev VA, Naumenko K (2015) A relationship between effective work of adhesion and peel force forthin hyperelastic films undergoing large deformation. Mech Res Commun 69:24–26
134. Gent AN, Kaang SY (1987) Effect of peel angle upon peel force. J Adhes 24(2–4):173–181
135. Jiang D, Feng X, Qu B, Wang Y, Fang D (2012) Rate-dependent interaction between thin films and interfaces during micro/nanoscale transfer printing. Soft Matter 8:418–423
136. Kaelble DH (1992) Theory and analysis of peel adhesion: adhesive thickness effects. J Adhes 37(1–3):205–214
137. Park Ju, Hardy M, Kang SJ, Barton K, Adair K, Mukhopadhyay DK, Lee CY, Strano MS, Alleyne AG, Georgiadis JG, Ferreira PM, Rogers JA (2007) High-resolution electrohydrodynamic jet printing. Nat Mater 6:782–789
138. Kovalchick C, Molinari A, Ravichandran G (2013) Rate dependent adhesion energy and nonsteady peeling of inextensible tapes. J Appl Mech 81(4):041016
139. Cheng H, Li M, Wu J, Carlson A, Kim S, Huang Y, Kang Z, Hwang K-C, Rogers JA (2013) A viscoelastic model for the rate effect in transfer printing. J Appl Mech 80(4):041015–041019
140. Peng Z, Wang C, Chen L, Chen S (2014) Peeling behavior of a viscoelastic thin-film on a rigid substrate. Int J Solids Struct 51:4596–4603
141. Xiong Z, Tay AA (2000) Modeling of viscoelastic effects on interfacial delamination in IC packages. Paper presented at the electronic components and technology conference, 21–24 May
142. Joe DJ, Kim S, Park JH, Park DY, Lee HE, Im TH, Choi I, Ruoff RS, Lee KJ (2017) Laser—material interactions for flexible applications. Adv Mater 29(26):1606586
143. Kim S, Son JH, Lee SH, You BK, Park KI, Lee HK, Byun M, Lee KJ (2014) Flexible crossbar-structured resistive memory arrays on plastic substrates via inorganic-based laser lift-off. Adv Mater 26(44):7480–7487

144. Park KI, Son JH, Hwang GT, Jeong CK, Ryu J, Koo M, Choi I, Lee SH, Byun M, Wang ZL (2014) Highly-efficient, flexible piezoelectric PZT thin film nanogenerator on plastic substrates. Adv Mater 26(16):2514–2520
145. Lee CH, Kim SJ, Oh Y, Kim MY, Yoon Y-J, Lee H-S (2010) Use of laser lift-off for flexible device applications. J Appl Phys 108(10):102814
146. Kim SJ, Lee HE, Choi H, Kim Y, We JH, Shin JS, Lee KJ, Cho BJ (2016) High-performance flexible thermoelectric power generator using laser multiscanning lift-off process. ACS Nano 10(12):10851–10857
147. Do YH, Kang MG, Kim JS, Kang CY, Yoon SJ (2012) Fabrication of flexible device based on PAN-PZT thin films by laser lift-off process. Sens Actuators A 184:124–127
148. Bian J, Zhou L, Wan X, Liu M, Zhu C, Huang Y, Yin Z (2019) Experimental study of laser lift-off of ultra-thin polyimide film for flexible electronics. Sci China Technol Sci 62(2):233–242
149. Delmdahl R, Pätzel R, Brune J (2013) Large-area laser-lift-off processing in microelectronics. Phys Proced 41:241–248
150. Doany FE, Narayan C (1997) Laser release process to obtain freestanding multilayer metal-polyimide circuits. IBM J Res Dev 41(1/2):151–157
151. Delmdahl R, Fricke M, Fechner B (2014) Laser lift-off systems for flexible-display production. J Inform Disp 15(1):1–4
152. MacCarthy N, Wood T, Ameri H, O'Connell D, Alderman J (2006) A laser release method for producing prototype flexible retinal implant devices. Sens Actuators A 132(1):296–301
153. Wong WS, Sands T, Cheung NW, Kneissl M, Bour DP, Mei P, Romano LT, Johnson NM (1999) Fabrication of thin-film InGaN light-emitting diode membranes by laser lift-off. Appl Phys Lett 75(10):1360–1362
154. Chu C-F, Lai F-I, Chu J-T, Yu C-C, Lin C-F, Kuo H-C, Wang SC (2004) Study of GaN light-emitting diodes fabricated by laser lift-off technique. J Appl Phys 95(8):3916–3922
155. Lin WY, Wuu DS, Pan KF, Huang SH, Lee CE, Wang WK, Hsu SC, Su YY, Huang SY, Horng RH (2005) High-power GaN-mirror-Cu light-emitting diodes for vertical current injection using laser liftoff and electroplating techniques. IEEE Photonics Technol Lett 17(9):1809–1811
156. Seo J-H, Li J, Lee J, Gong S, Lin J, Jiang H, Ma Z (2015) A simplified method of making flexible blue LEDs on a plastic substrate. IEEE Photonics J 7(2):1–7
157. Jung Y, Wang X, Kim J, Hyun Kim S, Ren F, Pearton SJ, Kim J (2012) GaN-based light-emitting diodes on origami substrates. Appl Phys Lett 100(23):231113
158. Jung Y, Wang X, Kim SH, Ren F, Kim J, Pearton SJ (2012) A facile method for flexible GaN-based light-emitting diodes. Phys Status Solidi (RRL)-Rapid Res Lett 6(11):421–423
159. Kim Ti, Jung YH, Song J, Kim D, Li Y, Kim Hs, Song IS, Wierer JJ, Pao HA, Huang Y (2012) High-efficiency, microscale GaN light-emitting diodes and their thermal properties on unusual substrates. Small 8(11):1643–1649
160. Kim RH, Tao H, Kim Ti, Zhang Y, Kim S, Panilaitis B, Yang M, Kim DH, Jung YH, Kim BH, Li Y, Huang Y, Omenetto FG, Rogers JA (2012) Materials and designs for wirelessly powered implantable light-emitting systems. Small 8(18):2812–2818
161. Cheung YF, Li KH, Choi HW (2016) Flexible free-standing III-nitride thin films for emitters and displays. ACS Appl Mater Interfaces 8(33):21440–21445
162. Lee SH, Park SY, Lee KJ (2012) Laser lift-off of GaN thin film and its application to the flexible light emitting diodes. In: Biosensing and nanomedicine V. 2012 International society for optics and photonics, p 846011
163. Chun J, Hwang Y, Choi YS, Jeong T, Baek JH, Ko HC, Park SJ (2012) Transfer of GaN LEDs from sapphire to flexible substrates by laser lift-off and contact printing. IEEE Photonics Technol Lett 24(23):2115–2118
164. Chun J, Hwang Y, Choi Y-S, Kim J-J, Jeong T, Baek JH, Ko HC, Park S-J (2014) Laser lift-off transfer printing of patterned GaN light-emitting diodes from sapphire to flexible substrates using a Cr/Au laser blocking layer. Scripta Mater 77:13–16
165. Choi WS, Park HJ, Park SH, Jeong T (2014) Flexible InGaN LEDs on a polyimide substrate fabricated using a simple direct-transfer method. IEEE Photonics Technol Lett 26(21):2115–2117

166. Tsakalakos L, Sands T (2000) Epitaxial ferroelectric (Pb,La)(Zr,Ti)O_3 thin films on stainless steel by excimer laser liftoff. Appl Phys Lett 76(2):227–229
167. Tsakalakos L, Sands T, Carleton E, Yu KM (2003) Modification of (Pb,La)(Zr,Ti)O_3 thin films during pulsed laser liftoff from MgO substrates. J Appl Phys 94(6):4047–4052
168. Li K-Y, Tai N-H, Lin IN (2005) Preparation of PNN-PZT thick film on Pt/Ti/SiO2/Si substrate by laser lift-off process. Integr Ferroelectr 69(1):135–141
169. Do YH, Jung WS, Kang MG, Kang CY, Yoon SJ (2013) Preparation on transparent flexible piezoelectric energy harvester based on PZT films by laser lift-off process. Sens Actuators A 200:51–55
170. Tsakalakos L (2000) Laser liftoff of epitaxial oxide films. The University of California
171. Noh M-S, Kim S, Hwang D-K, Kang C-Y (2017) Self-powered flexible touch sensors based on PZT thin films using laser lift-off. Sens Actuators A 261:288–294
172. Choi Y-Y, Choi K-H, Kim H-K (2011) Characteristics of a transparent GaON sacrificial layer for transfer of devices from sapphire to a flexible substrate. J Electrochem Soc 158(11):J349–J353
173. Kim K, Kim SY, Lee J-L (2014) Flexible organic light-emitting diodes using a laser lift-off method. J Mater Chem C 2(12):2144–2149
174. Jeong CK, Cho SB, Han JH, Park DY, Yang S, Park K-I, Ryu J, Sohn H, Chung Y-C, Lee KJ (2017) Flexible highly-effective energy harvester via crystallographic and computational control of nanointerfacial morphotropic piezoelectric thin film. Nano Res 10(2):437–455
175. Jeong CK, Han JH, Palneedi H, Park H, Hwang G-T, Joung B, Kim S-G, Shin HJ, Kang I-S, Ryu J, Lee KJ (2017) Comprehensive biocompatibility of nontoxic and high-output flexible energy harvester using lead-free piezoceramic thin film. APL Mater 5(7):074102
176. Geon-Tae H, Venkateswarlu A, Hyun HJ, J. JD, Changyeon B, Yong PD, Hyun KD, Hwan PJ, Kyu JC, Kwi-Il P, Jong-Jin C, Kyung KD, Jungho R, Jae LK (2016) Self-powered wireless sensor node enabled by an aerosol-deposited PZT flexible energy harvester. Adv Energy Mater 6(13):1600237
177. Jeong CK, Park K-I, Son JH, Hwang G-T, Lee SH, Park DY, Lee HE, Lee HK, Byun M, Lee KJ (2014) Self-powered fully-flexible light-emitting system enabled by flexible energy harvester. Energy Environ Sci 7(12):4035–4043
178. Lee HS, Chung J, Hwang GT, Jeong CK, Jung Y, Kwak JH, Kang H, Byun M, Kim WD, Hur S, Oh SH, Lee KJ (2014) Flexible inorganic piezoelectric acoustic nanosensors for biomimetic artificial hair cells. Adv Func Mater 24(44):6914–6921
179. Park DY, Joe DJ, Kim DH, Park H, Han JH, Jeong CK, Park H, Park JG, Joung B, Lee KJ (2017) Self-powered real-time arterial pulse monitoring using ultrathin epidermal piezoelectric sensors. Adv Mater 29(37):1702308
180. Lee HE, Kim S, Ko J, Yeom HI, Byun CW, Lee SH, Joe DJ, Im TH, Park SHK, Lee KJ (2016) Skin-like oxide thin-film transistors for transparent displays. Adv Func Mater 26(34):6170–6178
181. Delmdahl R (2010) The excimer laser: precision engineering. Nat Photonics 4:286
182. French I, George D, Kretz T, Templier F, Lifka H (2007) 58.4: Invited paper: flexible displays and electronics made in AM-LCD facilities by the EPLaR™ process. SID Symp Dig Tech Papers 38(1):1680–1683
183. Lee S-W, Lee T-H, Park J-W, Park C-H, Kim H-J, Kim S-M, Lee S-H, Song J-Y, Lee J-H (2015) The effect of laser irradiation on peel strength of temporary adhesives for wafer bonding. Int J Adhes Adhes 57:9–12
184. Dang B, Andry P, Tsang C, Maria J, Polastre R, Trzcinski R, Prabhakar A, Knickerbocker J (2010) CMOS compatible thin wafer processing using temporary mechanical wafer, adhesive and laser release of thin chips/wafers for 3D integration. In: 60th electronic components and technology conference (ECTC), 1–4 June 2010, pp 1393–1398
185. Webb BC, Andry P (2015) Simulation of thermal pulse evolution during laser debonding. In: IEEE 65th electronic components and technology conference (ECTC), 26–29 May 2015, pp 1423–1429

186. Phommahaxay A, Potoms G, Verbinnen G, Sleeckx E, Beyer G, Beyne E, Guerrero A, Bai D, Liu X, Yess K, Arnold K, Spiess W, Griesbach T, Rapps T, Lutter S (2016) Extremely low-force debonding of thinned CMOS substrate by laser release of a temporary bonding material. In: IEEE 66th electronic components and technology conference (ECTC), 31 May–3 June 2016, pp 1685–1690
187. Delmdahl R, Fechner B (2010) Large-area microprocessing with excimer lasers. Appl Phys A 101(2):283–286
188. Hwang S-W, Tao H, Kim D-H, Cheng H, Song J-K, Rill E, Brenckle MA, Panilaitis B, Won SM, Kim Y-S (2012) A physically transient form of silicon electronics. Science 337(6102):1640–1644
189. Rogers J, Lagally M, Nuzzo R (2011) Synthesis, assembly and applications of semiconductor nanomembranes. Nature 477(7362):45–53
190. Kim SJ, We JH, Cho BJ (2014) A wearable thermoelectric generator fabricated on a glass fabric. Energy Environ Sci 7(6):1959–1965
191. Chen J, Xu Z, Huang Y, Duan Y, Yin Z (2016) Analytical investigation on thermal-induced warpage behavior of ultrathin chip-on-flex (UTCOF) assembly. Sci China Technol Sci 59(11):1646–1655
192. Yousefsani SA, Tahani M (2013) Accurate determination of stress distributions in adhesively bonded homogeneous and heterogeneous double-lap joints. Eur J Mech A/Solids 39:197–208
193. Yousefsani SA, Tahani M (2013) Analytical solutions for adhesively bonded composite single-lap joints under mechanical loadings using full layerwise theory. Int J Adhes Adhes 43:32–41
194. Reddy J (1987) A generalization of two-dimensional theories of laminated composite plates. Commun Appl Numer Methods 3(3):173–180
195. Tahani M, Yousefsani S (2015) On thermomechanical stress analysis of adhesively bonded composite joints in presence of an interfacial void. Compos Struct 130:116–123
196. Cheng H-C, Ma C-H, Yu C-F, Lu S-T, Chen W-H, Jang KP, kuk Son J, Kwon SH, Yang Z, Cui J (2013) Process-dependent thermal-mechanical behaviors of an advanced thin-flip-chip-on-flex interconnect technology with anisotropic conductive film joints. CMC: Comput Mater Contin 38(3):129–154
197. Lu S-T, Chen W-H (2010) Experimental/numerical analysis of thermally induced warpage of ultrathin chip-on-flex (UTCOF) interconnects. IEEE Trans Compon Packag Technol 33(4):819–829
198. Cheng H-C, Chen W-H, Lin C-S, Hsu Y-Y, Uang R-H (2009) On the thermal–mechanical behaviors of a novel nanowire-based anisotropic conductive film technology. IEEE Trans Adv Packag 32(2):546–563
199. Cheng H-C, Huang H-H, Chen W-H, Lu S-T (2015) Hygro-thermo-mechanical behavior of adhesive-based flexible chip-on-flex packaging. J Electron Mater 44(4):1220–1237
200. Suhir E (2001) Analysis of interfacial thermal stresses in a trimaterial assembly. J Appl Phys 89(7):3685–3694
201. Tsai M, Hsu C, Han C (2004) A note on Suhir's solution of thermal stresses for a die-substrate assembly. J Electron Packag 126(1):115–119
202. Tsai M-Y, Chang H-Y, Pecht M (2009) Warpage analysis of flip-chip PBGA packages subject to thermal loading. IEEE Trans Device Mater Reliab 9(3):419–424
203. Xu Z, Liu Z, Liu H, Yin Z, Huang Y, Chen J (2015) Analytical evaluation of interfacial crack propagation in vacuum-based picking-up process. IEEE Trans Compon Packag Manuf Technol 5(11):1700–1708
204. Yeung B, Lee T-Y (2003) An overview of experimental methodologies and their applications for die strength measurement. IEEE Trans Compon Packag Technol 26(2):423–428
205. Barnat S, Frémont H, Gracia A, Cadalen E (2012) Evaluation by three-point-bend and ball-on-ring tests of thinning process on silicon die strength. Microelectron Reliab 52(9):2278–2282
206. Zhang T-Y, Su Y-J, Qian C-F, Zhao M-H, Chen L-Q (2000) Microbridge testing of silicon nitride thin films deposited on silicon wafers. Acta Mater 48(11):2843–2857
207. Wu J, Huang C, Liao C (2003) Fracture strength characterization and failure analysis of silicon dies. Microelectron Reliab 43(2):269–277

208. Paul I, Majeed B, Razeeb K, Barton J (2006) Statistical fracture modelling of silicon with varying thickness. Acta Mater 54(15):3991–4000
209. Majeed B, Paul I, Razeeb KM, Barton J, O'Mathuna SC (2006) Microstructural, mechanical, fractural and electrical characterization of thinned and singulated silicon test die. J Micromech Microeng 16(8):1519–1529
210. Tsai M, Chen C (2008) Evaluation of test methods for silicon die strength. Microelectron Reliab 48(6):933–941
211. Chae S-H, Zhao J-H, Edwards DR, Ho PS (2010) Effect of dicing technique on the fracture strength of Si dies with emphasis on multimodal failure distribution. IEEE Trans Device Mater Reliab 10(1):149–156
212. Liu Z, Huang Y, Xiao L, Tang P, Yin Z (2015) Nonlinear characteristics in fracture strength test of ultrathin silicon die. Semicond Sci Technol 30(4):045005
213. Thomas KR, Steiner U (2011) Direct stress measurements in thin polymer films. Soft Matter 7(17):7839–7842
214. Ericson F, Schweitz JÅ (1990) Micromechanical fracture strength of silicon. J Appl Phys 68(11):5840–5844
215. Lu S-T, Chen W-H (2010) Reliability and flexibility of ultra-thin chip-on-flex (UTCOF) interconnects with anisotropic conductive adhesive (ACA) joints. IEEE Trans Adv Packag 33(3):702–712
216. Chen L (2010) An integral approach for large deflection cantilever beams. Int J Non-Linear Mech 45(3):301–305
217. Holden J (1972) On the finite deflections of thin beams. Int J Solids Struct 8(8):1051–1055
218. Wang T, Lee S, Zienkiewicz O (1961) A numerical analysis of large deflections of beams. Int J Mech Sci 3(3):219–228
219. Wang C, Lam K, He X, Chucheepsakul S (1997) Large deflections of an end supported beam subjected to a point load. Int J Non-Linear Mech 32(1):63–72
220. Iyengar KSR, Rao SL (1955) Large deflections of simply supported beams. J Franklin Inst 259(6):523–528
221. Conway H (1947) XCIV. The large deflection of simply supported beams. Lond Edinb Dublin Philos Mag J Sci 38(287):905–911
222. Ohtsuki A (1986) An analysis of large deflection in a symmetrical three-point bending of beam. Bull JSME 29(253):1988–1995
223. Mohyeddin A, Fereidoon A (2014) An analytical solution for the large deflection problem of Timoshenko beams under three-point bending. Int J Mech Sci 78:135–139

Chapter 2
Interfacial Modeling of Flexible Multilayer Structures

2.1 Introduction

Advanced packaging is increasingly demanded in consumer electronics and communication devices, such as the three-dimensional chip stacking, system in package, and package on package. There are many challenges for the current electronic assembly technology. One of the major challenges is the capability of the pick and place of thin chips during the assembly process [1–6]. It is of significance to understand fundamentally the effects of thin chips on electronic packaging, such as how the geometric dimensions such as the length and thickness of the chip affect the peeling and assembling in advanced packaging [7–9].

The layout of this chapter is as follows: a mechanical model for the chip-on-substrate structure in the chip peeling-off through needle ejecting was first proposed to obtain the ERR of interfacial peeling. Then an analytical solution strategy for the chip-adhesive-substrate structure was performed to predict the adhesive stresses in adhesively bonded joints under balanced and unbalanced conditions.

2.2 Modeling of Chip-on-Substrate Structure

2.2.1 Mechanical Model

An illustration of the mechanical model of the chip-on-substrate structure is shown in Fig. 2.1. The silicon chip is attached to the top surface of the substrate layer with the center aligned with the origin of the coordinate system, as shown in Fig. 2.1a. Thicknesses of the chip and substrate layers are denoted by h_{chip} and h_{sub}, respectively, and the corresponding lengths are represented by l_{chip} and $l_{\text{sub}} = 2l_{\text{sub-half}}$, respectively. The substrate layer is subjected to clamped–clamped boundary conditions and a concentrated force P_{needle} resulting from the ejector needle. The bending moments, axial loads, and transverse shear loads applied on each segment of the chip-on-

Y. Huang et al., *Modeling and Application of Flexible Electronics Packaging*,
https://doi.org/10.1007/978-981-13-3627-0_2

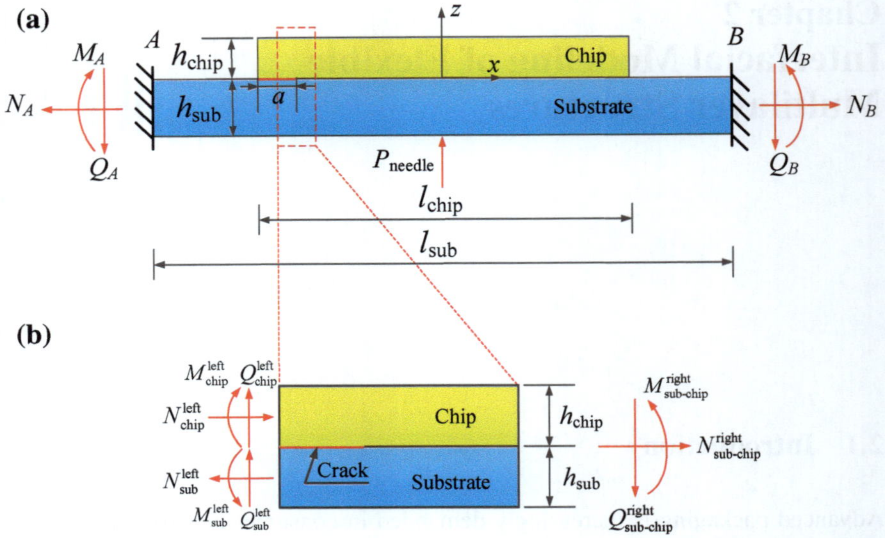

Fig. 2.1 Mechanical model: **a** the chip-on-substrate structure, and **b** the bending moments, axial and transverse loads acting on a segment of the delaminated chip-on-substrate structure at the front and back of the crack tip

substrate structure at the front and back of the crack tip are described in Fig. 2.1b. Both the chip and substrate layers are modeled as beams. The volume forces are neglected here since it is much smaller than other force loadings. Both the chip layer and substrate layer are supposed to be linear elastic and isotropic materials.

The chip-on-substrate structure in Fig. 2.1a is a statically indeterminate system; reaction forces at two ends of the substrate layer are as follows:

$$Q_A = Q_B = P_{\text{needle}}/2, N_A = N_B \approx 0, M_A = M_B = P_{\text{needle}} l_{\text{sub}}/8 \tag{2.1}$$

As depicted in Fig. 2.1b, the loads applied to a geometry consisting of two bonded layers can be generally represented by a set of bending moments ($M^{\text{left}}_{\text{chip}}$, $M^{\text{left}}_{\text{sub}}$, $M^{\text{right}}_{\text{sub-chip}}$), axial forces ($N^{\text{left}}_{\text{chip}}$, $N^{\text{left}}_{\text{sub}}$, $N^{\text{right}}_{\text{sub-chip}}$), and shear forces ($V^{\text{left}}_{\text{chip}}$, $V^{\text{left}}_{\text{sub}}$, $V^{\text{right}}_{\text{sub-chip}}$) acting on the structure at the front and back of the delamination crack tip. By solving the equilibrium equations, the following load configuration can be derived:

$$\begin{aligned} M^{\text{left}}_{\text{chip}} &= 0,\ M^{\text{left}}_{\text{sub}} = P_{\text{needle}}\left(l_{\text{sub-half}} - l_{\text{chip}} + 2a\right)/4, \\ M^{\text{right}}_{\text{sub-chip}} &= -P_{\text{needle}}\left(l_{\text{sub-half}} - l_{\text{chip}} + 2a\right)/4 \end{aligned} \tag{2.2}$$

$$N^{\text{left}}_{\text{chip}} = N^{\text{left}}_{\text{sub}} = N^{\text{right}}_{\text{sub-chip}} = 0 \tag{2.3}$$

$$Q_{\text{chip}}^{\text{left}} = 0,\ Q_{\text{sub}}^{\text{left}} = Q_{\text{sub-chip}}^{\text{right}} = -P_{\text{needle}}/2 \tag{2.4}$$

Under the hypothesis of isotropic and linear elastic deformation, a set of four equivalent loads can be obtained [10]: the bending moment, M_{eq}, the axial load, N_{eq}, the double transverse shear load, Q_{double}, and the single transverse shear load, Q_{single}. The equivalent loads can be represented by

$$M_{\text{eq}} = M_{\text{chip}}^{\text{left}} - C_3 M_{\text{sub-chip}}^{\text{right}} = C_3 P_{\text{needle}}\left(l_{\text{sub-half}} - l_{\text{chip}} + 2a\right)/4 \tag{2.5}$$

$$\begin{aligned} N_{\text{eq}} &= N_{\text{chip}}^{\text{left}} - C_1 N_{\text{sub-chip}}^{\text{right}} - C_2 M_{\text{sub-chip}}^{\text{right}}/h_{\text{chip}} \\ &= C_2 P_{\text{needle}}\left(l_{\text{sub-half}} - l_{\text{chip}} + 2a\right)/4h_{\text{chip}} \end{aligned} \tag{2.6}$$

$$Q_{\text{double}} = V_{\text{chip}}^{\text{left}} - V_{\text{sub-chip}}^{\text{right}} = P_{\text{needle}}/2 \tag{2.7}$$

$$Q_{\text{single}} = V_{\text{sub-chip}}^{\text{right}} = -P_{\text{needle}}/2 \tag{2.8}$$

where $C_1 = \Sigma H/(1+\Sigma H)$, $C_2 = (2 - 2\Delta + H)\Sigma H^2/2\tilde{I}_{(2)}$, $C_3 = \Sigma H^3/12\tilde{I}_{(2)}$, $\Sigma = \overline{E}_{\text{chip}}/\overline{E}_{\text{sub}}$, $H = h_{\text{chip}}/h_{\text{sub}}$, $\tilde{I}_{(2)} = \left\{\Sigma\left[(1 + H - \Delta) - (1 - \Delta)^3\right] + \Delta^3 + (1-\Delta)^3\right\}/3$, $\Delta = \left(1 + 2\Sigma H + \Sigma H^2\right)/2(1 + \Sigma H)$. E_{chip}, ν_{chip}, and h_{chip} are, respectively, the Young's modulus, Poisson's ratio, and thickness of the chip layer. E_{sub}, ν_{sub}, and h_{sub} are, respectively, the Young's modulus, Poisson's ratio, and thickness of the substrate layer. $\overline{E}_{\text{chip}} = E_{\text{chip}}/(1 - \nu_{\text{chip}}^2)$ and $\overline{E}_{\text{sub}} = E_{\text{sub}}/(1 - \nu_{\text{sub}}^2)$ are, respectively, the Young's moduli of the chip and substrate layers under the condition of plane strain.

2.2.2 *Analytical Evaluation on ERR of Interfacial Peeling*

The theory of fracture mechanics is often employed to predict the crack propagation in layered structures. The crack propagation can be predicted by comparing the ERR at the crack tip with the critical value [11–13]. Generally, an interfacial crack occurs involving the components of opening mode and sliding mode. The mode-mixity angle can be used to characterize the relative contribution of these two modes. The critical material parameter associated with the crack propagation is the adhesive fracture energy, which can be generally measured by experimental tests [14]. For layered structure, the adhesive fracture energy depends not only on the material properties but also on the mode-mixity angle [15].

The loads at the front and back of the crack tip are illustrated in Fig. 2.1b, where N's and Q's are loads per unit width, and M's are bending moments per unit width. Basic solutions of the equivalent crack-tip loads were derived by Suo and Hutchinson

[16], Li [10], and Thouless [17]. Subsequently, the total ERR of interfacial peeling and net phase angle can be deduced by superimposing these basic solutions under actual loading conditions [10, 16, 17].

$$\frac{\overline{E}_{\text{chip}}h_{\text{chip}}^3 G_M}{(l_{\text{sub-half}} - l_{\text{chip}} + 2a)^2 P_{\text{needle}}^2} = \frac{C_3^2}{16}\left[f_M(\alpha, H)\right]^2 \tag{2.9}$$

$$\frac{\overline{E}_{\text{chip}}h_{\text{chip}}^3 G_N}{(l_{\text{sub-half}} - l_{\text{chip}} + 2a)^2 P_{\text{needle}}^2} = \frac{C_2^2}{16}\left[f_N(\alpha, H)\right]^2 \tag{2.10}$$

$$\frac{\overline{E}_{\text{chip}}h_{\text{chip}} G_Q^{\text{double}}}{P_{\text{needle}}^2} = \frac{h_{\text{chip}}^2}{4}\left[f_Q^{\text{double}}(\alpha, H)\right]^2 \tag{2.11}$$

$$\frac{\overline{E}_{\text{chip}}h_{\text{chip}} G_Q^{\text{single}}}{P_{\text{needle}}^2} = \frac{h_{\text{chip}}^2}{4} f_Q^{\text{single}}(\alpha, H) \tag{2.12}$$

where G_M, G_N, G_Q^{double}, and G_Q^{single} are ERR components with respect to the bending moment, axial load, double transverse shear load, and single transverse shear load, respectively. Here, f_M, f_N, f_Q^{double}, and f_Q^{single} are relevant nondimensional functions, and $\alpha = (\overline{E}_{\text{chip}} - \overline{E}_{\text{sub}})/(\overline{E}_{\text{chip}} + \overline{E}_{\text{sub}})$ is the first Dundurs' parameter.

By superimposing the components in Eq. (2.9)–(2.12), the total ERR of interfacial peeling can be obtained as

$$G_{\text{crack-tip}} = G_{\text{I}} + G_{\text{II}} \tag{2.13}$$

where G_{I} and G_{II} are, respectively, the contributions of the mode I and mode II cracking to the total ERR of interfacial peeling, and there exist

$$\begin{aligned}\sqrt{\frac{\overline{E}_{\text{chip}}h_{\text{chip}}^3 G_{\text{I}}}{(l_{\text{sub-half}} - l_{\text{chip}} + 2a)^2 P_{\text{needle}}^2}} &= \frac{C_3}{4} f_M(\alpha, H)\cos(\psi_M) + \frac{C_2}{4} f_N(\alpha, H)\cos(\psi_N) \\ &+ \frac{h_{\text{chip}}}{2(l_{\text{sub-half}} - l_{\text{chip}} + 2a)} f_Q^{\text{double}}(\alpha, H)\cos(\psi_Q^{\text{double}}) \\ &+ \frac{h_{\text{chip}}}{2(l_{\text{sub-half}} - l_{\text{chip}} + 2a)} f_Q^{\text{single}}(\alpha, H)\cos(\psi_Q^{\text{single}})\end{aligned} \tag{2.14}$$

$$\sqrt{\frac{\overline{E}_{\text{chip}} h_{\text{chip}}^{3} G_{\text{II}}}{(l_{\text{sub-half}} - l_{\text{chip}} + 2a)^{2} P_{\text{needle}}^{2}}} = \frac{C_3}{4} f_M(\alpha, H)\sin(\psi_M) + \frac{C_2}{4} f_N(\alpha, H)\sin(\psi_N) + \frac{h_{\text{chip}}}{2(l_{\text{sub-half}} - l_{\text{chip}} + 2a)} f_Q^{\text{double}}(\alpha, H)\sin(\psi_Q^{\text{double}}) + \frac{h_{\text{chip}}}{2(l_{\text{sub-half}} - l_{\text{chip}} + 2a)} f_Q^{\text{single}}(\alpha, H)\sin(\psi_Q^{\text{single}}) \tag{2.15}$$

where ψ_M, ψ_N, ψ_Q^{double}, and ψ_Q^{single} are corresponding mode-mixity angles.

2.2.3 Virtual Crack Closure Technique (VCCT)

In the finite element model, the VCCT is applied to estimate the ERR at the midplane of the adhesive layer. The approach is related to the use of a very small virtual crack propagation increment Δa. An existing crack with length a located at the midplane of the adhesive layer is hypothesized and propagated by a small length Δa. Before the virtual crack propagation, nonzero peel stress and shear stress exist at the segment $C'C$ in the adhesive layer, as shown in Fig. 2.2a. The stresses are statically equivalent to the concentrated forces, $N_{\text{crack-tip}}$ and $Q_{\text{crack-tip}}$, and a couple, $M_{\text{crack-tip}}$, applied at the crack-tip, as depicted in Fig. 2.2b. When the virtual crack propagates from the point C to the point C', the previous crack-tip C is assumed to be split into the points A and B, as illustrated in Fig. 2.2c. In order to close the small virtual crack propagation increment, the crack-tip forces and couple have to be applied at the points A and B to move them back to the original locations. The ERR due to a small increase in the crack length is equivalent to the work required to close that small virtual crack propagation increment. The ERR of mode I and mode II cracking can be written as [18]

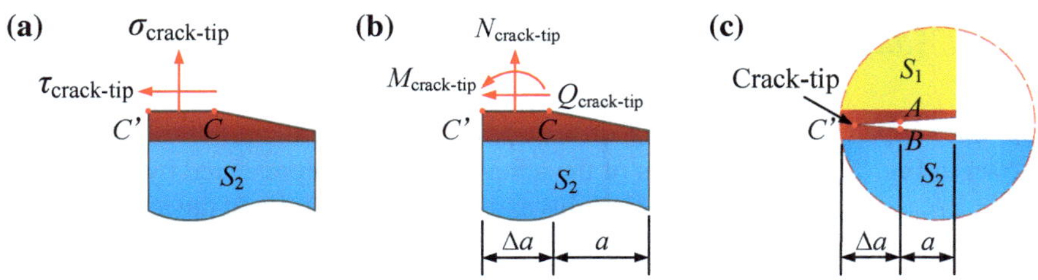

Fig. 2.2 A schematic illustration of the VCCT: **a** peel stress and shear stress at the bottom surface of the crack, **b** equivalent concentrated forces and couple at the crack tip, and **c** the adhesive layer with an initial crack length a and virtual crack propagation length Δa. Reprinted from Ref. [6], Copyright 2013, with permission from Elsevier

$$G_{\mathrm{I}} = \frac{Q_{\text{crack-tip}} \Delta w_{A-B}}{2\Delta a}, \; G_{\mathrm{II}} = \frac{N_{\text{crack-tip}} \Delta u_{A-B}}{2\Delta a} \tag{2.16}$$

where $\Delta u_{A\text{-}B}$ and $\Delta w_{A\text{-}B}$ are relative axial and transverse displacements of the points A and B, respectively.

2.2.4 Numerical Model

To verify the theoretical result of the interfacial peeling, Abaqus software is employed here to calculate the ERR of interfacial peeling using the VCCT with dummy nodes. The fracture interface element is implemented by the user-defined element subroutines. These special elements enable the calculation of ERR in combination with the finite element analysis. More details on the VCCT method with dummy nodes can be found in literatures [3, 5, 11, 19–21]. It should be noted that the crack path is assumed to be embedded in the midplane of the adhesive layer. Considering that the convergence of the ERR is related to the virtual crack propagation length Δa, a very small propagation length is used.

Both the chip and substrate layers are taken to be linear elastic materials. Because of symmetry, only one half of the layered structure depicted in Fig. 2.1a is constructed. The layered structure is considered to deform under plane strain condition. The plane strain element CPE4R is adopted. A concentrated force resulting from the ejector needle is applied at the bottom surface center of the substrate layer. Two boundary conditions are applied to the model, one is fixation at the left side end of the substrate layer, and the other is symmetric about z-axis.

2.3 Modeling of Chip-Adhesive-Substrate Adhesively Bonded Joints

2.3.1 Adhesive Model on Overlapped Joints

The overlapped joints consisting of a chip, a substrate, and an adhesive layer are shown in Fig. 2.3. The thickness of the adhesive layer is represented by h_{adh}, where $h_{\mathrm{adh}} << h_{\mathrm{chip}}, h_{\mathrm{sub}}$. The Young's modulus and the Poisson's ratio of the adhesive layer are represented by E_{adh} and ν_{adh}, respectively. Local coordinate systems are adopted with the origin at the midplanes of the chip and substrate layers. The coordinate x measures the distance in the axial direction, while local coordinates z_1 and z_2 denote the distances in the transverse direction from the midplanes of the chip and substrate layers, respectively. The midplane displacements of the chip layer along the axial and transverse directions are indicated by u_{chip} and w_{chip}, respectively. The midplane displacements of the substrate layer along the axial and transverse directions are

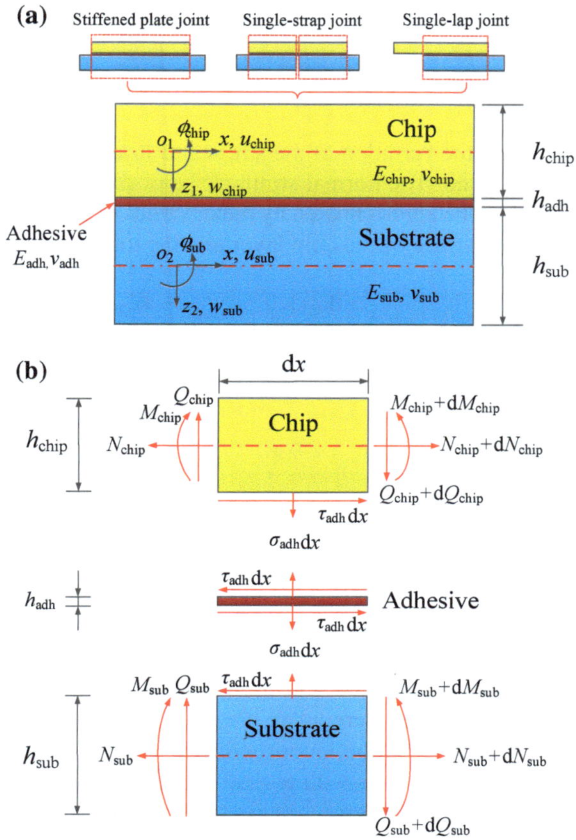

Fig. 2.3 Schematic illustration of overlapped joints: **a** typical adhesively bonded joints and corresponding overlapped region, and **b** free-body diagrams of the chip, substrate, and adhesive layers. Reprinted from Ref. [22], Copyright 2014, with permission from Elsevier

indicated by u_{sub} and w_{sub}, respectively. The cross-section rotations of the chip and substrate layers are, respectively, represented by ϕ_{chip} and ϕ_{sub}, which are assumed to be positive in the counterclockwise direction. The size in y-direction is defined as the unit one. The hypothesis of plane strain is adopted.

According to the theory of Timoshenko beam, the axial and the transverse displacements at a point inside the chip layer can be given as

$$U_{\text{chip}}(x, z_1) = u_{\text{chip}}(x) + z_1\phi_{\text{chip}}(x) \tag{2.17}$$

$$W_{\text{chip}}(x, z_1) = w_{\text{chip}}(x) \tag{2.18}$$

Similarly, the axial and the transverse displacements at a point inside the substrate layer can be given as

$$U_{\text{sub}}(x, z_2) = u_{\text{sub}}(x) + z_2\phi_{\text{sub}}(x) \tag{2.19}$$

$$W_{\text{sub}}(x, z_2) = w_{\text{sub}}(x) \tag{2.20}$$

Here, the variation of the stress and strain in the adhesive layer along z-direction is neglected. The normal strain $\varepsilon_{zz}^{\text{adh}}$ and shear strain γ_{xz}^{adh} at a point in the adhesive layer can be approximated by mean values of the relative displacements between the bottom surface of the chip layer and the top surface of the substrate layer. Hence:

$$\varepsilon_{zz}^{\text{adh}} = \frac{W_{\text{sub}}|_{z_2=-h_{\text{sub-half}}} - W_{\text{chip}}|_{z_1=h_{\text{chip-half}}}}{h_{\text{adh}}} = \frac{w_{\text{sub}} - w_{\text{chip}}}{h_{\text{adh}}} \tag{2.21}$$

$$\gamma_{xz}^{\text{adh}} = \frac{U_{\text{sub}}|_{z_2=-h_{\text{sub-half}}} - U_{\text{chip}}|_{z_1=h_{\text{chip-half}}}}{h_{\text{adh}}} = \frac{u_{\text{sub}} - u_{\text{chip}} - h_{\text{chip}}\phi_{\text{chip}} - h_{\text{sub}}\phi_{\text{sub}}}{h_{\text{adh}}} \tag{2.22}$$

where $h_{\text{chip-half}} = h_{\text{chip}}/2$ and $h_{\text{sub-half}} = h_{\text{sub}}/2$ are half thicknesses of the chip and the substrate layers, respectively.

The adhesive peel stress and shear stress can be obtained using Hooke's law as $\sigma_{\text{adh}} = \overline{E}_{\text{adh}}\varepsilon_{zz}^{\text{adh}}$ and $\tau_{\text{adh}} = \overline{G}_{\text{adh}}\gamma_{xz}^{\text{adh}}$, respectively, where $\overline{E}_{\text{adh}} = E_{\text{adh}}/(1 - \nu_{\text{adh}}^2)$ and $\overline{G}_{\text{adh}} = E_{\text{adh}}/2(1 + \nu_{\text{adh}})$ are the Young's modulus of the adhesive layer under plane strain condition and the shear modulus of the adhesive layer. The adhesive layer is assumed here as an elastic interface with zero thickness, which is treated as uniformly and continuously distributed linear elastic springs acting in the normal and shear directions with respect to the plane of interface. The peel stiffness and the shear stiffness are given, respectively, by $k_\sigma = \overline{E}_{\text{adh}}/h_{\text{adh}}$ and $k_\tau = \overline{G}_{\text{adh}}/h_{\text{adh}}$, and there exist

$$\sigma_{\text{adh}} = k_\sigma\left(w_{\text{sub}} - w_{\text{chip}}\right) \tag{2.23}$$

$$\tau_{\text{adh}} = k_\tau\left(u_{\text{sub}} - u_{\text{chip}} - h_{\text{chip-half}}\phi_{\text{chip}} - h_{\text{sub-half}}\phi_{\text{sub}}\right) \tag{2.24}$$

2.3.2 Equilibrium Equations

The free-body diagrams of the chip, substrate, and adhesive layers are shown in Fig. 2.3b, including the forces and moments as well as the adhesive peel stress and shear stress. Considering the equilibrium condition, following differential equations can be obtained:

$$\frac{\mathrm{d}N_{\text{chip}}}{\mathrm{d}x} + \tau_{\text{adh}} = 0, \quad \frac{\mathrm{d}Q_{\text{chip}}}{\mathrm{d}x} + \sigma_{\text{adh}} = 0, \quad \frac{\mathrm{d}M_{\text{chip}}}{\mathrm{d}x} + h_{\text{chip-half}}\tau_{\text{adh}} - Q_{\text{chip}} = 0 \tag{2.25}$$

$$\frac{dN_{sub}}{dx} - \tau_{adh} = 0, \quad \frac{dQ_{sub}}{dx} - \sigma_{adh} = 0, \quad \frac{dM_{sub}}{dx} + h_{sub\text{-}half}\tau_{adh} - Q_{sub} = 0 \tag{2.26}$$

where N_{chip}, N_{sub}, Q_{chip}, Q_{sub}, M_{chip}, and M_{sub} denote the internal axial force, shear force, and bending moment per unit width of the chip and substrate layers, respectively.

The internal forces can be given by the constitutive laws of Timoshenko beam:

$$N_{chip} = A_{chip}\frac{du_{chip}}{dx}, \; Q_{chip} = C_{chip}\left(\frac{dw_{chip}}{dx} + \phi_{chip}\right), \; M_{chip} = D_{chip}\frac{d\phi_{chip}}{dx} \tag{2.27}$$

$$N_{sub} = A_{sub}\frac{du_{sub}}{dx}, \; Q_{sub} = C_{sub}\left(\frac{dw_{sub}}{dx} + \phi_{sub}\right), \; M_{sub} = D_{sub}\frac{d\phi_{sub}}{dx} \tag{2.28}$$

where $A_{chip} = \overline{E}_{chip}h_{chip}$, $A_{sub} = \overline{E}_{sub}h_{sub}$, $C_{chip} = 2\overline{G}_{chip}k_{shear}h_{chip\text{-}half}$, $C_{sub} = 2\overline{G}_{sub}k_{shear}h_{sub\text{-}half}$, $D_{chip} = \overline{E}_{chip}h_{chip}^3/12$, and $D_{sub} = \overline{E}_{sub}h_{sub}^3/12$ are extensional stiffness, shear stiffness, and bending stiffness of the isotropic chip and substrate layers, respectively. Here, $\overline{G}_{chip} = E_{chip}/2(1+\nu_{chip})$ and $\overline{G}_{sub} = E_{sub}/2(1+\nu_{sub})$ are shear moduli of the chip and substrate layers, respectively. Here, k_{shear} is the shear correction factor, which is assumed to be 5/6 [23]. In addition, the corresponding compliances are defined as $a_{chip} = 1/A_{chip}$, $a_{sub} = 1/A_{sub}$, $c_{chip} = 1/C_{chip}$, $c_{sub} = 1/C_{sub}$, $d_{chip} = 1/D_{chip}$, and $d_{sub} = 1/D_{sub}$.

By substituting Eq. (2.28) into Eqs. (2.25)–(2.26), the following differential governing equations can be derived, which establish the relationship between the stresses of the adhesive layer and the displacements of the chip and substrate layers:

$$\begin{cases} \dfrac{d^2u_{chip}}{dx^2} = -a_{chip}\tau_{adh} \\ \dfrac{d\phi_{chip}}{dx} + \dfrac{d^2w_{chip}}{dx^2} = -c_{chip}\sigma_{adh} \\ \dfrac{d^3\phi_{chip}}{dx^3} = -d_{chip}\sigma_{adh} - d_{chip}h_{chip\text{-}half}\dfrac{d\tau_{adh}}{dx} \end{cases} \tag{2.29}$$

$$\begin{cases} \dfrac{d^2u_{sub}}{dx^2} = a_{sub}\tau_{adh} \\ \dfrac{d\phi_{sub}}{dx} + \dfrac{d^2w_{sub}}{dx^2} = c_{sub}\sigma_{adh} \\ \dfrac{d^3\phi_{sub}}{dx^3} = d_{sub}\sigma_{adh} - d_{sub}h_{sub\text{-}half}\dfrac{d\tau_{adh}}{dx} \end{cases} \tag{2.30}$$

2.3.3 Establishment of Differential Equation Set

As indicated by da Silva et al. [24], it is not straightforward to obtain a closed-form solution for the differential problem formulated by Eqs. (2.25)–(2.26) or Eqs. (2.29)–(2.30). As the model or boundary conditions gets more generalized, the governing equations become increasingly complicated and more computing capacity is demanded to solve the problem. Generally speaking, there are two classes of computer-based solution methods. One way is to directly solve the differential equations by the numerical method [18, 25–27]. The other way is to calculate theoretically the values of selected parameters (integration constants, roots of the characteristic equation, etc.), and derive an analytical expression for the differential problem. The latter is adopted here to calculate the distribution of the adhesive peel stress and shear stress in the bonded region.

Referring to the solution strategy proposed for the delamination of composite structure [28–30], one can differentiate Eqs. (2.23) and (2.24) with respect to x for four and three times, respectively. After substituting into Eqs. (2.29) and (2.30), the following equation sets can be obtained:

$$\frac{\mathrm{d}^4\sigma_{\mathrm{adh}}}{\mathrm{d}x^4} - k_\sigma\left(c_{\mathrm{chip}} + c_{\mathrm{sub}}\right)\frac{\mathrm{d}^2\sigma_{\mathrm{adh}}}{\mathrm{d}x^2} + k_\sigma\left(d_{\mathrm{chip}} + d_{\mathrm{sub}}\right)\sigma_{\mathrm{adh}} + k_\sigma\left(d_{\mathrm{chip}}h_{\text{chip-half}} - d_{\mathrm{sub}}h_{\text{sub-half}}\right)\frac{\mathrm{d}\tau_{\mathrm{adh}}}{\mathrm{d}x} = 0 \tag{2.31}$$

$$\frac{\mathrm{d}^3\tau_{\mathrm{adh}}}{\mathrm{d}x^3} - k_\tau\left(a_{\mathrm{sub}} + a_{\mathrm{chip}} + d_{\mathrm{sub}}h^2_{\text{sub-half}} + d_{\mathrm{chip}}h^2_{\text{chip-half}}\right)\frac{\mathrm{d}\tau_{\mathrm{adh}}}{\mathrm{d}x} - k_\tau\left(d_{\mathrm{chip}}h_{\text{chip-half}} - d_{\mathrm{sub}}h_{\text{sub-half}}\right)\sigma_{\mathrm{adh}} = 0 \tag{2.32}$$

Two cases are considered here to solve the differential problem: $d_{\mathrm{chip}}h_{\mathrm{chip}} = d_{\mathrm{sub}}h_{\mathrm{sub}}$, or $\overline{E}_{\mathrm{chip}}h^2_{\mathrm{chip}} = \overline{E}_{\mathrm{sub}}h^2_{\mathrm{sub}}$, which is termed here as the "balanced" case; $d_{\mathrm{chip}}h_{\mathrm{chip}} \neq d_{\mathrm{sub}}h_{\mathrm{sub}}$, or $\overline{E}_{\mathrm{chip}}h^2_{\mathrm{chip}} \neq \overline{E}_{\mathrm{sub}}h^2_{\mathrm{sub}}$, which is termed here as the "unbalanced" case. The superscripts "B" and "U" will be used in the following analysis to denote the "balanced" and "unbalanced" cases, respectively.

In the balanced case, after uncoupling Eqs. (2.31)–(2.32), one can get

$$\frac{\mathrm{d}^4\sigma^{\mathrm{B}}_{\mathrm{adh}}}{\mathrm{d}x^4} + \eta^{\mathrm{B}}_1\frac{\mathrm{d}^2\sigma^{\mathrm{B}}_{\mathrm{adh}}}{\mathrm{d}x^2} + \eta^{\mathrm{B}}_2\sigma^{\mathrm{B}}_{\mathrm{adh}} = 0 \tag{2.33}$$

$$\frac{\mathrm{d}^3\tau^{\mathrm{B}}_{\mathrm{adh}}}{\mathrm{d}x^3} + \eta^{\mathrm{B}}_3\frac{\mathrm{d}\tau^{\mathrm{B}}_{\mathrm{adh}}}{\mathrm{d}x} = 0 \tag{2.34}$$

where $\eta^{\mathrm{B}}_1 = -k_\sigma\left(c_{\mathrm{chip}} + c_{\mathrm{sub}}\right)$, $\eta^{\mathrm{B}}_2 = k_\sigma\left(d_{\mathrm{chip}} + d_{\mathrm{sub}}\right)$, and $\eta^{\mathrm{B}}_3 = -k_\tau\left(a_{\mathrm{sub}} + a_{\mathrm{chip}} + d_{\mathrm{sub}}h^2_{\text{sub-half}} + d_{\mathrm{chip}}h^2_{\text{chip-half}}\right)$.

The characteristic equation for the adhesive peel stress:

$$\lambda^4 + \eta_1^{\mathrm{B}}\lambda^2 + \eta_2^{\mathrm{B}} = 0 \tag{2.35}$$

The adhesive peel stress and shear stress can be, respectively, obtained by solving Eqs. (2.33) and (2.34) as follows:

$$\sigma_{\mathrm{adh}}^{\mathrm{B}}(x) = \sum_{n=1}^{4} F_n^{\mathrm{B}} \exp\left(\lambda_n^{\mathrm{B}} x\right) \tag{2.36}$$

$$\tau_{\mathrm{adh}}^{\mathrm{B}}(x) = \sum_{n=5}^{6} F_n^{\mathrm{B}} \exp\left(\lambda_n^{\mathrm{B}} x\right) + F_7^{\mathrm{B}} \tag{2.37}$$

where $F_1^{\mathrm{B}}, F_2^{\mathrm{B}}, F_3^{\mathrm{B}}, F_4^{\mathrm{B}}, F_5^{\mathrm{B}}, F_6^{\mathrm{B}}$, and F_7^{B} are integration constants to be determined, $\lambda_1^{\mathrm{B}}, \lambda_2^{\mathrm{B}}, \lambda_3^{\mathrm{B}}$, and λ_4^{B} are roots of Eq. (2.35), $\lambda_5^{\mathrm{B}} = (-\eta_3^{\mathrm{B}})^{0.5}$ and $\lambda_6^{\mathrm{B}} = (-\eta_3^{\mathrm{B}})^{0.5}$ are another two real characteristic roots.

In the unbalanced case, uncoupling of Eqs. (2.31)–(2.32) can be achieved by substituting the result of $\mathrm{d}\tau_{\mathrm{adh}}/\mathrm{d}x$ expressed by σ_{adh} into Eq. (2.32). A sixth-order linear homogeneous differential equation for the adhesive peel stress can then be obtained

$$\frac{\mathrm{d}^6\sigma_{\mathrm{adh}}^{\mathrm{U}}}{\mathrm{d}x_1^6} + \eta_1^{\mathrm{U}}\frac{\mathrm{d}^4\sigma_{\mathrm{adh}}^{\mathrm{U}}}{\mathrm{d}x_1^4} + \eta_2^{\mathrm{U}}\frac{\mathrm{d}^2\sigma_{\mathrm{adh}}^{\mathrm{U}}}{\mathrm{d}x_1^2} + \eta_3^{\mathrm{U}}\sigma_{\mathrm{adh}}^{\mathrm{U}} = 0 \tag{2.38}$$

where $\eta_1^{\mathrm{U}} = -k_\tau\left(a_{\mathrm{chip}} + a_{\mathrm{sub}} + d_{\mathrm{chip}}h_{\mathrm{chip\text{-}half}}^2 + d_{\mathrm{sub}}h_{\mathrm{sub\text{-}half}}^2\right) - k_\sigma\left(c_{\mathrm{chip}} + c_{\mathrm{sub}}\right)$, $\eta_2^{\mathrm{U}} = k_\sigma k_\tau\left(a_{\mathrm{chip}} + a_{\mathrm{sub}} + d_{\mathrm{chip}}h_{\mathrm{chip\text{-}half}}^2 + d_{\mathrm{sub}}h_{\mathrm{sub\text{-}half}}^2\right)\left(c_{\mathrm{chip}} + c_{\mathrm{sub}}\right) + k_\sigma\left(d_{\mathrm{chip}} + d_{\mathrm{sub}}\right)$, $\eta_3^{\mathrm{U}} = -k_\sigma k_\tau\left[\left(a_{\mathrm{chip}} + a_{\mathrm{sub}}\right)\left(d_{\mathrm{chip}} + d_{\mathrm{sub}}\right) + d_{\mathrm{chip}}d_{\mathrm{sub}}\left(h_{\mathrm{chip\text{-}half}} + h_{\mathrm{sub\text{-}half}}\right)^2\right]$.

The characteristic equation for the adhesive peel stress:

$$\lambda^6 + \eta_1^{\mathrm{U}}\lambda^4 + \eta_2^{\mathrm{U}}\lambda^4 + \eta_3^{\mathrm{U}} = 0 \tag{2.39}$$

By treating λ^2 as a whole, one can convert Eq. (2.39) into a cubic equation about λ^2, the root properties of which can be determined easily. When the adhesive layer is relatively thick, the cubic equation has one real root and one pair of conjugate complex roots. When the adhesive layer is very thin, there exist three real roots, as described by Luo and Tong [31].

Similarly, the adhesive shear stress can be described by a seventh-order differential equation as follows:

$$\frac{\mathrm{d}^7\tau_{\mathrm{adh}}^{\mathrm{U}}}{\mathrm{d}x_1^7} + \eta_1^{\mathrm{U}}\frac{\mathrm{d}^5\tau_{\mathrm{adh}}^{\mathrm{U}}}{\mathrm{d}x_1^5} + \eta_2^{\mathrm{U}}\frac{\mathrm{d}^3\tau_{\mathrm{adh}}^{\mathrm{U}}}{\mathrm{d}x_1^3} + \eta_3^{\mathrm{U}}\frac{\mathrm{d}\tau_{\mathrm{adh}}^{\mathrm{U}}}{\mathrm{d}x_1} = 0 \tag{2.40}$$

It is obvious that Eqs. (2.38) and (2.40) are similar to some extent. In fact, in addition to a zero root, there exist the same six roots for both Eqs. (2.38) and (2.40). The general expressions for the adhesive peel stress and shear stress can be written as

$$\sigma_{\mathrm{adh}}^{\mathrm{U}}(x) = \sum_{n=1}^{6} F_n^{\mathrm{U}} \exp\left(\lambda_n^{\mathrm{U}} x\right) \tag{2.41}$$

$$\tau_{\mathrm{adh}}^{\mathrm{U}}(x) = -\frac{1}{d_{\mathrm{chip}} h_{\mathrm{chip\text{-}half}} - d_{\mathrm{sub}} h_{\mathrm{sub\text{-}half}}} \times \left\{ \sum_{n=1}^{6} F_n^{\mathrm{U}} \left[\frac{\left(\lambda_n^{\mathrm{U}}\right)^3}{k_\sigma} - \left(c_{\mathrm{chip}} + c_{\mathrm{sub}}\right)\lambda_n^{\mathrm{U}} + \frac{d_{\mathrm{chip}} + d_{\mathrm{sub}}}{\lambda_n^{\mathrm{U}}} \right] \exp\left(\lambda_n^{\mathrm{U}} x\right) + F_7^{\mathrm{U}} \right\} \tag{2.42}$$

where $F_1^{\mathrm{U}}, F_2^{\mathrm{U}}, F_3^{\mathrm{U}}, F_4^{\mathrm{U}}, F_5^{\mathrm{U}}, F_6^{\mathrm{U}}$, and F_7^{U} are integration constants to be determined, $\lambda_1^{\mathrm{U}}, \lambda_2^{\mathrm{U}}, \lambda_3^{\mathrm{U}}, \lambda_4^{\mathrm{U}}, \lambda_5^{\mathrm{U}}$, and λ_6^{U} are roots of Eq. (2.39).

2.3.4 *Solutions for Internal Forces and Displacements*

For the balanced case, the expressions of the internal forces and bending moments can be deduced by substituting Eqs. (2.36)–(2.37) into Eqs. (2.25)–(2.26) and integrating with respect to x. In turn, by substituting the internal forces into Eqs. (2.27)–(2.28) and integrating with respect to x, the expressions of the displacements and cross-section rotations can also be derived. The related general expressions in the chip and substrate layers can be obtained finally.

The axial forces:

$$N_{\mathrm{chip}}^{\mathrm{B}}(x) = -\left[\sum_{n=5}^{6} F_n^{\mathrm{B}} \frac{1}{\lambda_n^{\mathrm{B}}} \exp\left(\lambda_n^{\mathrm{B}} x\right) + F_7^{\mathrm{B}} x + F_8^{\mathrm{B}}\right] \tag{2.43}$$

$$N_{\mathrm{sub}}^{\mathrm{B}}(x) = \sum_{n=5}^{6} F_n^{\mathrm{B}} \frac{1}{\lambda_n^{\mathrm{B}}} \exp\left(\lambda_n^{\mathrm{B}} x\right) + F_7^{\mathrm{B}} x + F_{11}^{\mathrm{B}} \tag{2.44}$$

The shear forces:

$$Q_{\mathrm{chip}}^{\mathrm{B}}(x) = -\left[\sum_{n=1}^{4} F_n^{\mathrm{B}} \frac{1}{\lambda_n^{\mathrm{B}}} \exp\left(\lambda_n^{\mathrm{B}} x\right) + F_9^{\mathrm{B}}\right] \tag{2.45}$$

$$Q_{\mathrm{sub}}^{\mathrm{B}}(x) = \sum_{n=1}^{4} F_n^{\mathrm{B}} \frac{1}{\lambda_n^{\mathrm{B}}} \exp\left(\lambda_n^{\mathrm{B}} x\right) + F_{12}^{\mathrm{B}} \tag{2.46}$$

The bending moments:

$$M^{\mathrm{B}}_{\mathrm{chip}}(x) = -\left[\begin{array}{l}\sum_{n=1}^{4} F^{\mathrm{B}}_{n}\frac{1}{\left(\lambda^{\mathrm{B}}_{n}\right)^{2}}\exp\left(\lambda^{\mathrm{B}}_{n}x\right) + h_{\mathrm{chip\text{-}half}}\sum_{n=5}^{6} F^{\mathrm{B}}_{n}\frac{1}{\lambda^{\mathrm{B}}_{n}}\exp\left(\lambda^{\mathrm{B}}_{n}x\right)\\ +\left(F^{\mathrm{B}}_{9} + h_{\mathrm{chip\text{-}half}}F^{\mathrm{B}}_{7}\right)x + F^{\mathrm{B}}_{10}\end{array}\right] \tag{2.47}$$

$$\begin{aligned}M^{\mathrm{B}}_{\mathrm{sub}}(x) = {} & \sum_{n=1}^{4} F^{\mathrm{B}}_{n}\frac{1}{\left(\lambda^{\mathrm{B}}_{n}\right)^{2}}\exp\left(\lambda^{\mathrm{B}}_{n}x\right) - h_{\mathrm{sub\text{-}half}}\sum_{n=5}^{6} F^{\mathrm{B}}_{n}\frac{1}{\lambda^{\mathrm{B}}_{n}}\exp\left(\lambda^{\mathrm{B}}_{n}x\right)\\ & + \left(F^{\mathrm{B}}_{12} - h_{\mathrm{sub\text{-}half}}F^{\mathrm{B}}_{7}\right)x + F^{\mathrm{B}}_{13}\end{aligned} \tag{2.48}$$

The midplane axial displacements:

$$u^{\mathrm{B}}_{\mathrm{chip}}(x) = -a_{\mathrm{chip}}\left[\sum_{n=5}^{6} F^{\mathrm{B}}_{n}\frac{1}{\left(\lambda^{\mathrm{B}}_{n}\right)^{2}}\exp\left(\lambda^{\mathrm{B}}_{n}x\right) + \frac{1}{2}F^{\mathrm{B}}_{7}x^{2} + F^{\mathrm{B}}_{8}x + F^{\mathrm{B}}_{14}\right] \tag{2.49}$$

$$u^{\mathrm{B}}_{\mathrm{sub}}(x) = a_{\mathrm{sub}}\left[\sum_{n=5}^{6} F^{\mathrm{B}}_{n}\frac{1}{\left(\lambda^{\mathrm{B}}_{n}\right)^{2}}\exp\left(\lambda^{\mathrm{B}}_{n}x\right) + \frac{1}{2}F^{\mathrm{B}}_{7}x^{2} + F^{\mathrm{B}}_{11}x + F^{\mathrm{B}}_{17}\right] \tag{2.50}$$

The cross-section rotations:

$$\phi^{\mathrm{B}}_{\mathrm{chip}}(x) = -d_{\mathrm{chip}}\left[\begin{array}{l}\sum_{n=1}^{4} F^{\mathrm{B}}_{n}\frac{1}{\left(\lambda^{\mathrm{B}}_{n}\right)^{3}}\exp\left(\lambda^{\mathrm{B}}_{n}x\right) + h_{\mathrm{chip\text{-}half}}\sum_{n=5}^{6} F^{\mathrm{B}}_{n}\frac{1}{\left(\lambda^{\mathrm{B}}_{n}\right)^{2}}\exp\left(\lambda^{\mathrm{B}}_{n}x\right)\\ +\frac{1}{2}\left(F^{\mathrm{B}}_{9} + h_{\mathrm{chip\text{-}half}}F^{\mathrm{B}}_{7}\right)x^{2} + F^{\mathrm{B}}_{10}x + F^{\mathrm{B}}_{15}\end{array}\right] \tag{2.51}$$

$$\phi^{\mathrm{B}}_{\mathrm{sub}}(x) = d_{\mathrm{sub}}\left[\begin{array}{l}\sum_{n=1}^{4} F^{\mathrm{B}}_{n}\frac{1}{\left(\lambda^{\mathrm{B}}_{n}\right)^{3}}\exp\left(\lambda^{\mathrm{B}}_{n}x\right) - h_{\mathrm{sub\text{-}half}}\sum_{n=5}^{6} F^{\mathrm{B}}_{n}\frac{1}{\left(\lambda^{\mathrm{B}}_{n}\right)^{2}}\exp\left(\lambda^{\mathrm{B}}_{n}x\right)\\ +\frac{1}{2}\left(F^{\mathrm{B}}_{12} - h_{\mathrm{sub\text{-}half}}F^{\mathrm{B}}_{7}\right)x^{2} + F^{\mathrm{B}}_{13}x + F^{\mathrm{B}}_{18}\end{array}\right] \tag{2.52}$$

The midplane transverse displacements:

$$\begin{aligned}w^{\mathrm{B}}_{\mathrm{chip}}(x) = {} & \sum_{n=1}^{4} F^{\mathrm{B}}_{n}\left[\frac{d_{\mathrm{chip}}}{\left(\lambda^{\mathrm{B}}_{n}\right)^{4}} - \frac{c_{\mathrm{chip}}}{\left(\lambda^{\mathrm{B}}_{n}\right)^{2}}\right]\exp\left(\lambda^{\mathrm{B}}_{n}x\right)\\ & + d_{\mathrm{chip}}h_{\mathrm{chip\text{-}half}}\sum_{n=5}^{6} F^{\mathrm{B}}_{n}\frac{1}{\left(\lambda^{\mathrm{B}}_{n}\right)^{3}}\exp\left(\lambda^{\mathrm{B}}_{n}x\right)\\ & + \frac{1}{6}\left(d_{\mathrm{chip}}F^{\mathrm{B}}_{9} + d_{\mathrm{chip}}h_{\mathrm{chip\text{-}half}}F^{\mathrm{B}}_{7}\right)x^{3}\\ & + \frac{1}{2}d_{\mathrm{chip}}F^{\mathrm{B}}_{10}x^{2} + \left(d_{\mathrm{chip}}F^{\mathrm{B}}_{15} - c_{\mathrm{chip}}F^{\mathrm{B}}_{9}\right)x + F^{\mathrm{B}}_{16}\end{aligned} \tag{2.53}$$

$$
\begin{aligned}
w_{\text{sub}}^{\mathrm{B}}(x) = {} & \sum_{n=1}^{4} F_n^{\mathrm{B}}\left[\frac{c_{\text{sub}}}{\left(\lambda_n^{\mathrm{B}}\right)^2} - \frac{d_{\text{sub}}}{\left(\lambda_n^{\mathrm{B}}\right)^4}\right]\exp\left(\lambda_n^{\mathrm{B}} x\right) \\
& + d_{\text{sub}} h_{\text{sub-half}} \sum_{n=5}^{6} F_n^{\mathrm{B}} \frac{1}{\left(\lambda_n^{\mathrm{B}}\right)^3}\exp\left(\lambda_n^{\mathrm{B}} x\right) \\
& - \frac{1}{6}\left(d_{\text{sub}} F_{12}^{\mathrm{B}} - d_{\text{sub}} h_{\text{sub-half}} F_7^{\mathrm{B}}\right)x^3 - \frac{1}{2} d_{\text{sub}} F_{13}^{\mathrm{B}} x^2 \\
& + \left(c_{\text{sub}} F_{12}^{\mathrm{B}} - d_{\text{sub}} F_{18}^{\mathrm{B}}\right)x + F_{19}^{\mathrm{B}}
\end{aligned} \tag{2.54}
$$

where $F_8^{\mathrm{B}}, F_9^{\mathrm{B}}, F_{10}^{\mathrm{B}}, F_{11}^{\mathrm{B}}, F_{12}^{\mathrm{B}}, F_{13}^{\mathrm{B}}, F_{14}^{\mathrm{B}}, F_{15}^{\mathrm{B}}, F_{16}^{\mathrm{B}}, F_{17}^{\mathrm{B}}, F_{18}^{\mathrm{B}}$, and F_{19}^{B} are integration constants to be determined.

In the unbalanced case, the analytical solutions for the internal forces, bending moments, displacements, and cross-section rotations in the chip and substrate layers can also be obtained similarly.

The axial forces:

$$
N_{\text{chip}}^{\mathrm{U}}(x) = \frac{\sum_{n=1}^{6} F_n^{\mathrm{U}}\left[\frac{\left(\lambda_n^{\mathrm{U}}\right)^2}{k_\sigma} - \left(c_{\text{chip}} + c_{\text{sub}}\right) + \frac{d_{\text{chip}} + d_{\text{sub}}}{\left(\lambda_n^{\mathrm{U}}\right)^2}\right]\exp\left(\lambda_n^{\mathrm{U}} x\right) + F_7^{\mathrm{U}} x + F_8^{\mathrm{U}}}{d_{\text{chip}} h_{\text{chip-half}} - d_{\text{sub}} h_{\text{sub-half}}} \tag{2.55}
$$

$$
N_{\text{sub}}^{\mathrm{U}}(x) = -\frac{\sum_{n=1}^{6} F_n^{\mathrm{U}}\left[\frac{\left(\lambda_n^{\mathrm{U}}\right)^2}{k_\sigma} - \left(c_{\text{chip}} + c_{\text{sub}}\right) + \frac{d_{\text{chip}} + d_{\text{sub}}}{\left(\lambda_n^{\mathrm{U}}\right)^2}\right]\exp\left(\lambda_n^{\mathrm{U}} x\right) + F_7^{\mathrm{U}} x + F_{11}^{\mathrm{U}}}{d_{\text{chip}} h_{\text{chip-half}} - d_{\text{sub}} h_{\text{sub-half}}} \tag{2.56}
$$

The shear forces:

$$
Q_{\text{chip}}^{\mathrm{U}}(x) = -\left[\sum_{n=1}^{6} F_n^{\mathrm{U}} \frac{1}{\lambda_n^{\mathrm{U}}}\exp\left(\lambda_n^{\mathrm{U}} x\right) + F_9^{\mathrm{U}}\right] \tag{2.57}
$$

$$
Q_{\text{sub}}^{\mathrm{U}}(x) = \sum_{n=1}^{6} F_n^{\mathrm{U}} \frac{1}{\lambda_n^{\mathrm{U}}}\exp\left(\lambda_n^{\mathrm{U}} x\right) + F_{12}^{\mathrm{U}} \tag{2.58}
$$

The bending moments:

$$
\begin{aligned}
M_{\text{chip}}^{\mathrm{U}}(x) = {} & \frac{h_{\text{chip-half}}}{d_{\text{chip}} h_{\text{chip-half}} - d_{\text{sub}} h_{\text{sub-half}}} \\
& \times \left\{\begin{array}{l} \sum_{n=1}^{6} F_n^{\mathrm{U}}\left[\frac{\left(\lambda_n^{\mathrm{U}}\right)^2}{k_\sigma} - \left(c_{\text{chip}} + c_{\text{sub}}\right) + \frac{d_{\text{sub}}\left(h_{\text{chip-half}} + h_{\text{sub-half}}\right)}{h_{\text{chip-half}}\left(\lambda_n^{\mathrm{U}}\right)^2}\right]\exp\left(\lambda_n^{\mathrm{U}} x\right) \\ + \left(F_7^{\mathrm{U}} - \frac{d_{\text{chip}} h_{\text{chip-half}} - d_{\text{sub}} h_{\text{sub-half}}}{h_{\text{chip-half}}} F_9^{\mathrm{U}}\right)x + F_{10}^{\mathrm{U}} \end{array}\right\}
\end{aligned} \tag{2.59}
$$

$$M_{\text{sub}}^{\text{U}}(x) = \frac{h_{\text{sub-half}}}{d_{\text{chip}}h_{\text{chip-half}} - d_{\text{sub}}h_{\text{sub-half}}} \times \left\{ \begin{array}{l} \sum\limits_{n=1}^{6} F_n^{\text{U}} \left[\dfrac{\left(\lambda_n^{\text{U}}\right)^2}{k_\sigma} - \left(c_{\text{chip}} + c_{\text{sub}}\right) + \dfrac{d_{\text{chip}}\left(h_{\text{chip-half}} + h_{\text{sub-half}}\right)}{h_{\text{sub-half}}\left(\lambda_n^{\text{U}}\right)^2} \right] \exp\left(\lambda_n^{\text{U}} x\right) \\ + \left(F_7^{\text{U}} + \dfrac{d_{\text{chip}}h_{\text{chip-half}} - d_{\text{sub}}h_{\text{sub-half}}}{h_{\text{sub-half}}} F_{12}^{\text{U}} \right) x + F_{13}^{\text{U}} \end{array} \right\} \tag{2.60}$$

The midplane axial displacements:

$$u_{\text{chip}}^{\text{U}}(x) = \frac{a_{\text{chip}}}{d_{\text{chip}}h_{\text{chip-half}} - d_{\text{sub}}h_{\text{sub-half}}} \left\{ \begin{array}{l} \sum\limits_{n=1}^{6} F_n^{\text{U}} \left(\begin{array}{l} \dfrac{\lambda_n^{\text{U}}}{k_\sigma} - \dfrac{c_{\text{chip}} + c_{\text{sub}}}{\lambda_n^{\text{U}}} \\ + \dfrac{d_{\text{chip}} + d_{\text{sub}}}{\left(\lambda_n^{\text{U}}\right)^3} \end{array} \right) \exp\left(\lambda_n^{\text{U}} x\right) \\ + \dfrac{1}{2} F_7^{\text{U}} x^2 + F_8^{\text{U}} x + F_{14}^{\text{U}} \end{array} \right\} \tag{2.61}$$

$$u_{\text{sub}}^{\text{U}}(x) = -\frac{a_{\text{sub}}}{d_{\text{chip}}h_{\text{chip-half}} - d_{\text{sub}}h_{\text{sub-half}}} \left\{ \begin{array}{l} \sum\limits_{n=1}^{6} F_n^{\text{U}} \left(\begin{array}{l} \dfrac{\lambda_n^{\text{U}}}{k_\sigma} - \dfrac{c_{\text{chip}} + c_{\text{sub}}}{\lambda_n^{\text{U}}} \\ + \dfrac{d_{\text{chip}} + d_{\text{sub}}}{\left(\lambda_n^{\text{U}}\right)^3} \end{array} \right) \exp\left(\lambda_n^{\text{U}} x\right) \\ + \dfrac{1}{2} F_7^{\text{U}} x^2 + F_{11}^{\text{U}} x + F_{17}^{\text{U}} \end{array} \right\} \tag{2.62}$$

The cross-section rotations:

$$\phi_{\text{chip}}^{\text{U}}(x) = \frac{d_{\text{chip}}h_{\text{chip-half}}}{d_{\text{chip}}h_{\text{chip-half}} - d_{\text{sub}}h_{\text{sub-half}}} \times \left\{ \begin{array}{l} \sum\limits_{n=1}^{6} F_n^{\text{U}} \left[\dfrac{\lambda_n^{\text{U}}}{k_\sigma} - \dfrac{c_{\text{chip}} + c_{\text{sub}}}{\lambda_n^{\text{U}}} + \dfrac{d_{\text{sub}}\left(h_{\text{chip-half}} + h_{\text{sub-half}}\right)}{h_{\text{chip-half}}\left(\lambda_n^{\text{U}}\right)^3} \right] \exp\left(\lambda_n^{\text{U}} x\right) \\ + \dfrac{1}{2} \left(F_7^{\text{U}} - \dfrac{d_{\text{chip}}h_{\text{chip-half}} - d_{\text{sub}}h_{\text{sub-half}}}{h_{\text{chip-half}}} F_9^{\text{U}} \right) x^2 + F_{10}^{\text{U}} x + F_{15}^{\text{U}} \end{array} \right\} \tag{2.63}$$

$$\phi_{\mathrm{sub}}^{\mathrm{U}}(x) = \frac{d_{\mathrm{sub}} h_{\mathrm{sub\text{-}half}}}{d_{\mathrm{chip}} h_{\mathrm{chip\text{-}half}} - d_{\mathrm{sub}} h_{\mathrm{sub\text{-}half}}} \times \left\{ \begin{array}{l} \sum_{n=1}^{6} F_n^{\mathrm{U}} \left[\frac{\lambda_n^{\mathrm{U}}}{k_\sigma} - \frac{c_{\mathrm{chip}} + c_{\mathrm{sub}}}{\lambda_n^{\mathrm{U}}} + \frac{d_{\mathrm{chip}} \left(h_{\mathrm{chip\text{-}half}} + h_{\mathrm{sub\text{-}half}} \right)}{h_{\mathrm{sub\text{-}half}} \left(\lambda_n^{\mathrm{U}} \right)^3} \right] \exp\left(\lambda_n^{\mathrm{U}} x \right) \\ + \frac{1}{2} \left(F_7^{\mathrm{U}} + \frac{d_{\mathrm{chip}} h_{\mathrm{chip\text{-}half}} - d_{\mathrm{sub}} h_{\mathrm{sub\text{-}half}}}{h_{\mathrm{sub\text{-}half}}} F_{12}^{\mathrm{U}} \right) x^2 + F_{13}^{\mathrm{U}} x + F_{18}^{\mathrm{U}} \end{array} \right\} \tag{2.64}$$

The midplane transverse displacements:

$$w_{\mathrm{chip}}^{\mathrm{U}}(x) = -\frac{1}{d_{\mathrm{chip}} h_{\mathrm{chip\text{-}half}} - d_{\mathrm{sub}} h_{\mathrm{sub\text{-}half}}} \times \left\{ \begin{array}{l} \sum_{n=1}^{6} F_n^{\mathrm{U}} \left[\begin{array}{l} \frac{d_{\mathrm{chip}} h_{\mathrm{chip\text{-}half}}}{k_\sigma} - \frac{d_{\mathrm{sub}} h_{\mathrm{sub\text{-}half}} c_{\mathrm{chip}} + d_{\mathrm{chip}} h_{\mathrm{chip\text{-}half}} c_{\mathrm{sub}}}{\left(\lambda_n^{\mathrm{U}} \right)^2} \\ + \frac{d_{\mathrm{chip}} d_{\mathrm{sub}} \left(h_{\mathrm{chip\text{-}half}} + h_{\mathrm{sub\text{-}half}} \right)}{\left(\lambda_n^{\mathrm{U}} \right)^4} \end{array} \right] \exp\left(\lambda_n^{\mathrm{U}} x \right) \\ + \frac{1}{6} \left[d_{\mathrm{chip}} h_{\mathrm{chip\text{-}half}} F_7^{\mathrm{U}} - d_{\mathrm{chip}} F_9^{\mathrm{U}} \left(d_{\mathrm{chip}} h_{\mathrm{chip\text{-}half}} - d_{\mathrm{sub}} h_{\mathrm{sub\text{-}half}} \right) \right] x^3 \\ + \frac{1}{2} d_{\mathrm{chip}} h_{\mathrm{chip\text{-}half}} F_{10}^{\mathrm{U}} x^2 + \left[\begin{array}{l} d_{\mathrm{chip}} h_{\mathrm{chip\text{-}half}} F_{15}^{\mathrm{U}} \\ \quad + c_{\mathrm{chip}} F_9^{\mathrm{U}} \left(d_{\mathrm{chip}} h_{\mathrm{chip\text{-}half}} - d_{\mathrm{sub}} h_{\mathrm{sub\text{-}half}} \right) \end{array} \right] x + F_{16}^{\mathrm{U}} \end{array} \right\} \tag{2.65}$$

$$w_{\mathrm{sub}}^{\mathrm{U}}(x) = -\frac{1}{d_{\mathrm{chip}} h_{\mathrm{chip\text{-}half}} - d_{\mathrm{sub}} h_{\mathrm{sub\text{-}half}}} \times \left\{ \begin{array}{l} \sum_{n=1}^{6} F_n^{\mathrm{U}} \left[\begin{array}{l} \frac{d_{\mathrm{sub}} h_{\mathrm{sub\text{-}half}}}{k_\sigma} - \frac{d_{\mathrm{sub}} h_{\mathrm{sub\text{-}half}} c_{\mathrm{chip}} + d_{\mathrm{chip}} h_{\mathrm{chip\text{-}half}} c_{\mathrm{sub}}}{\left(\lambda_n^{\mathrm{U}} \right)^2} \\ + \frac{d_{\mathrm{chip}} d_{\mathrm{sub}} \left(h_{\mathrm{chip\text{-}half}} + h_{\mathrm{sub\text{-}half}} \right)}{\left(\lambda_n^{\mathrm{U}} \right)^4} \end{array} \right] \exp\left(\lambda_n^{\mathrm{U}} x \right) \\ + \frac{1}{6} \left[d_{\mathrm{sub}} h_{\mathrm{sub\text{-}half}} F_7^{\mathrm{U}} + d_{\mathrm{sub}} F_{12}^{\mathrm{U}} \left(d_{\mathrm{chip}} h_{\mathrm{chip\text{-}half}} - d_{\mathrm{sub}} h_{\mathrm{sub\text{-}half}} \right) \right] x^3 \\ + \frac{1}{2} d_{\mathrm{sub}} h_{\mathrm{sub\text{-}half}} F_{13}^{\mathrm{U}} x^2 + \left[\begin{array}{l} d_{\mathrm{sub}} h_{\mathrm{sub\text{-}half}} F_{18}^{\mathrm{U}} \\ \quad - c_{\mathrm{sub}} F_{12}^{\mathrm{U}} \left(d_{\mathrm{chip}} h_{\mathrm{chip\text{-}half}} - d_{\mathrm{sub}} h_{\mathrm{sub\text{-}half}} \right) \end{array} \right] x + F_{19}^{\mathrm{U}} \end{array} \right\} \tag{2.66}$$

where $F_8^{\mathrm{U}}, F_9^{\mathrm{U}}, F_{10}^{\mathrm{U}}, F_{11}^{\mathrm{U}}, F_{12}^{\mathrm{U}}, F_{13}^{\mathrm{U}}, F_{14}^{\mathrm{U}}, F_{15}^{\mathrm{U}}, F_{16}^{\mathrm{U}}, F_{17}^{\mathrm{U}}, F_{18}^{\mathrm{U}}$, and F_{19}^{U} are integration constants to be determined.

2.3.5 *Relationships Among Integration Constants*

The complete solution to the differential problem depends on the 19 integration constants for both the balanced and unbalanced cases. However, the integration constants are not all independent of each other. Actually, by introducing the expressions for the adhesive stresses and displacements into Eqs. (2.23)–(2.24), seven relationships can be established among the integration constants.

In the balanced case, there exist

$$d_{\text{chip}} F_9^{\text{B}} + d_{\text{sub}} F_{12}^{\text{B}} = 0 \tag{2.67}$$

$$d_{\text{chip}} F_{10}^{\text{B}} + d_{\text{sub}} F_{13}^{\text{B}} = 0 \tag{2.68}$$

$$c_{\text{chip}} F_9^{\text{B}} + c_{\text{sub}} F_{12}^{\text{B}} - d_{\text{chip}} F_{15}^{\text{B}} - d_{\text{sub}} F_{18}^{\text{B}} = 0 \tag{2.69}$$

$$\begin{aligned} & d_{\text{chip}} h_{\text{chip-half}} F_9^{\text{B}} - d_{\text{sub}} h_{\text{sub-half}} F_{12}^{\text{B}} + \left(a_{\text{chip}} + a_{\text{sub}} + d_{\text{chip}} h_{\text{chip-half}}^2 \right. \\ & \quad \left. + d_{\text{sub}} h_{\text{sub-half}}^2 \right) F_7^{\text{B}} = 0 \end{aligned} \tag{2.70}$$

$$a_{\text{chip}} F_8^{\text{B}} + a_{\text{sub}} F_{11}^{\text{B}} - d_{\text{chip}} h_{\text{chip-half}} F_{10}^{\text{B}} - d_{\text{sub}} h_{\text{sub-half}} F_{13}^{\text{B}} = 0 \tag{2.71}$$

$$a_{\text{chip}} F_{14}^{\text{B}} + a_{\text{sub}} F_{17}^{\text{B}} - d_{\text{chip}} h_{\text{chip-half}} F_{15}^{\text{B}} - d_{\text{sub}} h_{\text{sub-half}} F_{18}^{\text{B}} - F_7^{\text{B}} / k_\tau = 0 \tag{2.72}$$

$$F_{16}^{\text{B}} - F_{19}^{\text{B}} = 0 \tag{2.73}$$

In the unbalanced case, there exist

$$F_7^{\text{U}} - \left(d_{\text{chip}} F_9^{\text{U}} + d_{\text{sub}} F_{12}^{\text{U}}\right) = 0 \tag{2.74}$$

$$d_{\text{chip}} h_{\text{chip-half}} F_{10}^{\text{U}} - d_{\text{sub}} h_{\text{sub-half}} F_{13}^{\text{U}} = 0 \tag{2.75}$$

$$\begin{aligned} & \left(d_{\text{chip}} h_{\text{chip-half}} - d_{\text{sub}} h_{\text{sub-half}}\right)\left(c_{\text{chip}} F_9^{\text{U}} + c_{\text{sub}} F_{12}^{\text{U}}\right) \\ & \quad + \left(d_{\text{chip}} h_{\text{chip-half}} F_{15}^{\text{U}} - d_{\text{sub}} h_{\text{sub-half}} F_{18}^{\text{U}}\right) = 0 \end{aligned} \tag{2.76}$$

$$\begin{aligned} & \left(a_{\text{chip}} + a_{\text{sub}} + d_{\text{chip}} h_{\text{chip-half}}^2 + d_{\text{sub}} h_{\text{sub-half}}^2\right) F_7^{\text{U}} \\ & \quad - \left(d_{\text{chip}} h_{\text{chip-half}} - d_{\text{sub}} h_{\text{sub-half}}\right)\left(d_{\text{chip}} h_{\text{chip-half}} F_9^{\text{U}} - d_{\text{sub}} h_{\text{sub-half}} F_{12}^{\text{U}}\right) = 0 \end{aligned} \tag{2.77}$$

$$a_{\text{chip}} F_8^{\text{U}} + a_{\text{sub}} F_{11}^{\text{U}} + d_{\text{chip}} h_{\text{chip-half}}^2 F_{10}^{\text{U}} + d_{\text{sub}} h_{\text{sub-half}}^2 F_{13}^{\text{U}} = 0 \tag{2.78}$$

$$a_{\text{chip}} F_{14}^{\text{U}} + a_{\text{sub}} F_{17}^{\text{U}} + d_{\text{chip}} h_{\text{chip-half}}^2 F_{15}^{\text{U}} + d_{\text{sub}} h_{\text{sub-half}}^2 F_{18}^{\text{U}} - F_7^{\text{U}} / k_\tau = 0 \tag{2.79}$$

$$F_{16}^{\text{U}} - F_{19}^{\text{U}} = 0 \tag{2.80}$$

There remain 12 independent integration constants to be determined. By imposing suitable boundary conditions, these integration constants can be obtained by solving a linear equation set. It should be pointed out that not all of the integration constants can be determined due to the possible existence of rigid-body displacements of the whole structure. In any case, the main attention is paid to the first seven integration constants included in the expressions for the adhesive peel stress and shear stress. Detailed determination of the integration constants will be given in following chapters.

2.3.6 Theoretical Calculation of ERR of Interfacial Peeling

Based on the above analytical model, G_{I} and G_{II} can be approximated as [32, 33]

$$G_{\mathrm{I}} = \frac{\sigma^2_{\text{crack-tip}}}{2k_\sigma},\ G_{\mathrm{II}} = \frac{\tau^2_{\text{crack-tip}}}{2k_\tau} \tag{2.81}$$

where $\sigma_{\text{crack-tip}}$ and $\tau_{\text{crack-tip}}$ represent the adhesive peel stress and shear stress at the crack tip, respectively. It is worth noting that finite values can be obtained by Eq. (2.81) even without crack initiation. The analytical results predicted by Eq. (2.81) can also be compared with those obtained by the VCCT for the validation purpose. Besides, the mode-mixity angle characterizing the contribution of mode I and II cracking can be defined as $\psi = \arctan[(G_{\mathrm{II}}/G_{\mathrm{I}})^{0.5}]$, where $\psi = 0°$ and $90°$ correspond to the pure mode I and pure mode II cracking, respectively.

2.4 Summary

In this chapter, the mechanical models of the chip-on-substrate and the chip-adhesive-substrate structures were constructed. For the chip-on-substrate structure subjected to the transversely concentrated load through needle ejecting, the interfacial peeling was modeled analytically and numerically from the fracture mechanics point of view. For the chip-adhesive-substrate structure, the very thin adhesive layer was modeled as linear elastic springs, and the generalized adhesive stresses in different adhesively bonded joints under both balanced and unbalanced conditions were given theoretically.

The current solution strategy can be applied for the analysis of adhesively bonded joints with relatively small adhesive layer thickness. The approach is meaningful in many practical applications and can help to shed light on the underlying mechanical mechanism, which can be considered as a unified engineering tool for the design, structural optimization, and evaluation of adhesively bonded joints. The analytical solution proposed here can be conveniently extended to a variety of bonded structures and layered structures, such as the electronic packaging and flexible electronics structures, where it is technically desired to predict the adhesive stresses and ERR to understand the mechanism of damage evolution and failure. The proposed method will be applied in the following chapters to derive design rules for the peeling-off process in advanced IC packaging.

References

1. Chong ACM, Cheung YM (2003) Finite element stress analysis of thin die detachment process. In: Proceedings of fifth international conference on electronic packaging technology, 28–30 Oct 2003. pp 44–51
2. Feil M, Adler C, Hemmetzberger D, Konig M, Bock K (2004) The challenge of ultra thin chip assembly. In: 54th electronic components and technology conference (IEEE Cat. No. 04CH37546), 1–4 June 2004. pp 35–40
3. Peng B, Huang Y, Yin Z, Xiong Y (2011) Analysis of interfacial peeling in IC chip pick-up process. J Appl Phys 110(7):073508
4. Peng B, Huang Y, Yin Z, Xiong Y (2011) On the analysis of dynamic effect in the die pick-up process. In: 12th international conference on electronic packaging technology and high density packaging. IEEE, 8–11 Aug 2011. pp 1–4
5. Peng B, Huang Y, Yin Z, Xiong Y (2012) Competing fracture modeling of thin chip pick-up process. IEEE Trans Compon Packag Manuf Technol 2(7):1217–1225
6. Liu Z, Valvo PS, Huang Y, Yin Z (2013) Cohesive failure analysis of an array of IC chips bonded to a stretched substrate. Int J Solids Struct 50(22–23):3528–3538
7. Xu Z, Liu Z, Huang Y, Chen J, Liu H, Yin Z (2015) Vacuum-based picking-up of thin chip from adhesive tape. J Adhes Sci Technol 29(13):1315–1329
8. Xu Z, Liu Z, Liu H, Yin Z, Huang Y, Chen J (2015) Analytical evaluation of interfacial crack propagation in vacuum-based picking-up process. IEEE Trans Compon Packag Manuf Technol 5(11):1700–1708
9. Chen J, Xu Z, Huang Y, Duan Y, Yin Z (2016) Analytical investigation on thermal-induced warpage behavior of ultrathin chip-on-flex (UTCOF) assembly. Science China Technological Sciences 59(11):1646–1655
10. Li S, Wang J, Thouless M (2004) The effects of shear on delamination in layered materials. J Mech Phys Solids 52(1):193–214
11. Liu Z, Huang Y, Chen J, Yin Z (2014) Tunable peeling technique and mechanism of thin chip from compliant adhesive tapes. IEEE Trans Compon Packag Manuf Technol 4(4):560–568
12. Huang Y, Liu H, Xu Z, Chen J, Yin Z (2018) Conformal peeling of device-on-substrate system in flexible electronic assembly. IEEE Trans Compon Packag Manuf Technol 8(8):1496–1506
13. Liu Z, Wan X, Huang Y, Chen J, Yin Z (2018) Theoretical and experimental studies of competing fracture for flexible chip-adhesive-substrate composite structure. IEEE Trans Compon Packag Manuf Technol 8(1):57–64
14. Liu Z, Tang P, Huang Y, Yin Z (2014) Experimental estimation of adhesive fracture energy of compliant adhesive tape. In: 15th International Conference on Electronic Packaging Technology, 12–15 Aug 2014. pp 842–846
15. Hutchinson JW, Suo Z (1991) Mixed mode cracking in layered materials. In: Hutchinson JW, Wu TY (eds) Advances in applied mechanics, vol 29. Elsevier, pp 63–191. https://doi.org/10.1016/s0065-2156(08)70164-9
16. Suo Z, Hutchinson JW (1990) Interface crack between two elastic layers. Int J Fract 43(1):1–18
17. Thouless M (2009) The effects of transverse shear on the delamination of edge-notch flexure and 3-point bend geometries. Compos B Eng 40(4):305–312
18. Yang C, Chadegani A, Tomblin JS (2008) Strain energy release rate determination of prescribed cracks in adhesively-bonded single-lap composite joints with thick bondlines. Compos B Eng 39(5):863–873
19. Xie D, Biggers SB (2006) Progressive crack growth analysis using interface element based on the virtual crack closure technique. Finite Elem Anal Des 42(11):977–984
20. Xie D, Biggers SB (2007) Calculation of transient strain energy release rates under impact loading based on the virtual crack closure technique. Int J Impact Eng 34(6):1047–1060
21. Liu Z, Huang YA, Liu H, Chen J, Yin Z (2014) Reliable peeling of ultrathin die with multineedle ejector. IEEE Trans Compon Packag Manuf Technol 4(9):1545–1554

22. Liu Z, Huang Y, Yin Z, Bennati S, Valvo PS (2014) A general solution for the two-dimensional stress analysis of balanced and unbalanced adhesively bonded joints. Int J Adhes Adhes 54:112–123
23. Gruttmann F, Wagner W (2001) Shear correction factors in Timoshenko's beam theory for arbitrary shaped cross-sections. Comput Mech 27(3):199–207
24. da Silva LF, das Neves PJ, Adams R, Spelt J (2009) Analytical models of adhesively bonded joints—Part I: Literature survey. Int J Adhes Adhes 29 (3):319–330
25. Yang C, Pang S-S (1996) Stress-strain analysis of single-lap composite joints under tension. J Eng Mater Technol 118(2):247–255
26. Chadegani A, Batra RC (2011) Analysis of adhesive-bonded single-lap joint with an interfacial crack and a void. Int J Adhes Adhes 31(6):455–465
27. Chadegani A, Yang C, Smeltzer SS (2012) Adhesive-bonded composite joints analysis with delaminated surface ply using strain-energy release rate. Journal of Aircraft 49(2):503–520
28. Allix O, Ladevèze P (1992) Interlaminar interface modelling for the prediction of delamination. Compos Struct 22(4):235–242
29. Corigliano A (1993) Formulation, identification and use of interface models in the numerical analysis of composite delamination. Int J Solids Struct 30(20):2779–2811
30. Tay T (2003) Characterization and analysis of delamination fracture in composites: an overview of developments from 1990 to 2001. Appl Mech Rev 56(1):1–32
31. Luo Q, Tong L (2009) Energy release rates for interlaminar delamination in laminates considering transverse shear effects. Compos Struct 89(2):235–244
32. Edde F, Verreman Y (1992) On the fracture parameters in a clamped cracked lap shear adhesive joint. Int J Adhes Adhes 12(1):43–48
33. Krenk S (1992) Energy release rate of symmetric adhesive joints. Eng Fract Mech 43(4):549–559

Chapter 3
Measurement of Fracture Strength of Ultra-thin Silicon Chip and Adhesive Fracture Energy

3.1 Introduction

Experimental approaches like the ball-on-ring test [1, 2], three/four-point bending test [3–10], and cantilever beam method [11, 12] are generally adopted to estimate the fracture strength of silicon chip [13], among which the three-point bending test is more preferred for thin silicon devices due to the advantage in aspects of simplicity and reliability. In the traditional three-point bending test, the bending theory of linear elastic beam can give a good approximation of the fracture strength under the condition of small deflection. Attentions were mostly focused on thick silicon chips (>100 μm-thick) [4, 6], in which the experimental load-deflection curve remained a linear response. However, the response may behave nonlinearly for ultra-thin samples. The traditional strategy of fracture strength estimation may induce extra errors. In recent years, the nonlinear behavior in estimation of the fracture strength of ultra-thin chip has been investigated by researchers like Schoenfelder et al. [7], Huang and Tsai [14], and Endler et al. [15]. The nonlinear load-deflection relationship was established through nonlinear finite element analysis. However, related theoretical explanations and experimental descriptions were seldom reported. Thus, it is essential to derive a nonlinear solution of the three-point bending test to get a more precise estimation of the fracture strength.

In addition, from the viewpoint of fracture mechanics, the chip peeling-off from the adhesive tape can be treated as the crack propagation along the adhesive tape [16, 17], which is determined by Griffith fracture criterion, namely, $G_{\text{adh}}^{\text{crack-tip}} \geq G_{\text{adh}}^{c}$ [18–20], in which $G_{\text{adh}}^{\text{crack-tip}}$ denotes the ERR of interfacial peeling at the crack-tip, and the G_{adh}^{c} indicates the adhesive fracture energy, namely, the minimum energy required to peel a chip off the adhesive tape. Investigations have found that both the peel angle and peel speed play important roles in the chip peeling-off from the adhesive tape [21]. An improved peeling apparatus with automatically controlled peel angle/speed is desired to investigate the underlying peeling mechanism.

Y. Huang et al., *Modeling and Application of Flexible Electronics Packaging*,
https://doi.org/10.1007/978-981-13-3627-0_3

The chapter is outlined as follows: The fracture strength of ultra-thin silicon chip samples was estimated through the three-point bending test incorporating the geometric nonlinearity. A novel angle/speed controlled peeling apparatus was designed for the estimation of the adhesive fracture energy.

3.2 Fracture Strength of Ultra-thin Silicon Chip

3.2.1 Experimental Procedure

The present work starts with the specimen preparation, where the ultra-thin silicon chip samples with different thicknesses are fabricated using a novel dicing before thinning technique [1, 22]. Both the thinning and separation of an ultra-thin wafer are still a challenging task, especially for a wafer thinner than 50 μm. Here, the dicing before thinning is adopted to obtain such thin silicon sheet, and the process can be summarized as follows: The predesigned dicing grooves with a desired thickness on the active side of the wafer are first prepared using a diamond saw as depicted in Fig. 3.1a. The diced active side is bonded to a protective wafer, as seen in Fig. 3.1b, and then thinned from the opposite side until the chip grooves are opened, as shown in Fig. 3.1c, d, respectively. The thinning process started with a mechanical grinding to remove the bulk of silicon and followed by a special fine grinding, such as chemical mechanical polishing, to attain the target thickness. The desired chip samples can be detached from the protective wafer with water bath heating method, as described in Fig. 3.1e. The specimens are then cleaned with acetone to remove contamination like the residual adhesive before experimental tests, as shown in Fig. 3.1f.

As stated in literatures [1, 6, 10], the advantages of the dicing before thinning process exist in minimizing the edge crack and dicing marks as much as possible, so that the corresponding effect on the chip strength can be reduced. A good kerf and cross section are, respectively, shown in Fig. 3.2a, b, where the edges are kept neat with very little crack and marks. In addition, a good surface state with an average roughness of 1.47 nm can be detected in Fig. 3.2c, which benefits from the fine chemical mechanical polishing for a long time. The fake silicon chip fabricated possesses excellent mechanical properties, especially in the aspect of the bending flexibility, as depicted in Fig. 3.2d.

The testing chips prepared from a 4-in. silicon wafer are in the size of 5 mm × 10 mm with different thicknesses of 30, 40, 50, and 100 μm. The traditional three-point bending test is performed via the XYZTEC Condor classic-70 multifunction tester with a specialized fixture to control over the span length. The experimental setup with a close-up view of the three-point bending fixture and silicon chip is shown in Fig. 3.3. All samples are cleaned with acetone before experimental tests. To avoid contamination when handling the silicon chip, the clean rubber gloves are used during

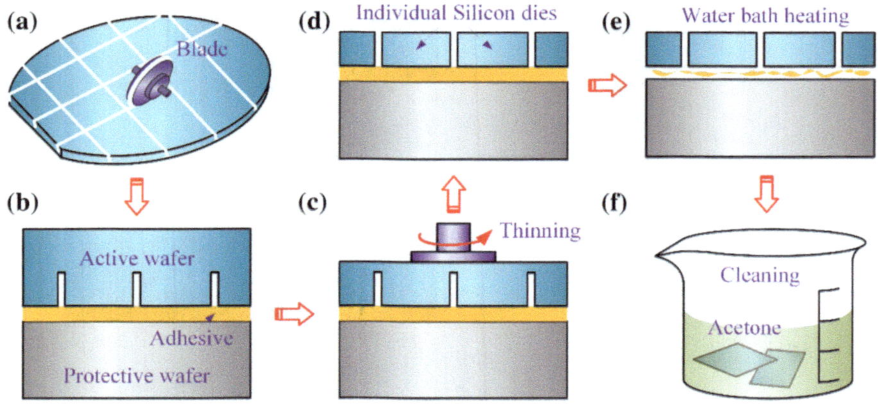

Fig. 3.1 Fabrication process of the ultra-thin chips with dicing before thinning technique: **a** dicing with a diamond composite blade, **b** bonded to a protective wafer, **c** thinning, **d** individual silicon chips, **e** detaching the silicon chips from protective wafer using the water bath heating method, and **f** cleaning the silicon chips with acetone.

the whole experiment. Each specimen is carefully placed in the fixture with a tweezer. The whole test process is automatically accomplished and the corresponding testing data can be read conveniently.

3.2.2 Nonlinear Mechanical Characterization in Bending Test

3.2.2.1 Three-Point Bending Test

A schematic illustration of the three-point bending test using a flexible specimen supported on two frictionless supports with the surface radius r_{sup} is depicted in Fig. 3.4. The initial span length, width, and thickness of the specimen are, respectively, represented by $2l_{\text{TPB-half}}$, b_{TPB}, and h_{TPB}, as shown in Fig. 3.4a. When the specimen deflects, it can slide freely between supports from the initial support points A and B to new points A' and B' with a new span length $2l'_{\text{TPB-half}}$, as shown in Fig. 3.4b. An illustration of the geometric deformation is given in Fig. 3.4c. If the point o is defined as the origin of the coordinate system, the horizontal distance from the origin, the transverse displacement, and the deflection angle of an arbitrary point D on the specimen are represented by x_{TPB}, w_{TPB}, and θ_{TPB}, respectively. The vertical load P_{TPB} leads to the formation of the reaction force F_{TPB} and deflection angle $\theta_{0\text{-TPB}}$, and there exists $F_{\text{TPB}} = P_{\text{TPB}}/[2\cos(\theta_{0\text{-TPB}})]$. Generally, the relationship between the bending moment M_{TPB} and curvature $1/\rho_{\text{TPB}}$ can be written as $1/\rho_{\text{TPB}} = M_{\text{TPB}}/E_{\text{TPB}}I_{\text{TPB}}$

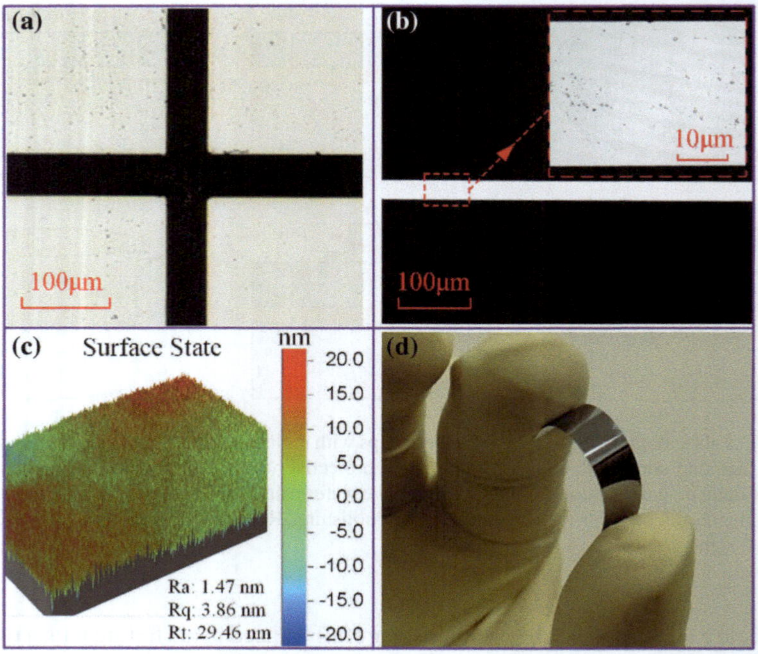

Fig. 3.2 The kerf and cross section of the ultra-thin silicon specimen fabricated by the dicing before thinning technique: **a** the photograph of a good kerf in the completed wafer, **b** the smooth cross section, **c** the good surface quality, and **d** an ultra-thin silicon specimen with the thickness of 30 μm exhibiting excellent flexibility.

in an elastic range, where $E_{\text{TPB}} I_{\text{TPB}}$ is the flexural stiffness. In addition, the curvature can also be expressed as $1/\rho_{\text{TPB}} = (\mathrm{d}^2 w_{\text{TPB}}/\mathrm{d}x_{\text{TPB}}^2)/\left[1 + (\mathrm{d}w_{\text{TPB}}/\mathrm{d}x_{\text{TPB}})^2\right]^{3/2}$. Considering the symmetric condition, the bending moment can be calculated by

$$M_{\text{TPB}} = \frac{P_{\text{TPB}}}{2}[x_{\text{TPB}} + w_{\text{TPB}} \tan(\theta_{0\text{-TPB}})], \quad 0 \le x_{\text{TPB}} \le l'_{\text{TPB}} \tag{3.1}$$

It should be noted that due to the effect of large deflection, the slippage of the specimen and change of support points cannot be neglected. Then the actual deflection w_{absolute} and span length $l'_{\text{TPB-half}}$ can be modified as $w_{\text{absolute}} = w_{\text{TPB}} + \Delta w_{\text{TPB}}$ and $l'_{\text{TPB-half}} = l_{\text{TPB-half}} - \Delta l_{\text{TPB-half}}$, where Δw_{TPB} and $\Delta l_{\text{TPB-half}}$ can be calculated by the following equations $\Delta w_{\text{TPB}} = \left(r_{\text{sup}} + h_{\text{TPB}}/2\right)[1 - \cos(\theta_{0\text{-TPB}})]$ and $\Delta l_{\text{TPB-half}} = \left(r_{\text{sup}} + h_{\text{TPB}}/2\right) \sin(\theta_{0\text{-TPB}})$.

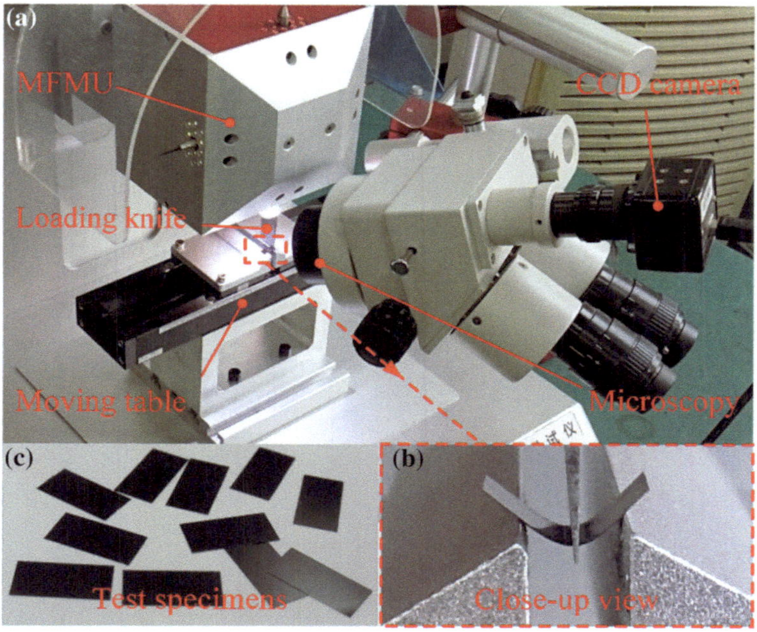

Fig. 3.3 The experimental setup: **a** load–deflection testing configuration with a multifunction measurement unit and a real-time digital microscopy device, and **b** a close-up view of the test fixture and silicon chip, **c** part of the test specimens.

3.2.2.2 Classical Linear Solution

Based on the small deflection theory, it can be approximated that $(\mathrm{d}w_{\mathrm{TPB}}/\mathrm{d}x_{\mathrm{TPB}})^2 \ll 1$ and the reaction force F_{TPB} is vertical. The curvature and bending moment can thus be written as $1/\rho_{\mathrm{TPB}} = \mathrm{d}^2 w_{\mathrm{TPB}}/\mathrm{d}x_{\mathrm{TPB}}^2$ and $M_{\mathrm{TPB}} = P_{\mathrm{TPB}} x_{\mathrm{TPB}}/2$. Therefore, the differential equation $\mathrm{d}^2 w_{\mathrm{TPB}}/\mathrm{d}x_{\mathrm{TPB}}^2 = P_{\mathrm{TPB}} x_{\mathrm{TPB}}/(2E_{\mathrm{TPB}} I_{\mathrm{TPB}})$ in the linear theory can be obtained. The deflection angle, nondimensional maximum deflection, and bending stress can be expressed as follows:

$$\theta_{0\text{-TPB}} = \lambda_{\mathrm{TPB}}, \quad \delta_{\max\text{-TPB}} = \frac{2}{3}\lambda_{\mathrm{TPB}}, \quad \sigma_{\max\text{-TPB}} = \left(\frac{E_{\mathrm{TPB}} h_{\mathrm{TPB}}}{l_{\text{TPB-half}}}\right)\lambda_{\mathrm{TPB}} \tag{3.2}$$

where $\lambda_{\mathrm{TPB}} = P_{\mathrm{TPB}} l_{\mathrm{TPB}}^2/(4E_{\mathrm{TPB}} I_{\mathrm{TPB}})$ and $\delta_{\mathrm{TPB}} = w_{\mathrm{TPB}}/l_{\text{TPB-half}}$ are, respectively, defined as the deflection parameter and nondimensional deflection.

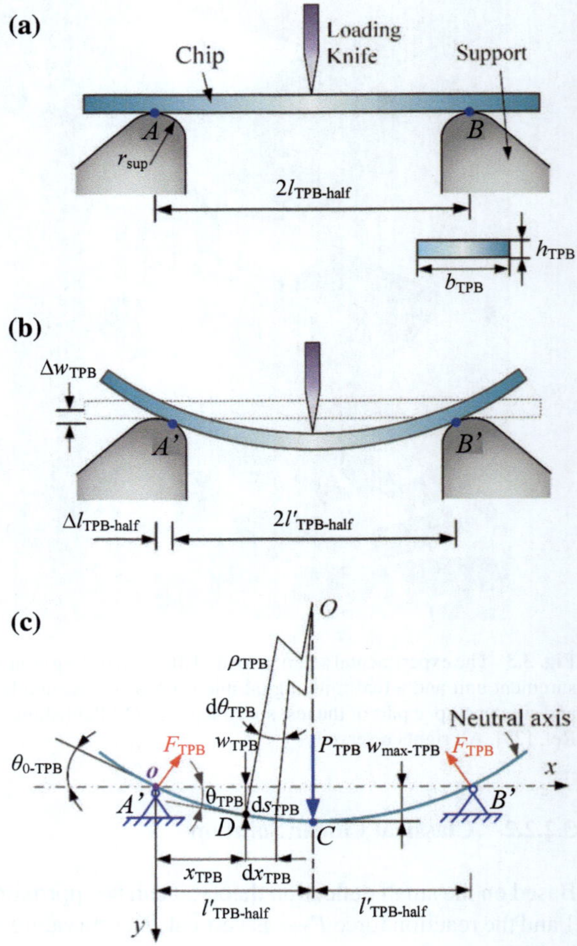

Fig. 3.4 Schematic illustration of the three-point bending test: **a** initial state, **b** bending state, and **c** description of large deflection in a symmetric three-point bending test. © IOP Publishing. Reproduced with permission from Ref. [23]. All rights reserved

3.2.2.3 Nonlinear Solution Based on Elliptic Integrals

As to the large deflection of ultra-thin silicon chips, the linear solution of Eq. (3.2) is no longer applicable to the strength evaluation. Therefore, the geometric nonlinearity should be considered. To analytically solve the nonlinear differential equation governing large deflection, it is promising to introduce the elliptic integrals [24, 25]. If the arc length of curved specimen is represented by s_{TPB}, the constitutive geometric relationship can be given by

$$\frac{d\theta_{TPB}}{ds_{TPB}} = -\frac{M_{TPB}}{E_{TPB}I_{TPB}}, \quad \frac{dx_{TPB}}{ds_{TPB}} = \cos(\theta_{TPB}), \quad \frac{dw_{TPB}}{ds_{TPB}} = \sin(\theta_{TPB}) \tag{3.3}$$

In view of the equilibrium equation of Eq. (3.1), one can rewrite Eq. (3.3) as

$$\frac{\mathrm{d}\theta_{\mathrm{TPB}}}{\mathrm{d}s_{\mathrm{TPB}}} = -\sqrt{\frac{P_{\mathrm{TPB}}\sin(\theta_{0\text{-TPB}} - \theta_{\mathrm{TPB}})}{E_{\mathrm{TPB}}I_{\mathrm{TPB}}\cos(\theta_{0\text{-TPB}})} + C} \tag{3.4}$$

where C is an integral constant. The curvature at support ends is zero, namely, $\theta_{\mathrm{A}'} = \theta_{0\text{-TPB}}$ and $\mathrm{d}\theta_{\mathrm{TPB}}/\mathrm{d}s_{\mathrm{TPB}} = 0$, which confirms that the constant $C = 0$. To simplify the derivation, here $\chi_{\mathrm{TPB}} = [E_{\mathrm{TPB}}I_{\mathrm{TPB}}\cos(\theta_{0\text{-TPB}})/P_{\mathrm{TPB}}]^{0.5}$ and $\sin(\varphi_{\mathrm{TPB}}) = [\sin(\theta_{0\text{-TPB}} - \theta_{\mathrm{TPB}})]^{0.5}$ are defined, where φ_{TPB} is varied from $\varphi_{\mathrm{A}'\text{-TPB}} = 0$ to $\varphi_{\mathrm{C\text{-}TPB}} = \arcsin\{[\sin(\theta_{0\text{-TPB}})]^{0.5}\}$ with the deflection angle varied from $\theta_{\mathrm{A}'\text{-TPB}} = \theta_{0\text{-TPB}}$ to $\theta_{\mathrm{C\text{-}TPB}} = 0$. Then, the differential relationship between θ_{TPB} and φ_{TPB} can be determined as $\mathrm{d}\theta_{\mathrm{TPB}} = -2\sin(\varphi_{\mathrm{TPB}})/\left[1+\sin^2(\varphi_{\mathrm{TPB}})\right]^{0.5}\mathrm{d}\varphi_{\mathrm{TPB}}$. If it is denoted that $\phi(\varphi_{\mathrm{TPB}}) = \left[1+\sin^2(\varphi_{\mathrm{TPB}})\right]^{0.5} - 1/[\sin(\theta_{0\text{-TPB}} - \theta_{\mathrm{TPB}})]^{0.5}$, one can rewrite Eqs. (3.3) and (3.4) as

$$\mathrm{d}s_{\mathrm{TPB}} = 2\chi_{\mathrm{TPB}}/\sqrt{1+\sin^2(\varphi_{\mathrm{TPB}})}\mathrm{d}\varphi_{\mathrm{TPB}} \tag{3.5}$$

$$\mathrm{d}x_{\mathrm{TPB}} = 2\chi_{\mathrm{TPB}}\cos(\theta_{0\text{-TPB}})[\cos(\varphi_{\mathrm{TPB}}) + \tan(\theta_{0\text{-TPB}})\phi(\varphi_{\mathrm{TPB}})]\mathrm{d}\varphi_{\mathrm{TPB}} \tag{3.6}$$

$$\mathrm{d}w_{\mathrm{TPB}} = 2\chi_{\mathrm{TPB}}\cos(\theta_{0\text{-TPB}})[\tan(\theta_{0\text{-TPB}})\cos(\varphi_{\mathrm{TPB}}) - \phi(\varphi_{\mathrm{TPB}})]\mathrm{d}\varphi_{\mathrm{TPB}} \tag{3.7}$$

The arc length, horizontal distance, and deflection of an arbitrary point D can be determined by applying Eqs. (3.8)–(3.9), which are, respectively, the incomplete elliptic integrals of the first kind and the second kind, into Eqs. (3.5)–(3.7) from $\varphi_{\mathrm{A\text{-}TPB}} = 0$ to $\varphi_{\mathrm{D\text{-}TPB}}$.

$$F(\Phi, m) = \int_0^{\Phi} 1/\sqrt{1+m\sin^2(\varphi_{\mathrm{TPB}})}\mathrm{d}\varphi_{\mathrm{TPB}} \tag{3.8}$$

$$E(\Phi, m) = \int_0^{\Phi} \sqrt{1+m\sin^2(\varphi_{\mathrm{TPB}})}\mathrm{d}\varphi_{\mathrm{TPB}} \tag{3.9}$$

In particular, for the midspan C, $\varphi_{\mathrm{C\text{-}TPB}} = \arcsin[\sin(\theta_{0\text{-TPB}})]^{0.5}\}$, the half span of the specimen $l'_{\mathrm{TPB\text{-}half}} = l_{\mathrm{TPB\text{-}half}} - \Delta l_{\mathrm{TPB\text{-}half}}$ can be obtained by the integral operation through Eq. (3.10),

$$\int_0^{l_{\mathrm{TPB\text{-}half}} - \Delta l_{\mathrm{TPB\text{-}half}}} \mathrm{d}x_{\mathrm{TPB}} = 2\chi_{\mathrm{TPB}}\cos(\theta_{0\text{-TPB}}) \int_0^{\varphi_{\mathrm{C\text{-}TPB}}} \begin{bmatrix} \cos(\varphi_{\mathrm{TPB}}) \\ + \tan(\theta_{0\text{-TPB}})\phi(\varphi_{\mathrm{TPB}}) \end{bmatrix} \mathrm{d}\varphi_{\mathrm{TPB}} \tag{3.10}$$

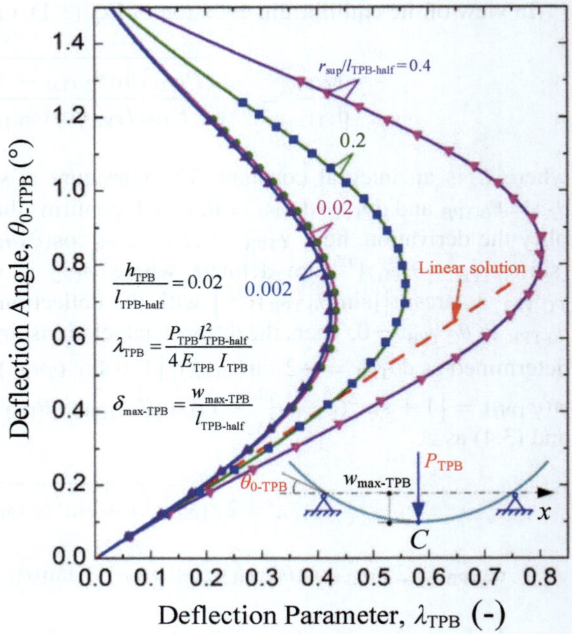

Fig. 3.5 The deflection angle $\theta_{0\text{-TPB}}$ versus the deflection parameter λ_{TPB} under conditions of $r_{sup}/l_{TPB\text{-half}} = 0.002$, 0.02, 0.2, and 0.4, where the analytical predictions and finite element results are represented by solid lines and scatter plots, respectively. © IOP Publishing. Reproduced with permission from Ref. [23]. All rights reserved

Finally, the nondimensional result can be described as follows:

$$\left(1 - \frac{\Delta l_{TPB\text{-half}}}{l_{TPB\text{-half}}}\right)^2 \lambda_{TPB} = \cos^3(\theta_{0\text{-TPB}})[\sin(\varphi_{C\text{-TPB}}) + \tan(\theta_{0\text{-TPB}})\phi(\varphi_{C\text{-TPB}})]^2 \tag{3.11}$$

where $\sin(\varphi_{C\text{-TPB}}) = [\sin(\theta_{0\text{-TPB}})]^{0.5}$, $\phi(\varphi_{C\text{-TPB}}) = E(\varphi_{C\text{-TPB}}, 1) - F(\varphi_{C\text{-TPB}}, 1)$, and $\Delta l_{TPB\text{-half}}/l_{TPB\text{-half}} = \left[r_{sup}/l_{TPB\text{-half}} + h_{TPB}/(2l_{TPB\text{-half}})\right]\sin(\theta_{0\text{-TPB}})$.

It can be implied from Eq. (3.11) that the initial deflection angle $\theta_{0\text{-TPB}}$ is a function about λ_{TPB}, $r_{sup}/l_{TPB\text{-half}}$, and $h_{TPB}/l_{TPB\text{-half}}$, namely, $\theta_{0\text{-TPB}} = f\left(\lambda_{TPB}, r_{sup}/l_{TPB\text{-half}}, h_{TPB}/l_{TPB\text{-half}}\right)$. However, the function is not simply a linear relationship as that in Eq. (3.2). $r_{sup}/l_{TPB\text{-half}}$ can be designed before the experimental test, but $h_{TPB}/l_{TPB\text{-half}}$ is generally determined by the sample itself. Here, attentions are focused on varying the ratio $r_{sup}/l_{TPB\text{-half}}$ while maintaining a constant $h_{TPB}/l_{TPB\text{-half}}$, e.g., $h_{TPB}/l_{TPB\text{-half}} = 0.02$. The result of $\theta_{0\text{-TPB}}$ versus λ_{TPB} shows a high consistency between analytical and finite element predictions, and exhibits significant nonlinearity, as depicted in Fig. 3.5. It is shown that with the increase of $\theta_{0\text{-TPB}}$, λ_{TPB} can attain a maximum value, and then decreases as $\theta_{0\text{-TPB}}$ further increases. When $\theta_{0\text{-TPB}}$ is at the maximum level, the corresponding λ_{TPB} of nonlinear solution is greater than that of linear solution. Furthermore, the nonlinear results are insensitive to $r_{sup}/l_{TPB\text{-half}}$ when $r_{sup}/l_{TPB\text{-half}} < 2\%$.

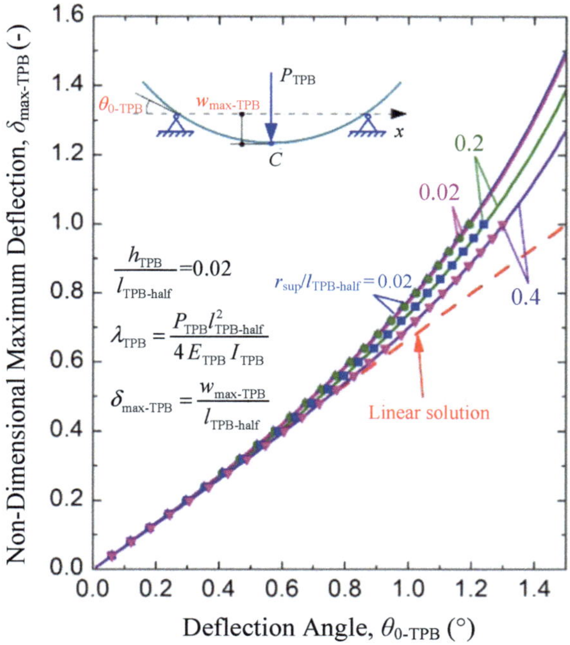

Fig. 3.6 The nondimensional maximum deflection versus the deflection angle under conditions of $r_{\text{sup}}/l_{\text{TPB-half}} =$ 0.002, 0.02, 0.2, and 0.4, where the analytical predictions and finite element results are represented by solid lines and scatter plots, respectively.

The nondimensional maximum deflection $\delta_{\text{max-TPB}} = (w_{\text{max-TPB}} + \Delta w_{\text{TPB}})/l_{\text{TPB-half}}$ at the midspan C can be given by

$$\delta_{\text{max-TPB}} = \left(1 - \frac{\Delta l_{\text{TPB-half}}}{l_{\text{TPB-half}}}\right) \frac{\tan(\theta_{\text{0-TPB}})\sin(\varphi_{\text{C-TPB}}) - \phi(\varphi_{\text{C-TPB}})}{\sin(\varphi_{\text{C-TPB}}) + \tan(\theta_{\text{0-TPB}})\phi(\varphi_{\text{C-TPB}})} + \frac{\Delta w_{\text{TPB}}}{l_{\text{TPB-half}}} \tag{3.12}$$

where $w_{\text{max-TPB}}$ is the maximum value of deflection, $\Delta w_{\text{TPB}}/l_{\text{TPB-half}} = \left(r_{\text{sup}}/l_{\text{TPB-half}} + h_{\text{TPB}}/2l_{\text{TPB-half}}\right)[1 - \cos(\theta_{\text{0-TPB}})]$, and $\delta_{\text{max-TPB}} = g\left(\theta_{\text{0-TPB}}, r_{\text{sup}}/l_{\text{TPB-half}}, h_{\text{TPB}}/l_{\text{TPB-half}}\right)$.

The nondimensional maximum deflection versus the deflection angle under conditions of $r_{\text{sup}}/l_{\text{TPB-half}} = 0.002$, 0.02, 0.2, and 0.4 is shown in Fig. 3.6. The prediction of Eq. (3.12) tends to change curvilinearly but the linear solution still keeps a linear trend. A good agreement can be found between the finite element simulation and nonlinear solution. Based on the results of Eqs. (3.11) and (3.12), it can be concluded that when $r_{\text{sup}}/l_{\text{TPB-half}}$ and $h_{\text{TPB}}/l_{\text{TPB-half}}$ are both small enough, e.g., $r_{\text{sup}}/l_{\text{TPB-half}} < 2\%$ and $h_{\text{TPB}}/l_{\text{TPB-half}} < 2\%$, the effect of nonlinearity on the analytical results can be neglected. The engineering precision of formulae regardless of both $r_{\text{sup}}/l_{\text{TPB-half}}$ and $h_{\text{TPB}}/l_{\text{TPB-half}}$ is adequate to meet the experimental requirement as long as they are small enough.

3.2.2.4 Discussion of the Experiment

Combining both Eqs. (3.11) and (3.12), the normalized load-deflection relationship $\delta_{\text{max-TPB}} = \delta(\lambda_{\text{TPB}})$ can be obtained. To further ascertain the nonlinear analytical predictions, a series of tests are conducted 30, 40, 50, and 100 μm-thick silicon chips, separately. There are 26 samples prepared for each group. The corresponding experimental procedure has been described in the previous section. In the bending test, the loading force is applied on backside of the chip at a speed of about 0.6 mm/min.

Average values of experimental results of unit width load-deflection relationships and corresponding analytical predictions for 30, 40, 50, and 100 μm-thick silicon chips are shown in Fig. 3.7. The nonlinear characteristics for samples thinner than 50 μm can be clearly observed from Fig. 3.7a. A linear trend is detected for thick ones. It can be indicated by the curve slope that higher flexibility exists for the thinner chip. Obviously, the nonlinearity can no longer be neglected when the test specimen is changed from normal to ultra-thin condition. The classical linear solution is insufficient to deal with such a change. The nonlinear predictions presented here match experimental results very well.

The nondimensional maximum deflection $\delta_{\text{max-TPB}}$ versus the deflection parameter λ_{TPB} is shown in Fig. 3.7b. It can be clearly seen that the nonlinearity increases with the decrease in the chip thickness. The thicker chip corresponds to a smaller $\delta_{\text{max-TPB}}$. Furthermore, the deflection parameter λ_{TPB} reaches the limit value of about 0.431 when $\delta_{\text{max-TPB}} \approx 0.484$, which implies that the maximum vertical reaction force is attained. The specimen begins to slip between support ends if the deflection further increases. Small deviations between experimental and analytical results can be found, especially for 30 μm-thick samples. The experimental error can hardly be avoided although each step is carefully operated including the measurement of load span, the fixation of load and specimen, etc. It should also be pointed out that the abscissa errors are inevitably magnified several times when transforming the dimensional load P_{TPB} into the nondimensional parameter λ_{TPB}. The nonlinear solution strategy presented here can be used to investigate the physical fundamental of the three-point bending test of an ultra-thin chip.

3.2.3 Estimation of Fracture Strength

The maximum bending stress generally occurs at the bottom surface center of the silicon chip, which can be obtained by substituting Eq. (3.4) into $\sigma_{\text{max-TPB}} = (E_{\text{TPB}} w_{\text{TPB}} / \rho_{\text{TPB}})\big|_{\theta_{\text{TPB}}=0} = [E_{\text{TPB}} h_{\text{TPB}} (\mathrm{d}\theta_{\text{TPB}} / \mathrm{d}s_{\text{TPB}})]\big|_{\theta_{\text{TPB}}=0}$, and finally:

$$\sigma_{\text{max-TPB}} = \left(\frac{E_{\text{TPB}} h_{\text{TPB}}}{l_{\text{TPB-half}}}\right)\sqrt{\lambda_{\text{TPB}} \tan(\theta_{0\text{-TPB}})} \tag{3.13}$$

Recalling the linear results involved in Eq. (3.2), the deflection angle $\theta_{0\text{-TPB}}$ is very small. It can thus be approximated that $\theta_{0\text{-TPB}} = \tan(\theta_{0\text{-TPB}}) = \lambda_{\text{TPB}}$. The linear

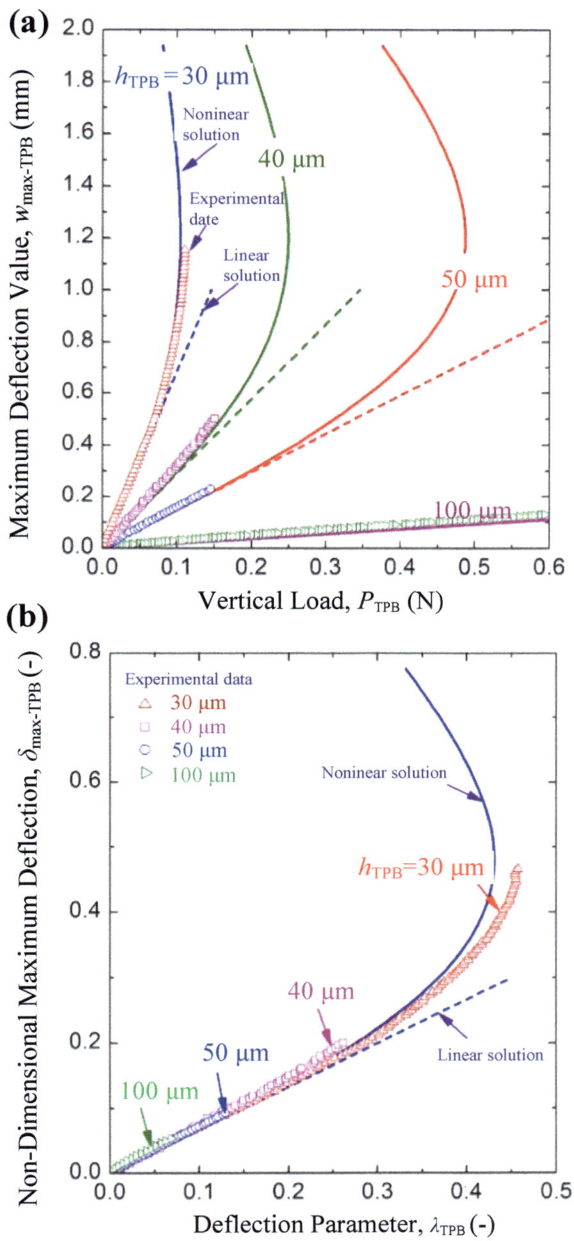

Fig. 3.7 A comparison of the theoretical prediction and experimental result on the relationship between the deflection and load under different thickness conditions: **a** original results and **b** nondimensional results involving both the deflection parameter λ_{TPB} and nondimensional deflection $\delta_{max\text{-}TPB}$. Nonlinear analytical results, linear estimations, and experimental results are represented by solid lines, dash lines, and scatter plots, respectively. © IOP Publishing. Reproduced with permission from Ref. [23]. All rights reserved

result can then be included in the nonlinear solution of Eq. (3.13) in this way. In order to apply the nonlinear evaluation more conveniently into the practical test, here, a correction factor η_{TPB} is defined after compared with the linear solution:

$$\eta_{\text{TPB}} = \sqrt{\frac{\tan(\theta_{0\text{-TPB}})}{\lambda_{\text{TPB}}}} \tag{3.14}$$

To validate the analytical estimation of the fracture strength, the nonlinear finite element analysis is first implemented, from which the data of both the loading force P_{TPB} and the maximum bending stress $\sigma_{\text{max-TPB}}$ versus the corresponding deflection value $w_{\text{max-TPB}}$ can be obtained. The simulated P_{TPB} and $\sigma_{\text{max-TPB}}$ are then substituted into Eqs. (3.2) and (3.13) for a theoretical prediction of the linear and nonlinear bending stress, respectively. The corresponding results are shown in Fig. 3.8. It can be found that the maximum bending stresses obtained from the nonlinear solution and finite element analysis both gradually get larger with the increase in $\delta_{\text{max-TPB}}$ and match each other very well, while those from the linear theory tend to be parabolic with a peak appeared at $\delta_{\text{max-TPB}} \approx 0.484$ coinciding with those in Fig. 3.7b. The correction factor η_{TPB} versus $\delta_{\text{max-TPB}}$ is shown in Fig. 3.8b. η_{TPB} represents the scale factor converting the bending stress derived from the linear theory into the corresponding nonlinear prediction. $\eta_{\text{TPB}} \approx 1.0$ when $\delta_{\text{max-TPB}} < 0.1$, under which condition the linear relationship between the bending stress and the deflection maintains very well. The $\eta_{\text{TPB}} \approx 1.4$ when $\delta_{\text{max-TPB}} = 0.5$, which means that the corresponding bending stress predicted by the nonlinear model is 140% of that by the linear estimation.

In general, the brittle material shows a large scattering behavior in the distribution of the fracture strength. It is therefore difficult to determine the fracture strength simply by several sample values. The statistical evaluation is used more frequently to derive reliable fracture strength from the experimental data. Here, the Weibull theory is introduced to statistically analyze the fracture strength of ultra-thin silicon chip. The typical two-parameter Weibull distribution [5, 10] is plotted in Fig. 3.9. Corresponding Weibull parameters determined from the experimental data, including the Weibull modulus and characteristic strength, are, respectively, 2.53, 3.41, 4.51, 6.25, and 440 MPa, 621 MPa, 802 MPa, 1433 MPa for silicon chips with thicknesses of 100 μm, 50 μm, 40 μm, 30 μm. Weibull modulus is in an increasing trend with the decrease in sample thickness. A higher value of Weibull modulus indicates a lower strength dispersion, which shows that the fracture mechanism of silicon chip transits from brittle to ductile. It can also be found that the thinner specimen possesses a higher characteristic fracture strength. It may be implied that a high stress can be withstood for an ultra-thin chip during the device packaging and subsequent application [26].

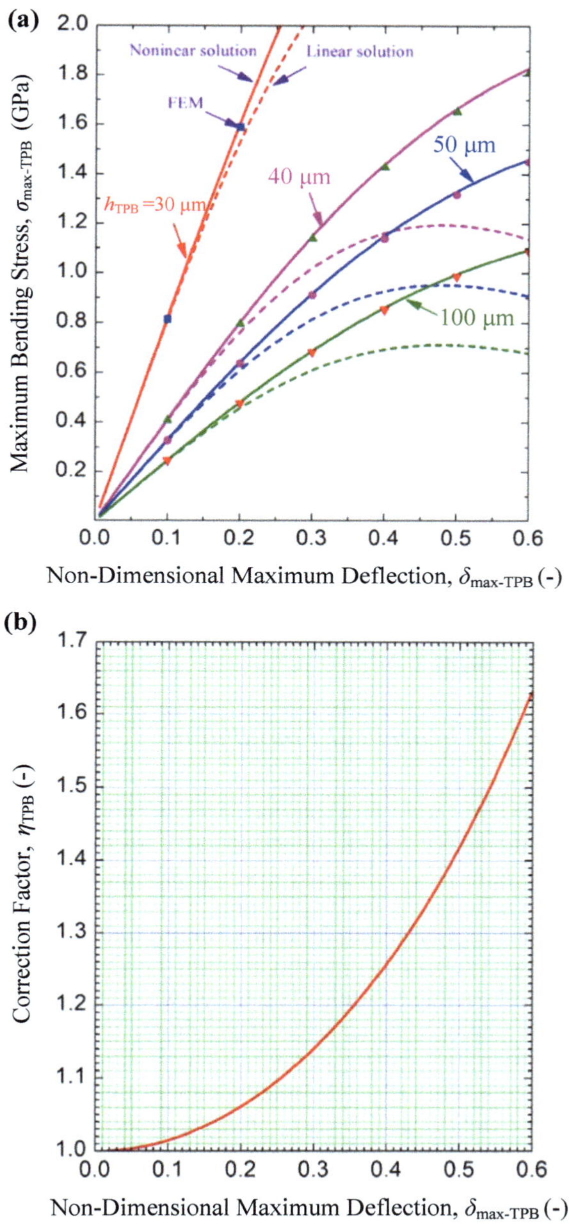

Fig. 3.8 Results of the maximum bending stress $\sigma_{max\text{-}TPB}$ at the midspan: **a** the sample thicknesses are selected as 30 μm, 40 μm, 50 μm, and 100 μm separately, where the nonlinear analytical results, linear estimations, and experimental results are represented by solid lines, dash lines, and scatter plots, respectively; **b** the nomograph of the correction factor η_{TPB} versus the nondimensional maximum deflection $\delta_{max\text{-}TPB}$. © IOP Publishing. Reproduced with permission from Ref. [23]. All rights reserved

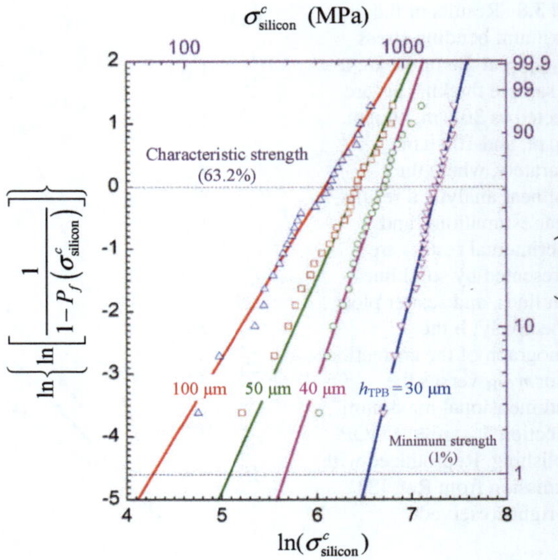

Fig. 3.9 Weibull plot and failure probability in the three-point bending test. The Weibull fitting and experimental data are represented by solid lines and scatter plots, respectively. © IOP Publishing. Reproduced with permission from Ref. [23]. All rights reserved

3.3 Estimation of Adhesive Fracture Energy of Adhesive Tape

3.3.1 *Theoretical Foundation*

In general, the ERR of interfacial peeling can be calculated by analytical and numerical methods [26, 27], and the adhesive fracture energy can be estimated through the peeling tests [28, 29]. The essence of detachment between an adhesive strip and a rigid adherend is the propagation of a crack along the adhesive interface. The total system energy in a strained and bonded length l_{strip} is denoted here as U_{total}. Based on Griffith fracture criterion, the initial peeling occurs when $\mathrm{d}U_{\text{total}}/\mathrm{d}l_{\text{strip}} \geq 0$. The critical peeling condition is satisfied when $\mathrm{d}U_{\text{total}}/\mathrm{d}l_{\text{strip}} = 0$.

In practice, an adhesive strip with the width w_{strip} and thickness h_{strip} is pulled by the peel force P_{peel} at a peel angle θ_{peel} from a rigid silicate glass surface, as shown in Fig. 3.10. The total energy of the peeling system is related to the adhesion energy $U_{\text{adhesion}} = -l_{\text{strip}} b_{\text{strip}} G^c_{\text{adh}}$ (here, G^c_{adh} is experimental adhesive fracture energy required to peel unit area of adhesion), external work of the load U_{external}, stored elastic energy in the peel region U_{elastic}, kinetic energy U_{kinetic}, and dissipated energy during bending/stretching of the peel arm $U_{\text{dissipated}}$, where the basic energy balance equation gives

$$U_{\text{total}} = U_{\text{adhesion}} + U_{\text{external}} + U_{\text{elastic}} + U_{\text{kinetic}} + U_{\text{dissipated}} \qquad (3.15)$$

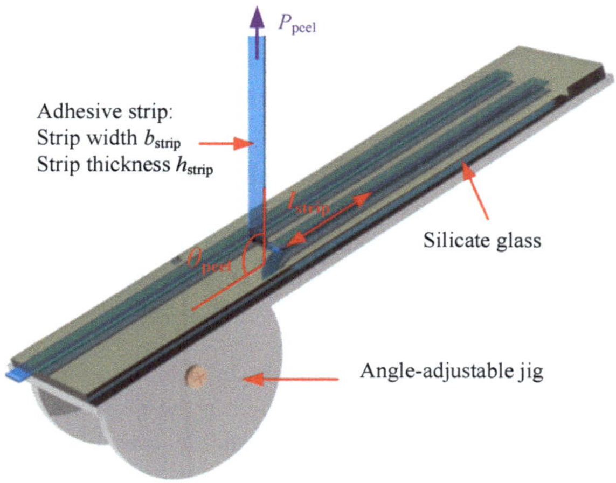

Fig. 3.10 A schematic diagram of the peeling test: an adhesive strip with the width b_{strip} and thickness h_{strip} is peeled off from the silicate glass. © 2014 IEEE. Reprinted, with permission, from Ref. [30]

When peeling an inextensible tape at a certain peel speed, $U_{elastic}$, $U_{kinetic}$, and $U_{dissipated}$ in Eq. (3.15) are usually insignificant compared with other components, which means that $\mathrm{d}U_{elastic} = \mathrm{d}U_{kinetic} = \mathrm{d}U_{dissipated} \approx 0$. Recalling Eq. (3.15), the critical condition of initial peeling can be described as $\mathrm{d}U_{total}/\mathrm{d}l_{strip} = \mathrm{d}U_{adhesion}/\mathrm{d}l_{strip} + \mathrm{d}U_{external}/\mathrm{d}l_{strip} = 0$. In addition, $\mathrm{d}U_{external} = P_{peel}[1 - \cos(\theta_{peel})]\mathrm{d}l_{strip}$. Therefore, the peel angle can be written as

$$1 - \cos(\theta_{peel}) = \frac{G^c_{adh}}{P_{peel}/b_{strip}} \tag{3.16}$$

The above equation is in accordance with that derived by Rivlin [28]. The work needed to peel the strip from the substrate is exactly compensated by the gain in the system adhesion energy. It should be noted that the peel force is independent of the Young's moduli of both the strip and substrate.

The system elastic energy $U_{elastic}$ cannot be neglected when peeling an extensible strip. Considering that the strip is assumed as linear elastic in the peeling test, and the variation of the strip width and strip thickness is neglected, the strain energy density can thus be calculated as $P^2_{peel}/(2E_{strip}b^2_{strip}\,h^2_{strip})$, in which E_{strip} is the Young's modulus of the strip. The stored elastic energy per unit crack propagation is $\mathrm{d}U_{elastic} = -P^2_{peel}\,/(2E_{strip}b_{strip}h_{strip})\mathrm{d}l_{strip}$. It should be noted that the external work of the load is $\mathrm{d}U_{external} = P_{peel}\left[1 - \cos\left(\theta_{peel}\right) + P_{peel}/\left(2E_{strip}b_{strip}h_{strip}\right)\right]\mathrm{d}l_{strip}$. By substituting the energy terms into Eq. (3.15), one can obtain

$$\frac{1}{2E_{strip}h_{strip}}\left(\frac{P_{peel}}{b_{strip}}\right)^2 + \left[1 - \cos\left(\theta_{peel}\right)\right]\frac{P_{peel}}{b_{strip}} - G^c_{adh} = 0 \tag{3.17}$$

where $P^2_{peel}/(2E_{strip}b^2_{strip}\,h^2_{strip})$ is the so-called elastic term [29].

One can rewrite Eq. (3.17) as

$$1 - \cos(\theta_{\text{peel}}) = \frac{G_{\text{adh}}^{c}}{P_{\text{peel}}/b_{\text{strip}}} - \frac{P_{\text{peel}}}{2E_{\text{strip}}b_{\text{strip}}h_{\text{strip}}} \tag{3.18}$$

Comparing Eq. (3.18) with Eq. (3.16), if $P_{\text{peel}}/(b_{\text{strip}}h_{\text{strip}})$ is much smaller than E_{strip}, e.g., in a special case of peeling an inextensible strip, $P_{\text{peel}}/(E_{\text{strip}}b_{\text{strip}}h_{\text{strip}})$ is close to zero, and one can transform Eq. (3.18) into Eq. (3.16). The elastic term becomes obvious when the peel angle is very small. Here, $G_{\text{adh}}^{c} = P_{\text{peel}}^{2}/(2E_{\text{strip}}b_{\text{strip}}^{2})$ can be obtained under pure stretch condition, which corresponds to the zero peel angle.

Based on the above analysis, in order to obtain reliable data of the experimental adhesive fracture energy in peeling the inextensible/extensible tape, the following criteria should be satisfied as much as possible. First, the adhesive crack propagation should be kept at a steady speed to avoid extra energy dissipation. Second, a reasonable peel angle should be selected to satisfy $P_{\text{peel}}/(2E_{\text{strip}}b_{\text{strip}}h_{\text{strip}}) \ll [1 - \cos(\theta_{\text{peel}})]$ so that the effect of the elastic term can be neglected.

3.3.2 Experimental Procedure

All experiments are conducted via a novel displacement/angle/speed controlled platform, as shown in Fig. 3.11, including the peel displacement that varied from 0 to 350 mm, the peel angle that varied from 0° to 180°, and the peel speed that varied from 50 μm/s to 100 mm/s. In addition, the peel force can be recorded using the 222-N LCM300 load cell (FUTEK, USA) at an accuracy of 1 mN.

Peeling tests are conducted for an inextensible strip (a typical electronic packaging tape) with the width of 15 mm, the thickness 55 μm, and the Young's modulus of 7.5 GPa, and an extensible strip (a typical adhesive/blue tape) with the width of 15 mm, the thickness of 75 μm, and the Young's modulus of 110 MPa, separately. The silicate glass is used here as the peel substrate due to similar material properties between the silicon wafer and silicate glass. Prior to testing, the glass surface is cleaned with acetone and left to dry. Then, the strip is adhered to the substrate surface with a roller and left for 30 min before the peeling test.

3.3.3 Estimation of Adhesive Fracture Energy

One can rewrite Eq. (3.16) as

$$\lg(P_{\text{peel}}/b_{\text{strip}}) = -\lg[1 - \cos(\theta_{\text{peel}})] + \lg(G_{\text{adh}}^{c}) \tag{3.19}$$

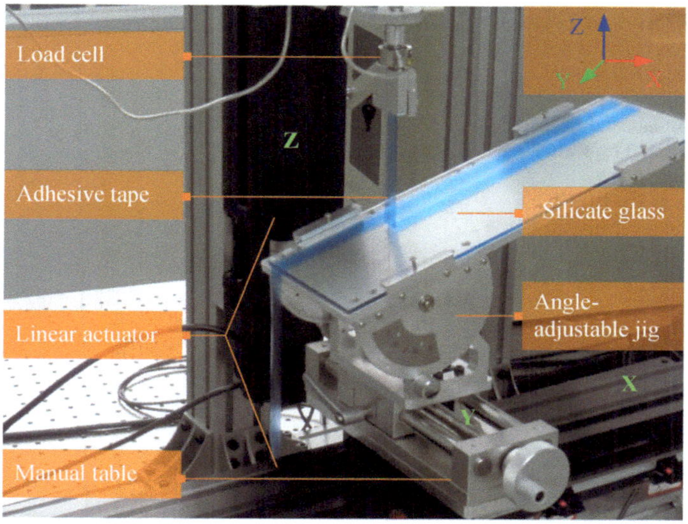

Fig. 3.11 A novel angle-adjustable experimental peeling apparatus, where X- or Z-motion can be driven by two independent linear/electric actuators, and the Y-motion depends on a manual table slide with single axis. © 2014 IEEE. Reprinted, with permission, from Ref. [30]

Experimental results of $\lg(P_{\text{peel}}/b_{\text{strip}})$ versus $\lg\left[1-\cos\left(\theta_{\text{peel}}\right)\right]$ for inextensible and extensible tapes are shown in Fig. 3.12. The intercept at the ordinate axis theoretically reflects the adhesive fracture energy. The experimental peel angles are selected as 40°, 50°, 60°, 70°, 90°, 120°, 150°, and 180°, separately. The peel speeds are selected as 1, 5, and 10 mm/s, separately.

3.3.4 Angle Dependence of Peel Force

The adhesive fracture energy can be derived from Fig. 3.12, e.g., $G^c_{\text{inexten}} = 0.0284\,\text{N/mm}$ and $G^c_{\text{exten}} = 0.0173\,\text{N/mm}$ correspond, respectively, to the inextensible and extensible strips at a peel speed of 1 mm/s. The theoretical relationship between the peel force and peel angle for an inextensible strip can be written as

$$\frac{P_{\text{peel}}}{E_{\text{strip}} b_{\text{strip}} h_{\text{strip}}} = \frac{G^c_{\text{adh}}}{E_{\text{strip}} h_{\text{strip}}} \times \frac{1}{1-\cos\left(\theta_{\text{peel}}\right)} \tag{3.20}$$

and for an extensible strip, there exists

$$\frac{P_{\text{peel}}}{E_{\text{strip}} b_{\text{strip}} h_{\text{strip}}} = \sqrt{\left[1-\cos\left(\theta_{\text{peel}}\right)\right]^2 + \frac{2G^c_{\text{adh}}}{E_{\text{strip}} h_{\text{strip}}}} - \left[1-\cos\left(\theta_{\text{peel}}\right)\right] \tag{3.21}$$

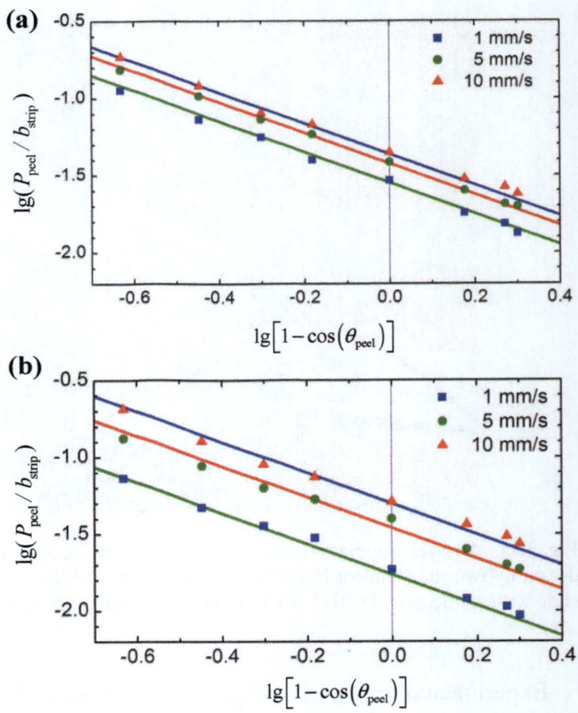

Fig. 3.12 Experimental data of the peel force versus peel angle: **a** inextensible strip and **b** extensible strip. The experimental data and linear fit are represented by scatter plots and solid lines, respectively. © 2014 IEEE. Reprinted, with permission, from Ref. [30]

It is obvious that there are three normalized parameters involved in Eqs. (3.20) and (3.21) including $P_{peel}/(E_{strip}b_{strip}h_{strip})$, $G^c_{adh}/(E_{strip}h_{strip})$, and $1-\cos(\theta_{peel})$. The strip strain during the tests is kept less than 10% so that the assumption of linear elasticity can be satisfied. The peel force is measured at different peel angles at a constant peel speed of 1 mm/s. The theoretical predictions and experimental results are shown in Fig. 3.13. A good agreement can be found between them.

It is shown that for inextensible and extensible strip, when the peel angle exceeds a certain value, e.g., $\theta_{inexten} > 10°$ and $\theta_{exten} > 25°$, namely, $1-\cos(\theta_{peel}) > 0.012$ and 0.094, the peel force can be given by Eq. (3.20), that is, the elastic term is essentially negligible. However, as the peel angle decreases, the peel force of extensible material approaches to Eq. (3.21). The upper limit approaches to $\left[2G^c_{adh}/(E_{strip}h_{strip})\right]^{1/2}$, which corresponds to the state of pure extension. As to the inextensible adhesive tape, the peel force is less sensitive to the elastic term, which can be deduced from the theoretical prediction and experimental result. However, in both cases, a smaller peel angle corresponds to a larger peel force. In order to improve the initial peeling of a chip from the adhesive tape, an effective way is to reasonably increase the initial peel angle.

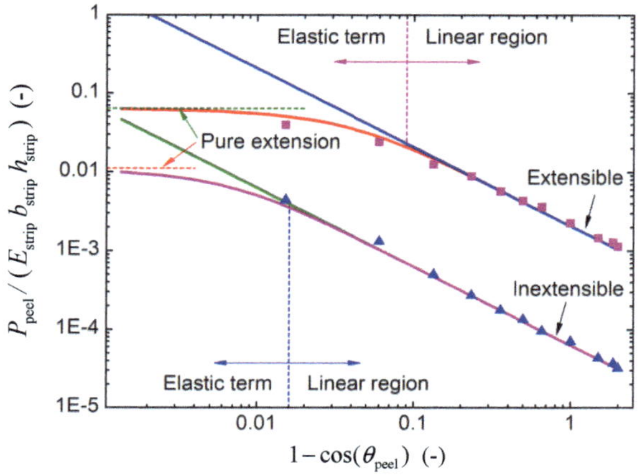

Fig. 3.13 Angle dependence of the peel force at a constant crack propagation speed of 1 mm/s, $G^c_{adh}/(E_{strip}h_{strip}) = 6.8848 \times 10^{-5}$ and 0.0021 for inextensible and extensible strip, respectively. The straight and curved lines are, respectively, derived from Eqs. (3.20) and (3.21), and the discrete points represent the experimental data. © 2014 IEEE. Reprinted, with permission, from Ref. [30]

3.3.5 Rate Dependence of Adhesive Fracture Energy

The dependence of experimental adhesive fracture energy on the peel speed can be observed in Fig. 3.12. A number of tests are implemented at several peel speeds with a constant peel angle to develop a rate-dependent law of experimental adhesive fracture energy. Based on above analysis, the peel angle is selected here as 90°, under which condition the measured force per unit width is exactly equal to the adhesive fracture energy. The adhesive fracture energy versus peel speeds of $v_{peel} = 1, 3, 5, 7,$ 10, 15, 20, 30, 40, 50, and 60 mm/s is given in Fig. 3.14, which can be empirically described by a power law [31] as follows:

$$G^c_{adh}(v_{peel}) = G_0\left[1+\left(\frac{v_{peel}}{v_0}\right)^n\right] \tag{3.22}$$

where G_0 is adhesive fracture energy corresponding to $v_{peel} \approx 0$, v_0 is the reference peel speed, and n is a scaling parameter. Clearly, $G^c_{adh} = 2G_0$ when $v_{peel} = v_0$. If it is denoted that $G^0_{inexten} = 0.0133\,\text{N/mm}$, $G^0_{exten} = 0.0077\,\text{N/mm}$, $v^0_{inexten} = 0.66\,\text{mm/s}$, and $v^0_{exten} = 0.7\,\text{mm/s}$, the same experimental data can be replotted in logarithmic form, as shown in Fig. 3.14b. An obvious linear trend is found between $\lg\left[G^c_{adh}(v_{peel})/G_0 - 1\right]$ and $\lg(v_{peel}/v_0)$. The linear fit can be described by

$$\lg\left[\frac{G^c_{adh}(v_{peel}) - G_0}{G_0}\right] = n\lg\left(\frac{v_{peel}}{v_0}\right) \tag{3.23}$$

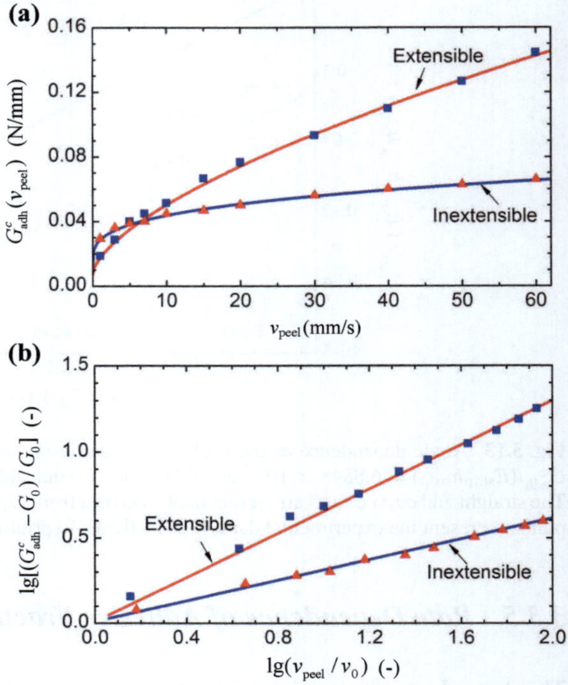

Fig. 3.14 Adhesive fracture energy versus the peel speed with a fitting curve: **a** the power-law form and **b** the logarithmic form. © 2014 IEEE. Reprinted, with permission, from Ref. [30]

As indicated by Eq. (3.23), the scaling factor can be determined by the slope of linear fit. The scaling factors for inextensible and extensible strip obtained from Fig. 3.14b are 0.302 and 0.644, respectively. The relationship between the adhesive fracture energy and the peel speed for an inextensible tape can be described by

$$G^c_{\text{inexten}}\left(v^{\text{peel}}_{\text{inexten}}\right) = 0.0133\left[1 + \left(\frac{v^{\text{peel}}_{\text{inexten}}}{0.66}\right)^{0.302}\right], \tag{3.24}$$

and for an extensible tape, there exists

$$G^c_{\text{exten}}\left(v^{\text{peel}}_{\text{exten}}\right) = 0.0077\left[1 + \left(\frac{v^{\text{peel}}_{\text{exten}}}{0.7}\right)^{0.644}\right] \tag{3.25}$$

Results of Eqs. (3.24) and (3.25) are also presented in Fig. 3.14a, which shows that the rate-dependent law matches well with the experimental data. Both the experimental results and theoretical predictions depict that the softer adhesive tape is more susceptible to the peel speed. The rate-dependent law is of significance to predict the adhesive fracture energy without related experimental certification.

It can be concluded that when peeling a relatively large and thin (e.g., 10 mm × 10 mm × 60 μm) chip from the adhesive tape, the peeling success rate reduces with the increase of the peel speed, which is mainly attributed to the rate dependence of the adhesive fracture energy. Sufficient peeling energy should be provided to achieve an initial chip peeling. The peel speed should be increased to improve the productivity. However, the corresponding adhesive fracture energy will be increased meanwhile. The predetermined peeling criterion may thus be failed due to the higher adhesive fracture energy. It can thus be concluded that the reliable peeling and fast peeling, which corresponds, respectively, to the peeling success rate and production rate, are essentially conflicting.

3.4 Summary

The efficient chip peeling-off from the adhesive tape plays an important role in the flip-chip packaging process, especially with the increasing demand for ultra-thin chip. Whether the chip can be peeled off successfully or not is dependent on the peeling fracture criterion, that is, the maximum tensile stress of the chip layer should be smaller than the corresponding fracture strength, and the ERR of interfacial peeling should exceed the adhesive fracture energy of the adhesive tape. Estimation of the fracture strength of ultra-thin silicon chip and adhesive fracture energy of the adhesive tape is essential for the establishment of a reliable chip peeling-off process.

(1) An analytical model based on the three-point bending test was presented to determine the fracture strength of ultra-thin silicon chip considering the geometric nonlinearity. The formula of nonlinear load-deflection relationship and bending stress were given in terms of nondimensional deflection parameters. The analytical solution agreed well with both the finite element simulation and experimental result. The classical linear solution showed an obvious departure from the experimental data with the decrease in the sample thickness. The nonlinear effect on the fracture strength of ultra-thin chip cannot be neglected anymore. Comparison results showed that the linear theory depicted a smaller estimation of the fracture strength than that of the nonlinear approach. A Weibull model was finally proposed to statistically analyze characteristics of the fracture strength. A higher Weibull modulus and fracture strength could be detected for a thinner chip, which demonstrated that the fracture mechanism transited from brittle to ductile with the decrease in the chip thickness.
(2) The peeling behavior of inextensible/extensible strip bonded to a silicate glass substrate was investigated using a novel angle/speed controlled peeling apparatus. Effects of the peel angle and peel speed on the peel force and adhesive fracture energy were analyzed. The angle dependence of the peel force was revealed through experiments. The peel speed was found to be important for the determination of the adhesive fracture energy. An effective power law incorpo-

rating the rate dependence of the adhesive fracture energy was finally established by fitting the test data. The proposed approach is beneficial for exploring the peeling mechanism of the chip peeling-off process.

References

1. Jeon E-B, Park J-D, Song JH, Lee HJ, Kim H-S (2012) Bi-axial fracture strength characteristic of an ultra-thin flash memory chip. J Micromech Microeng 22(10):105014
2. Barnat S, Frémont H, Gracia A, Cadalen E (2012) Evaluation by three-point-bend and ball-on-ring tests of thinning process on silicon die strength. Microelectron Reliab 52(9):2278–2282
3. Zhang T-Y, Su Y-J, Qian C-F, Zhao M-H, Chen L-Q (2000) Microbridge testing of silicon nitride thin films deposited on silicon wafers. Acta Mater 48(11):2843–2857
4. Wu J, Huang C, Liao C (2003) Fracture strength characterization and failure analysis of silicon dies. Microelectron Reliab 43(2):269–277
5. Paul I, Majeed B, Razeeb K, Barton J (2006) Statistical fracture modelling of silicon with varying thickness. Acta Mater 54(15):3991–4000
6. Majeed B, Paul I, Razeeb KM, Barton J, O'Mathuna SC (2006) Microstructural, mechanical, fractural and electrical characterization of thinned and singulated silicon test die. J Micromech Microeng 16(8):1519–1529
7. Schoenfelder S, Ebert M, Landesberger C, Bock K, Bagdahn J (2007) Investigations of the influence of dicing techniques on the strength properties of thin silicon. Microelectron Reliab 47(2):168–178
8. Tsai M, Chen C (2008) Evaluation of test methods for silicon die strength. Microelectron Reliab 48(6):933–941
9. Chae S-H, Zhao J-H, Edwards DR, Ho PS (2010) Effect of dicing technique on the fracture strength of Si dies with emphasis on multimodal failure distribution. IEEE Trans Device Mater Reliab 10(1):149–156
10. Porter D, Berfield T (2013) Die separation and rupture strength for deep reactive ion etched silicon wafers. J Micromech Microeng 23(8):085020
11. Thomas KR, Steiner U (2011) Direct stress measurements in thin polymer films. Soft Matter 7(17):7839–7842
12. Ericson F, Schweitz JÅ (1990) Micromechanical fracture strength of silicon. J Appl Phys 68(11):5840–5844
13. Yeung B, Lee T-Y (2003) An overview of experimental methodologies and their applications for die strength measurement. IEEE Trans Compon Packag Technol 26(2):423–428
14. Huang PS, Tsai MY (2011) Nonlinearities in thin-silicon die strength tests. In: 6th international microsystems, packaging, assembly and circuits technology conference (IMPACT), 19–21 Oct 2011. IEEE, pp 91–95
15. Endler S, Hoang T, Angelopoulos EA, Rempp H, Harendt C, Burghart JN (2011) Mechanical characterisation of ultra-thin chips. In: Semiconductor conference Dresden, 27–28 Sept 2011. IEEE, pp 1–4
16. Huang Y, Liu H, Xu Z, Chen J, Yin Z (2018) Conformal peeling of device-on-substrate system in flexible electronic assembly. IEEE Trans Compon Packag Manuf Technol 8(8):1496–1506
17. Liu Z, Wan X, Huang Y, Chen J, Yin Z (2018) Theoretical and experimental studies of competing fracture for flexible chip-adhesive-substrate composite structure. IEEE Trans Compon Packag Manuf Technol 8(1):57–64
18. Liu Z, Valvo PS, Huang Y, Yin Z (2013) Cohesive failure analysis of an array of IC chips bonded to a stretched substrate. Int J Solids Struct 50(22–23):3528–3538
19. Peng B, Huang Y, Yin Z, Xiong Y (2012) Competing fracture modeling of thin chip pick-up process. IEEE Trans Compon Packag Manuf Technol 2(7):1217–1225

20. Liu Z, Huang YA, Liu H, Chen J, Yin Z (2014) Reliable peeling of ultrathin die with multineedle ejector. IEEE Trans Compon Packag Manuf Technol 4(9):1545–1554
21. Medding J, Stalder R, Niederhauser M, Stoessel P (2004) Thin die bonding techniques. In: IEEE/CPMT/SEMI 29th international electronics manufacturing technology symposium (IEEE Cat. No. 04CH37585), 14–16 July 2004. pp 68–73
22. Burghartz JN, Appel W, Harendt C, Rempp H, Richter H, Zimmermann M (2010) Ultra-thin chip technology and applications, a new paradigm in silicon technology. Solid-State Electron 54(9):818–829
23. Liu Z, Huang Y, Xiao L, Tang P, Yin Z (2015) Nonlinear characteristics in fracture strength test of ultrathin silicon die. Semicond Sci Technol 30(4):045005
24. Conway H (1947) XCIV. The large deflection of simply supported beams. Lond Edinb Dublin Philos Mag J Sci 38(287):905–911
25. Ohtsuki A (1986) An analysis of large deflection in a symmetrical three-point bending of beam. Bull JSME 29(253):1988–1995
26. Liu Z, Huang Y, Chen J, Yin Z (2014) Tunable peeling technique and mechanism of thin chip from compliant adhesive tapes. IEEE Trans Compon Packag Manuf Technol 4(4):560–568
27. Peng B, Huang Y, Yin Z, Xiong Y (2011) Analysis of interfacial peeling in IC chip pick-up process. J Appl Phys 110(7):073508
28. Rivlin R (1997) The effective work of adhesion. In: Collected papers of RS Rivlin. Springer, pp 2611–2614
29. Kendall K (1975) Thin-film peeling-the elastic term. J Phys D Appl Phys 8(13):1449
30. Liu Z, Tang P, Huang Y, Yin Z (2014) Experimental estimation of adhesive fracture energy of compliant adhesive tape. In: 15th international conference on electronic packaging technology, 12–15 Aug 2014. pp 842–846
31. Feng X, Meitl MA, Bowen AM, Huang Y, Nuzzo RG, Rogers JA (2007) Competing fracture in kinetically controlled transfer printing. Langmuir 23(25):12555–12560

Chapter 4
Tension-Assisted Peeling

4.1 Introduction

Adhesively bonded joints have found extensive applications for load transfer, element connection, and structural repair in the field of aerospace, mechanical, and civil engineering [1–3]. Recently, they play increasingly important roles in the design and packaging of microelectronics and flexible electronics [4–7]. Adhesive joints like the stiffened plate, single-strap, and single-lap joints generally consist of three layers: two adherends and one adhesive layer, which can be utilized to typically denote the structure of flexible electronics including the chip, adhesive, and substrate layers.

Generally, the chip-adhesive-substrate structure is put into tension during typical manufacturing processes, e.g., the single/multiple needle peeling-off process [8, 9], where the substrate layer is generally subjected to a fixed prestrain [10, 11]. Due to the prestrain in substrate layer and discontinuity of layers at the edge, complicated stress fields and high stress concentrations usually occur in the vicinity of corners of the adhesive layer, which may lead to the overall failure of the adhesively bonded joint [12–14]. An efficient and accurate estimation of the adhesive stresses and interfacial peeling behavior is beneficial for the design of the chip-adhesive-substrate structure.

The layout of this chapter is as follows: First, a theoretical model was constructed to predict the interfacial behavior of both balanced and unbalanced adhesively bonded joints under different boundary conditions. Then, the adhesive stresses and debonding behavior of a layered structure with chips periodically bonded to a stretched substrate were investigated.

4.2 Theoretical Determination of Integration Constants

In Chap. 2, the mechanical model and theoretical framework have been developed for general adhesively bonded joints. The analytical expressions for internal forces, displacements, strain, and stress components were obtained by assuming the adhesive

Y. Huang et al., *Modeling and Application of Flexible Electronics Packaging*,
https://doi.org/10.1007/978-981-13-3627-0_4

stresses as main unknowns. This section will discuss how to determine the integration constants for various balanced and unbalanced adhesively bonded joints under mixed force and displacement boundary conditions.

4.2.1 Stiffened Plate Joint Under Axial Tension and Bending Moment

A typical stiffened plate joint under axial tension and bending moment is depicted in Fig. 4.1a [15]. Due to the symmetry of the structure and external loading, the analysis can be limited to the right-side half portion, as shown in Fig. 4.1b. The boundary conditions are summarized as follows:

(a) Symmetric conditions at the left-side end sections:

$$u_{\text{chip}}\big|_{x=0} = 0,\ \phi_{\text{chip}}\big|_{x=0} = 0,\ Q_{\text{chip}}\big|_{x=0} = 0 \tag{4.1}$$

$$u_{\text{sub}}\big|_{x=0} = 0,\quad \phi_{\text{sub}}\big|_{x=0} = 0,\quad Q_{\text{sub}}\big|_{x=0} = 0 \tag{4.2}$$

(b) Free end and applied load conditions at the right-side end sections of the chip and substrate layers, respectively:

$$N_{\text{chip}}\big|_{x=l_{\text{chip-half}}} = 0,\quad Q_{\text{chip}}\big|_{x=l_{\text{chip-half}}} = 0,\quad M_{\text{chip}}\big|_{x=l_{\text{chip-half}}} = 0 \tag{4.3}$$

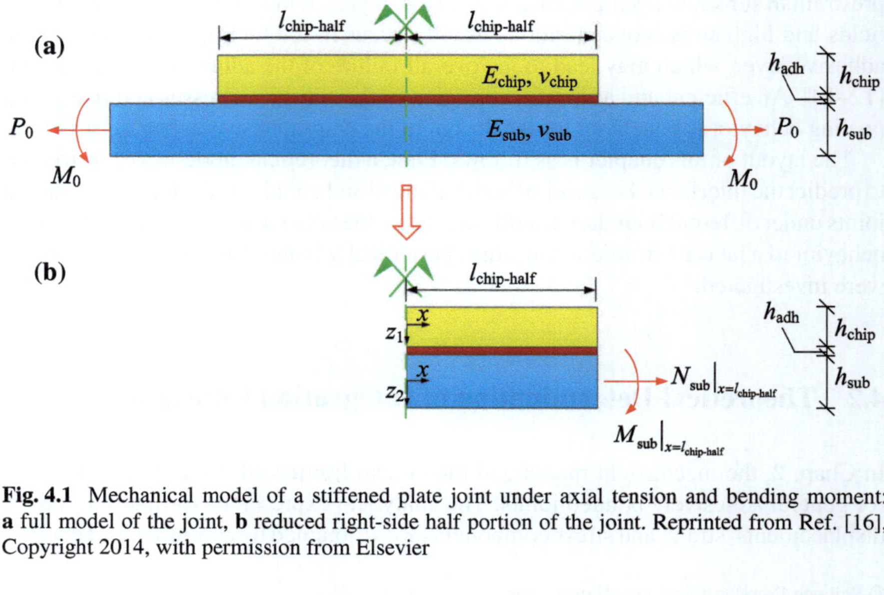

Fig. 4.1 Mechanical model of a stiffened plate joint under axial tension and bending moment: **a** full model of the joint, **b** reduced right-side half portion of the joint. Reprinted from Ref. [16], Copyright 2014, with permission from Elsevier

$$N_{\text{sub}}|_{x=l_{\text{chip-half}}} = P_0,\ Q_{\text{sub}}|_{x=l_{\text{chip-half}}} = 0,\ M_{\text{sub}}|_{x=l_{\text{chip-half}}} = -M_0 \tag{4.4}$$

The integration constants can be obtained by substituting the expressions for the internal forces and displacements into Eqs. (4.1)–(4.4), and combining the relationship among the integration constants as listed in Eqs. (2.67)–(2.80).

In the balanced case, the first four integration constants included in the expressions of the adhesive peel stress can be obtained through the following linear equation set:

$$\sum_{n=1}^{4} F_n^{\text{B}} \frac{1}{\lambda_n^{\text{B}}} = 0 \tag{4.5}$$

$$\sum_{n=1}^{4} F_n^{\text{B}} \frac{1}{\left(\lambda_n^{\text{B}}\right)^3} = 0 \tag{4.6}$$

$$\sum_{n=1}^{4} F_n^{\text{B}} \frac{\exp\left(\lambda_n^{\text{B}} l_{\text{chip-half}}\right)}{\lambda_n^{\text{B}}} = 0 \tag{4.7}$$

$$\sum_{n=1}^{4} F_n^{\text{B}} \frac{\exp\left(\lambda_n^{\text{B}} l_{\text{chip-half}}\right)}{\left(\lambda_n^{\text{B}}\right)^2} = h_{\text{chip-half}} F_8^{\text{B}} - F_{10}^{\text{B}} \tag{4.8}$$

The extra two integration constants included in the expressions of the adhesive shear stress are

$$F_5^{\text{B}} = -\frac{\lambda_5^{\text{B}}}{2\cosh\left(\lambda_5^{\text{B}} l_{\text{chip-half}}\right)} F_8 \tag{4.9}$$

$$F_6^{\text{B}} = \frac{\lambda_5^{\text{B}}}{2\cosh\left(\lambda_5^{\text{B}} l_{\text{chip-half}}\right)} F_8 \tag{4.10}$$

The integration constants of F_7^{B}, F_9^{B}, and F_{12}^{B} are found equal to zero. The remaining integration constants of F_8^{B}, F_{10}^{B}, F_{11}^{B}, and F_{13}^{B} can be expressed as follows:

$$F_8^{\text{B}} = -\frac{a_{\text{sub}}\left(d_{\text{chip}} + d_{\text{sub}}\right) P_0 + d_{\text{chip}} d_{\text{sub}}\left(h_{\text{chip-half}} + h_{\text{sub-half}}\right) M_0}{\left(a_{\text{chip}} + a_{\text{sub}}\right)\left(d_{\text{chip}} + d_{\text{sub}}\right) + d_{\text{chip}} d_{\text{sub}}\left(h_{\text{chip-half}} + h_{\text{sub-half}}\right)^2} \tag{4.11}$$

$$F_{10}^{\text{B}} = -\frac{a_{\text{sub}} d_{\text{sub}}\left(h_{\text{chip-half}} + h_{\text{sub-half}}\right) P_0 - d_{\text{sub}}\left(a_{\text{chip}} + a_{\text{sub}}\right) M_0}{\left(a_{\text{chip}} + a_{\text{sub}}\right)\left(d_{\text{chip}} + d_{\text{sub}}\right) + d_{\text{chip}} d_{\text{sub}}\left(h_{\text{chip-half}} + h_{\text{sub-half}}\right)^2} \tag{4.12}$$

$$F_{11}^{\text{B}} = \frac{\begin{bmatrix} a_{\text{chip}}\left(d_{\text{chip}} + d_{\text{sub}}\right) \\ + d_{\text{chip}} d_{\text{sub}}\left(h_{\text{chip-half}} + h_{\text{sub-half}}\right)^2 \end{bmatrix} P_0 - d_{\text{chip}} d_{\text{sub}}\left(h_{\text{chip-half}} + h_{\text{sub-half}}\right) M_0}{\left(a_{\text{chip}} + a_{\text{sub}}\right)\left(d_{\text{chip}} + d_{\text{sub}}\right) + d_{\text{chip}} d_{\text{sub}}\left(h_{\text{chip-half}} + h_{\text{sub-half}}\right)^2} \tag{4.13}$$

$$F_{13}^{\mathrm{B}} = \frac{a_{\mathrm{sub}} d_{\mathrm{chip}} \left(h_{\text{chip-half}} + h_{\text{sub-half}} \right) P_0 - d_{\mathrm{chip}} \left(a_{\mathrm{chip}} + a_{\mathrm{sub}} \right) M_0}{\left(a_{\mathrm{chip}} + a_{\mathrm{sub}} \right) \left(d_{\mathrm{chip}} + d_{\mathrm{sub}} \right) + d_{\mathrm{chip}} d_{\mathrm{sub}} \left(h_{\text{chip-half}} + h_{\text{sub-half}} \right)^2} \tag{4.14}$$

Similarly, in the unbalanced case, the first six constants can be obtained through the following linear equation set:

$$\sum_{n=1}^{6} F_n^{\mathrm{U}} \lambda_n^{\mathrm{U}} = 0 \tag{4.15}$$

$$\sum_{n=1}^{6} F_n^{\mathrm{U}} \frac{1}{\lambda_n^{\mathrm{U}}} = 0 \tag{4.16}$$

$$\sum_{n=1}^{6} F_n^{\mathrm{U}} \frac{1}{\left(\lambda_n^{\mathrm{U}} \right)^3} = 0 \tag{4.17}$$

$$\sum_{n=1}^{6} F_n^{\mathrm{U}} \frac{\exp\left(\lambda_n^{\mathrm{U}} l_{\text{chip-half}} \right)}{\lambda_n^{\mathrm{U}}} = 0 \tag{4.18}$$

$$\sum_{n=1}^{6} F_n^{\mathrm{U}} \frac{\exp\left(\lambda_n^{\mathrm{U}} l_{\text{chip-half}} \right)}{\left(\lambda_n^{\mathrm{U}} \right)^2} = -\frac{F_8^{\mathrm{U}} + d_{\mathrm{sub}} M_0}{d_{\mathrm{chip}} + d_{\mathrm{sub}}} \tag{4.19}$$

$$\sum_{n=1}^{6} F_n^{\mathrm{U}} \left[\frac{\left(\lambda_n^{\mathrm{U}} \right)^2}{k_\sigma} - \left(c_{\mathrm{chip}} + c_{\mathrm{sub}} \right) \right] \exp\left(\lambda_n^{\mathrm{U}} l_{\text{chip-half}} \right) = d_{\mathrm{sub}} M_0 \tag{4.20}$$

The integration constants of F_7^{U}, F_9^{U}, and F_{12}^{U} are found equal to zero. The remaining integration constants of F_8^{U}, F_{10}^{U}, F_{11}^{U}, and F_{13}^{U} can be expressed by the corresponding integration constants in the balanced case

$$F_8^{\mathrm{U}} = \left(d_{\mathrm{sub}} h_{\text{sub-half}} - d_{\mathrm{chip}} h_{\text{chip-half}} \right) F_8^{\mathrm{B}} \tag{4.21}$$

$$F_{10}^{\mathrm{U}} = \frac{d_{\mathrm{sub}} h_{\text{sub-half}} - d_{\mathrm{chip}} h_{\text{chip-half}}}{h_{\text{chip-half}}} F_{10}^{\mathrm{B}} \tag{4.22}$$

$$F_{11}^{\mathrm{U}} = \left(d_{\mathrm{sub}} h_{\text{sub-half}} - d_{\mathrm{chip}} h_{\text{chip-half}} \right) F_{11}^{\mathrm{B}} \tag{4.23}$$

$$F_{13}^{\mathrm{U}} = -\frac{d_{\mathrm{sub}} h_{\text{sub-half}} - d_{\mathrm{chip}} h_{\text{chip-half}}}{h_{\text{sub-half}}} F_{13}^{\mathrm{B}} \tag{4.24}$$

4.2.2 Single-Strap Joint Under Tension

A single-strap joint under tensile loading is illustrated in Fig. 4.2a [17]. Due to the symmetry, only the right-side half portion of the structure is considered, as shown in Fig. 4.2b. The boundary conditions are summarized as follows:

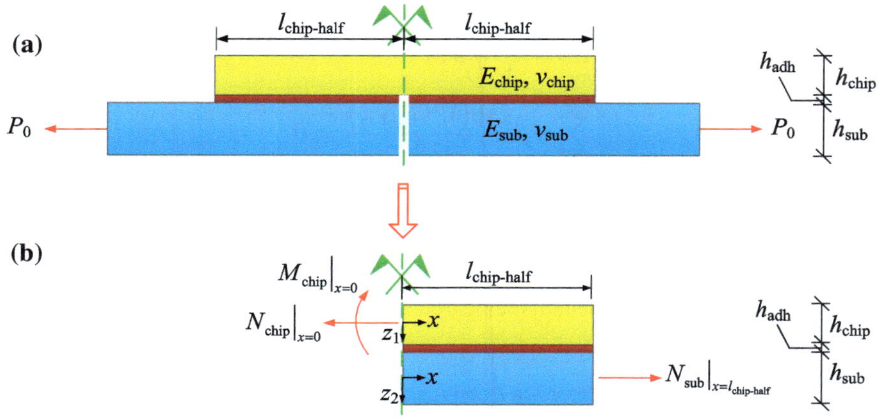

Fig. 4.2 Mechanical model of a single-strap joint under axial tensile loading: **a** full model of the joint, **b** reduced right-side half portion of the joint. Reprinted from Ref. [16], Copyright 2014, with permission from Elsevier

(a) Applied load and free end conditions at the left-side end sections of the chip and substrate layers, respectively

$$N_{\text{chip}}\big|_{x=0} = P_0, \quad Q_{\text{chip}}\big|_{x=0} = 0, \ M_{\text{chip}}\big|_{x=0} = M_0 \tag{4.25}$$

$$N_{\text{sub}}|_{x=0} = 0, \ Q_{\text{sub}}|_{x=0} = 0, \ M_{\text{sub}}|_{x=0} = 0 \tag{4.26}$$

where $M_0 = (h_{\text{chip-half}} + h_{\text{sub-half}})P$.

(b) Free end and applied load conditions at the right-side end sections of the chip and substrate layers, respectively

$$N_{\text{chip}}\big|_{x=l_{\text{chip-half}}} = 0, \ Q_{\text{chip}}\big|_{x=l_{\text{chip-half}}} = 0, \ M_{\text{chip}}\big|_{x=l_{\text{chip-half}}} = 0 \tag{4.27}$$

$$N_{\text{sub}}|_{x=l_{\text{chip-half}}} = P_0, \ Q_{\text{sub}}|_{x=l_{\text{chip-half}}} = 0, \ M_{\text{sub}}|_{x=l_{\text{chip-half}}} = 0 \tag{4.28}$$

The integration constants can be obtained by implementing the boundary conditions. In the balanced case, the first four integration constants included in the expression of the adhesive peel stress can be obtained through the following linear equation set:

$$\sum_{n=1}^{4} F_n^{\text{B}} \frac{1}{\lambda_n^{\text{B}}} = 0 \tag{4.29}$$

$$\sum_{n=1}^{4} F_n^{\mathrm{B}} \frac{1}{\left(\lambda_n^{\mathrm{B}}\right)^2} = -h_{\text{sub-half}} P_0 \tag{4.30}$$

$$\sum_{n=1}^{4} F_n^{\mathrm{B}} \frac{\exp\left(\lambda_n^{\mathrm{B}} l_{\text{chip-half}}\right)}{\lambda_n^{\mathrm{B}}} = 0 \tag{4.31}$$

$$\sum_{n=1}^{4} F_n^{\mathrm{B}} \frac{\exp\left(\lambda_n^{\mathrm{B}} l_{\text{chip-half}}\right)}{\left(\lambda_n^{\mathrm{B}}\right)^2} = 0 \tag{4.32}$$

The extra integration constants appeared in the expression of the adhesive shear stress are

$$F_5^{\mathrm{B}} = \frac{\lambda_5^{\mathrm{B}}\left(e^{-\lambda_5^{\mathrm{B}} l_{\text{chip-half}}} F_{11}^{\mathrm{B}} - F_8^{\mathrm{B}}\right)}{2\cosh\left(\lambda_5^{\mathrm{B}} l_{\text{chip-half}}\right)} \tag{4.33}$$

$$F_6^{\mathrm{B}} = \frac{\lambda_5^{\mathrm{B}}\left(e^{\lambda_5^{\mathrm{B}} l_{\text{chip-half}}} F_{11}^{\mathrm{B}} - F_8^{\mathrm{B}}\right)}{2\cosh\left(\lambda_5^{\mathrm{B}} l_{\text{chip-half}}\right)} \tag{4.34}$$

The integration constants of F_7^{B}, F_9^{B}, and F_{12}^{B} are found equal to zero. The remaining integration constants F_8^{B}, F_{10}^{B}, F_{11}^{B}, and F_{13}^{B} can be expressed as follows:

$$F_8^{\mathrm{B}} = -\frac{a_{\text{sub}}\left(d_{\text{chip}} + d_{\text{sub}}\right)}{\left(a_{\text{chip}} + a_{\text{sub}}\right)\left(d_{\text{chip}} + d_{\text{sub}}\right) + d_{\text{chip}} d_{\text{sub}}\left(h_{\text{chip-half}} + h_{\text{sub-half}}\right)^2} P_0 \tag{4.35}$$

$$F_{10}^{\mathrm{B}} = -\frac{a_{\text{sub}} d_{\text{sub}}\left(h_{\text{chip-half}} + h_{\text{sub-half}}\right)}{\left(a_{\text{chip}} + a_{\text{sub}}\right)\left(d_{\text{chip}} + d_{\text{sub}}\right) + d_{\text{chip}} d_{\text{sub}}\left(h_{\text{chip-half}} + h_{\text{sub-half}}\right)^2} P_0 \tag{4.36}$$

$$F_{11}^{\mathrm{B}} = \frac{a_{\text{chip}}\left(d_{\text{chip}} + d_{\text{sub}}\right) + d_{\text{chip}} d_{\text{sub}}\left(h_{\text{chip-half}} + h_{\text{sub-half}}\right)^2}{\left(a_{\text{chip}} + a_{\text{sub}}\right)\left(d_{\text{chip}} + d_{\text{sub}}\right) + d_{\text{chip}} d_{\text{sub}}\left(h_{\text{chip-half}} + h_{\text{sub-half}}\right)^2} P_0 \tag{4.37}$$

$$F_{13}^{\mathrm{B}} = \frac{a_{\text{sub}} d_{\text{chip}}\left(h_{\text{chip-half}} + h_{\text{sub-half}}\right)}{\left(a_{\text{chip}} + a_{\text{sub}}\right)\left(d_{\text{chip}} + d_{\text{sub}}\right) + d_{\text{chip}} d_{\text{sub}}\left(h_{\text{chip-half}} + h_{\text{sub-half}}\right)^2} P_0 \tag{4.38}$$

In the unbalanced case, the first six integration constants can be obtained through the following linear equation set:

$$\sum_{n=1}^{6} F_n^{\mathrm{U}} \frac{1}{\lambda_n^{\mathrm{U}}} = 0 \tag{4.39}$$

$$\sum_{n=1}^{6} F_n^{\mathrm{U}} \frac{1}{\left(\lambda_n^{\mathrm{U}}\right)^2} = -\frac{F_{11}^{\mathrm{U}} + d_{\text{chip}} M_0}{d_{\text{chip}} + d_{\text{sub}}} \tag{4.40}$$

$$\sum_{n=1}^{6} F_n^{\mathrm{U}} \left[\frac{\left(\lambda_n^{\mathrm{U}}\right)^2}{k_\sigma} - \left(c_{\text{chip}} + c_{\text{sub}}\right)\right] = d_{\text{chip}} M_0 \tag{4.41}$$

$$\sum_{n=1}^{6} F_n^{\mathrm{U}} \frac{\exp\left(\lambda_n^{\mathrm{U}} l_{\text{chip-half}}\right)}{\lambda_n^{\mathrm{U}}} = 0 \tag{4.42}$$

$$\sum_{n=1}^{6} F_n^{\mathrm{U}} \frac{\exp\left(\lambda_n^{\mathrm{U}} l_{\text{chip-half}}\right)}{\left(\lambda_n^{\mathrm{U}}\right)^2} = -\frac{F_8^{\mathrm{U}}}{d_{\text{chip}} + d_{\text{sub}}} \tag{4.43}$$

$$\sum_{n=1}^{6} F_n^{\mathrm{U}} \left[\frac{\left(\lambda_n^{\mathrm{U}}\right)^2}{k_\sigma} - \left(c_{\text{chip}} + c_{\text{sub}}\right) \right] \exp\left(\lambda_n^{\mathrm{U}} l_{\text{chip-half}}\right) = 0 \tag{4.44}$$

The integration constants of F_7^{U}, F_9^{U}, and F_{12}^{U} are found equal to zero. The remaining integration constants F_8^{U}, F_{10}^{U}, F_{11}^{U}, and F_{13}^{U} can be obtained through Eqs. (4.21)–(4.24).

4.2.3 Single-Lap Joint Under Tension

A single-lap joint under tension loading is shown in Fig. 4.3a [18], the boundary conditions can be summarized as follows:

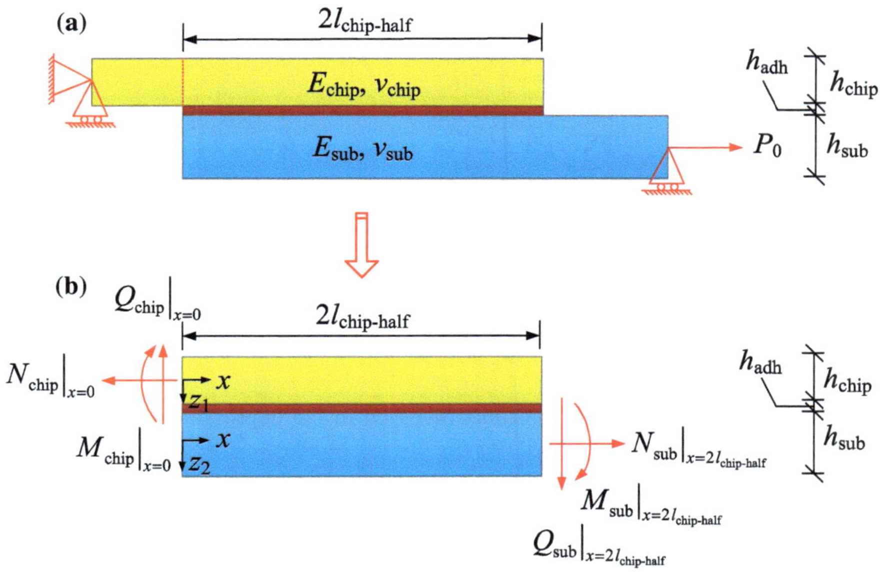

Fig. 4.3 Mechanical model of a single-lap joint under axial tensile loading: **a** full model of the joint, **b** overlapped area with boundary loadings. Reprinted from Ref. [16], Copyright 2014, with permission from Elsevier

(a) Applied load and free end conditions at the left-side end sections of the chip and substrate layers, respectively

$$N_{\text{chip}}\big|_{x=0} = P_0,\ Q_{\text{chip}}\big|_{x=0} = Q_0,\ M_{\text{chip}}\big|_{x=0} = M_1 \tag{4.45}$$

$$N_{\text{sub}}|_{x=0} = 0,\ Q_{\text{sub}}|_{x=0} = 0,\ M_{\text{sub}}|_{x=0} = 0 \tag{4.46}$$

(b) Free end and applied load conditions at the right-side end sections of the chip and substrate layers, respectively

$$N_{\text{chip}}\big|_{x=2l_{\text{chip-half}}} = 0,\ Q_{\text{chip}}\big|_{x=2l_{\text{chip-half}}} = 0,\ M_{\text{chip}}\big|_{x=2l_{\text{chip-half}}} = 0 \tag{4.47}$$

$$N_{\text{sub}}|_{x=2l_{\text{chip-half}}} = P_0,\ Q_{\text{sub}}|_{x=2l_{\text{chip-half}}} = Q_0,\ M_{\text{sub}}|_{x=2l_{\text{chip-half}}} = -M_2 \tag{4.48}$$

where $M_1 = k_1(h_{\text{chip-half}} + h_{\text{sub-half}})P_0$, $M_2 = k_2(h_{\text{chip-half}} + h_{\text{sub-half}})\,P_0$, and $Q_0 = (1 - k_1 - k_2)\,(h_{\text{chip-half}} + h_{\text{sub-half}})\,P_0/2l_{\text{chip-half}}$ for equilibrium, in which k_1 and k_2 are the bending moment factor [19, 20].

In the balanced case, the first four integration constants included in the expression of the adhesive peel stress can be obtained through the following linear equation set:

$$\sum_{n=1}^{4} F_n^{\text{B}} \frac{1}{\lambda_n^{\text{B}}} = -F_{12}^{\text{B}} \tag{4.49}$$

$$\sum_{n=1}^{4} F_n^{\text{B}} \frac{1}{\left(\lambda_n^{\text{B}}\right)^2} = -k_1 h_{\text{sub-half}} P_0 \tag{4.50}$$

$$\sum_{n=1}^{4} F_n^{\text{B}} \frac{\exp\left(2l_{\text{chip-half}}\lambda_n^{\text{B}}\right)}{\lambda_n^{\text{B}}} = -F_9^{\text{B}} \tag{4.51}$$

$$\sum_{n=1}^{4} F_n^{\text{B}} \frac{\exp\left(2l_{\text{chip-half}}\lambda_n^{\text{B}}\right)}{\left(\lambda_n^{\text{B}}\right)^2} = -k_2 h_{\text{chip-half}} P_0 \tag{4.52}$$

The extra two integration constants involved in the expression of the adhesive shear stress are

$$F_5^{\text{B}} = \frac{\lambda_5^{\text{B}}\left(e^{-2\lambda_5^{\text{B}} l_{\text{chip-half}}} F_{11}^{\text{B}} - F_8^{\text{B}} - 2l_{\text{chip-half}} F_7^{\text{B}}\right)}{2\cosh\left(2\lambda_5^{\text{B}} l_{\text{chip-half}}\right)} \tag{4.53}$$

$$F_6^{\text{B}} = \frac{\lambda_5^{\text{B}}\left(e^{2\lambda_5^{\text{B}} l_{\text{chip-half}}} F_{11}^{\text{B}} - F_8^{\text{B}} - 2l_{\text{chip-half}} F_7^{\text{B}}\right)}{2\cosh\left(2\lambda_5^{\text{B}} l_{\text{chip-half}}\right)} \tag{4.54}$$

The integration constants of F_7^{B}, F_9^{B}, and F_{12}^{B} can be expressed as follows:

$$F_7^{\mathrm{B}} = \frac{d_{\mathrm{chip}} d_{\mathrm{sub}} \left(h_{\text{chip-half}} + h_{\text{sub-half}}\right)}{\left(d_{\mathrm{chip}} + d_{\mathrm{sub}}\right)\left(a_{\mathrm{chip}} + a_{\mathrm{sub}} + d_{\mathrm{chip}} h_{\mathrm{chip}}^2 + d_{\mathrm{sub}} h_{\mathrm{sub}}^2\right)} Q_0 \tag{4.55}$$

$$F_9^{\mathrm{B}} = -\frac{d_{\mathrm{sub}}}{d_{\mathrm{chip}} + d_{\mathrm{sub}}} Q_0 \tag{4.56}$$

$$F_{12}^{\mathrm{B}} = \frac{d_{\mathrm{chip}}}{d_{\mathrm{chip}} + d_{\mathrm{sub}}} Q_0 \tag{4.57}$$

The remaining integration constants of F_8^{B}, F_{10}^{B}, F_{11}^{B}, and F_{13}^{B} can be expressed as follows:

$$F_8^{\mathrm{B}} = \frac{d_{\mathrm{chip}} d_{\mathrm{sub}} \left(h_{\text{chip-half}} + h_{\text{sub-half}}\right)^2 (k_1 - 1) - a_{\mathrm{sub}} \left(d_{\mathrm{chip}} + d_{\mathrm{sub}}\right)}{\left(a_{\mathrm{chip}} + a_{\mathrm{sub}}\right)\left(d_{\mathrm{chip}} + d_{\mathrm{sub}}\right) + d_{\mathrm{chip}} d_{\mathrm{sub}} \left(h_{\text{chip-half}} + h_{\text{sub-half}}\right)^2} P_0 \tag{4.58}$$

$$F_{10}^{\mathrm{B}} = \frac{d_{\mathrm{sub}} \left(h_{\text{chip-half}} + h_{\text{sub-half}}\right)\left(a_{\mathrm{chip}} - a_{\mathrm{chip}} k_1 - a_{\mathrm{sub}} k_1\right)}{\left(a_{\mathrm{chip}} + a_{\mathrm{sub}}\right)\left(d_{\mathrm{chip}} + d_{\mathrm{sub}}\right) + d_{\mathrm{chip}} d_{\mathrm{sub}} \left(h_{\text{chip-half}} + h_{\text{sub-half}}\right)^2} P_0 \tag{4.59}$$

$$F_{11}^{\mathrm{B}} = \frac{a_{\mathrm{chip}} \left(d_{\mathrm{chip}} + d_{\mathrm{sub}}\right) + d_{\mathrm{chip}} d_{\mathrm{sub}} k_1 \left(h_{\text{chip-half}} + h_{\text{sub-half}}\right)^2}{\left(a_{\mathrm{chip}} + a_{\mathrm{sub}}\right)\left(d_{\mathrm{chip}} + d_{\mathrm{sub}}\right) + d_{\mathrm{chip}} d_{\mathrm{sub}} \left(h_{\text{chip-half}} + h_{\text{sub-half}}\right)^2} P_0 \tag{4.60}$$

$$F_{13}^{\mathrm{B}} = \frac{d_{\mathrm{chip}} \left(h_{\text{chip-half}} + h_{\text{sub-half}}\right)\left(a_{\mathrm{chip}} k_1 + a_{\mathrm{sub}} k_1 - a_{\mathrm{chip}}\right)}{\left(a_{\mathrm{chip}} + a_{\mathrm{sub}}\right)\left(d_{\mathrm{chip}} + d_{\mathrm{sub}}\right) + d_{\mathrm{chip}} d_{\mathrm{sub}} \left(h_{\text{chip-half}} + h_{\text{sub-half}}\right)^2} P_0 \tag{4.61}$$

In the unbalanced case, the first six integration constants can be obtained through the following linear equation set:

$$\sum_{n=1}^{6} F_n^{\mathrm{U}} \frac{1}{\lambda_n^{\mathrm{U}}} = -F_{12}^{\mathrm{U}} \tag{4.62}$$

$$\sum_{n=1}^{6} F_n^{\mathrm{U}} \frac{1}{\left(\lambda_n^{\mathrm{U}}\right)^2} = -\frac{F_{11}^{\mathrm{U}} + d_{\mathrm{chip}} M_1}{d_{\mathrm{chip}} + d_{\mathrm{sub}}} \tag{4.63}$$

$$\sum_{n=1}^{6} F_n^{\mathrm{U}} \left[\frac{\left(\lambda_n^{\mathrm{U}}\right)^2}{k_\sigma} - \left(c_{\mathrm{chip}} + c_{\mathrm{sub}}\right)\right] = d_{\mathrm{chip}} M_1 \tag{4.64}$$

$$\sum_{n=1}^{6} F_n^{\mathrm{U}} \frac{\exp\left(2 l_{\text{chip-half}} \lambda_n^{\mathrm{U}}\right)}{\lambda_n^{\mathrm{U}}} = -F_9^{\mathrm{U}} \tag{4.65}$$

$$\sum_{n=1}^{6} F_n^{\mathrm{U}} \frac{\exp\left(2 l_{\text{chip-half}} \lambda_n^{\mathrm{U}}\right)}{\left(\lambda_n^{\mathrm{U}}\right)^2} = -\frac{F_8^{\mathrm{U}} + d_{\mathrm{sub}} M_2 + 2 l_{\text{chip-half}} F_7^{\mathrm{U}}}{d_{\mathrm{chip}} + d_{\mathrm{sub}}} \tag{4.66}$$

$$\sum_{n=1}^{6} F_n^{\mathrm{U}} \left[\frac{\left(\lambda_n^{\mathrm{U}}\right)^2}{k_\sigma} - \left(c_{\mathrm{chip}} + c_{\mathrm{sub}}\right)\right] \exp\left(2 l_{\text{chip-half}} \lambda_n^{\mathrm{U}}\right) = d_{\mathrm{sub}} M_2 \tag{4.67}$$

The integration constants of F_7^{U}, F_9^{U}, and F_{12}^{U} can be expressed as follows:

$$F_7^{\mathrm{U}} = -\frac{d_{\text{chip}} d_{\text{sub}} \left(h_{\text{chip-half}} + h_{\text{sub-half}}\right)\left(d_{\text{chip}} h_{\text{chip-half}} - d_{\text{sub}} h_{\text{sub-half}}\right)}{\left(a_{\text{chip}} + a_{\text{sub}}\right)\left(d_{\text{chip}} + d_{\text{sub}}\right) + d_{\text{chip}} d_{\text{sub}} \left(h_{\text{chip-half}} + h_{\text{sub-half}}\right)^2} Q_0 \quad (4.68)$$

$$F_9^{\mathrm{U}} = -\frac{d_{\text{sub}} \left[d_{\text{chip}} h_{\text{chip-half}} \left(h_{\text{chip-half}} + h_{\text{sub-half}}\right) + \left(a_{\text{chip}} + a_{\text{sub}}\right)\right]}{\left(a_{\text{chip}} + a_{\text{sub}}\right)\left(d_{\text{chip}} + d_{\text{sub}}\right) + d_{\text{chip}} d_{\text{sub}} \left(h_{\text{chip-half}} + h_{\text{sub-half}}\right)^2} Q_0 \quad (4.69)$$

$$F_{12}^{\mathrm{U}} = \frac{d_{\text{chip}} \left[d_{\text{sub}} h_{\text{sub}} \left(h_{\text{chip-half}} + h_{\text{sub-half}}\right) + \left(a_{\text{chip}} + a_{\text{sub}}\right)\right]}{\left(a_{\text{chip}} + a_{\text{sub}}\right)\left(d_{\text{chip}} + d_{\text{sub}}\right) + d_{\text{chip}} d_{\text{sub}} \left(h_{\text{chip-half}} + h_{\text{sub-half}}\right)^2} Q_0 \quad (4.70)$$

The remaining integration constants of F_8^{U}, F_{10}^{U}, F_{11}^{U}, and F_{13}^{U} can be obtained through Eqs. (4.21)–(4.24).

4.3 Modeling of Adhesively Bonded Chip Array

4.3.1 Mechanical Model Description

In general, arrays of chips adhesively bonded to a stretched substrate are commonly applied in the industry. The chip-adhesive-substrate structure consisting of a periodic array of chips adhesively bonded onto an elastomer substrate layer under the stretching condition is modeled here, as illustrated in Fig. 4.4a. In consideration of the periodicity, the mechanical model is restricted to a unit cell with length $2(l_{\text{chip-half}} + l_{\text{adja-half}})$, where $l_{\text{chip-half}}$ is the half length of a single chip and $l_{\text{adja-half}}$ is the half distance between two adjacent chips, as shown in Fig. 4.4b. Due to the symmetry of the unit cell and external loading, only the right-side half portion of the unit cell is considered, as seen in Fig. 4.4c. A tensile load per unit area along the axial direction, P_{tension}, is applied to the right-side end section of the substrate layer. It is assumed that the crack will initiate from the free end of the adhesive layer and propagate along the mid-plane of the adhesive layer.

As depicted in Fig. 4.4c, the structure is divided into Region 1 and Region 2 starting from the symmetry axis with lengths $l_{\text{chip-half}}$ and $l_{\text{adja-half}}$ respectively. Three segments are defined: S_1 corresponds to the chip layer, S_2 and S_3 correspond respectively to the portions of the substrate layer belonged to Region 1 and Region 2. Local reference axes x_1 and x_2 measure the distance in the axial direction from the left-side end sections of Region 1 and Region 2, respectively. Likewise, local axes z_1, z_2, and z_3 denote the distances in the transverse direction from the mid-planes of segments S_1, S_2, and S_3, respectively. Thus, for the segment S_i ($i = 1, 2, 3$), a local coordinate system x_j, y_i, z_i (here, $j = 1, 2$ represents Region 1 and Region 2, respectively) is established, with the origin at the mid-point of the left-side edge of the segment. Each segment is modeled as an elastic beam based on Timoshenko theory. Accordingly, the governing equations are derived with suitable continuity conditions, using the so-

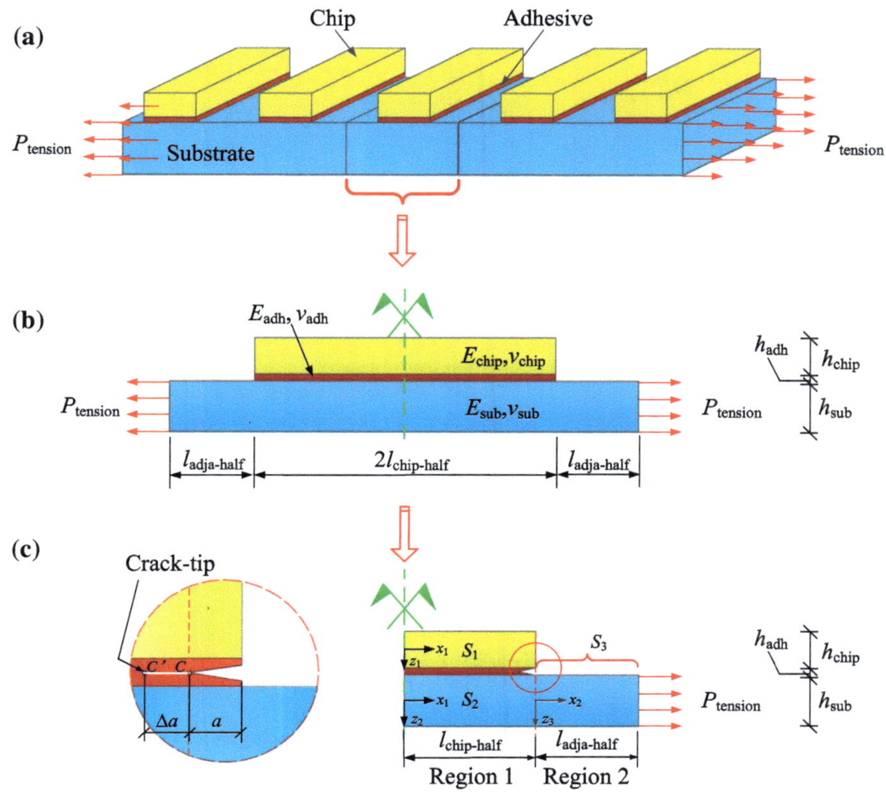

Fig. 4.4 Mechanical model of a periodic array of chips bonded to the substrate layer, subjected to uniaxial uniform traction: **a** schematic description of the model, **b** plane strain model of the unit cell, **c** reduced right-side half portion of the unit cell with a local magnification of the crack tip. Reprinted from Ref. [25], Copyright 2013, with permission from Elsevier

called multi-segment analysis method which has been adopted by other researchers [21–24]. The construction of the adhesive model for Region 1 can be referred to Chap. 2.

The free-body diagram of an elementary segment of the substrate layer in Region 2 ($x_2 \in [0, l_{\text{adja-half}}]$) is shown in Fig. 4.5. Following equilibrium equations can be deduced:

$$\frac{\mathrm{d}N_{\text{sub-reg-2}}^{\text{stretch}}}{\mathrm{d}x_2} = 0, \quad \frac{\mathrm{d}Q_{\text{sub-reg-2}}^{\text{stretch}}}{\mathrm{d}x_2} = 0, \quad \frac{\mathrm{d}M_{\text{sub-reg-2}}^{\text{stretch}}}{\mathrm{d}x_2} - Q_{\text{sub-reg-2}}^{\text{stretch}} = 0 \tag{4.71}$$

By combining the constitutive laws of Timoshenko theory, the following equation set can be obtained:

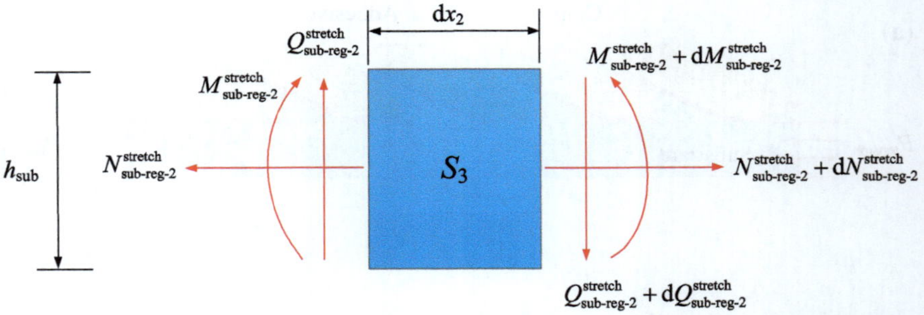

Fig. 4.5 Free-body diagram of an elementary segment of the substrate layer in region 2. Reprinted from Ref. [25], Copyright 2013, with permission from Elsevier

$$\frac{d^2 u^{stretch}_{sub\text{-}reg\text{-}2}}{dx_2^2} = 0, \quad \frac{d\phi^{stretch}_{sub\text{-}reg\text{-}2}}{dx_2} + \frac{d^2 w^{stretch}_{sub\text{-}reg\text{-}2}}{dx_2^2} = 0, \quad \frac{d^3 \phi^{stretch}_{sub\text{-}reg\text{-}2}}{dx_2^3} = 0 \tag{4.72}$$

4.3.2 Boundary and Continuity Conditions

(a) Symmetric conditions at the left-side end section of Region 1

$$u^{stretch}_{chip\text{-}reg\text{-}1}\Big|_{x_1=0} = 0, \quad \phi^{stretch}_{chip\text{-}reg\text{-}1}\Big|_{x_1=0} = 0, \quad Q^{stretch}_{chip\text{-}reg\text{-}1}\Big|_{x_1=0} = 0 \tag{4.73}$$

$$u^{stretch}_{sub\text{-}reg\text{-}1}\Big|_{x_1=0} = 0, \quad \phi^{stretch}_{sub\text{-}reg\text{-}1}\Big|_{x_1=0} = 0, \quad Q^{stretch}_{sub\text{-}reg\text{-}1}\Big|_{x_1=0} = 0 \tag{4.74}$$

(b) Free end conditions at the right-side end section of the chip layer in Region 1

$$N^{stretch}_{chip\text{-}reg\text{-}1}\Big|_{x_1=l_{chip\text{-}half}} = 0, \quad Q^{stretch}_{chip\text{-}reg\text{-}1}\Big|_{x_1=l_{chip\text{-}half}} = 0, \quad M^{stretch}_{chip\text{-}reg\text{-}1}\Big|_{x_1=l_{chip\text{-}half}} = 0 \tag{4.75}$$

(c) Continuity conditions at the cross section connecting segments S_2 and S_3 of the substrate layer

$$\begin{cases} u^{stretch}_{sub\text{-}reg\text{-}1}\Big|_{x_1=l_{chip\text{-}half}} = u^{stretch}_{sub\text{-}reg\text{-}2}\Big|_{x_2=0} \\ \phi^{stretch}_{sub\text{-}reg\text{-}1}\Big|_{x_1=l_{chip\text{-}half}} = \phi^{stretch}_{sub\text{-}reg\text{-}2}\Big|_{x_2=0} \\ w^{stretch}_{sub\text{-}reg\text{-}1}\Big|_{x_1=l_{chip\text{-}half}} = w^{stretch}_{sub\text{-}reg\text{-}2}\Big|_{x_2=0} \end{cases} \tag{4.76}$$

$$\begin{cases} N^{\text{stretch}}_{\text{sub-reg-1}}\Big|_{x_1=l_{\text{chip-half}}} = N^{\text{stretch}}_{\text{sub-reg-2}}\Big|_{x_2=0} \\ Q^{\text{stretch}}_{\text{sub-reg-1}}\Big|_{x_1=l_{\text{chip-half}}} = Q^{\text{stretch}}_{\text{sub-reg-2}}\Big|_{x_2=0} \\ M^{\text{stretch}}_{\text{sub-reg-1}}\Big|_{x_1=l_{\text{chip-half}}} = M^{\text{stretch}}_{\text{sub-reg-2}}\Big|_{x_2=0} \end{cases} \tag{4.77}$$

(d) Periodicity conditions at the right-side end section of the substrate layer in Region 2

$$N^{\text{stretch}}_{\text{sub-reg-2}}\Big|_{x_2=l_{\text{adj-half}}} = P_{\text{tension}} h_{\text{sub}}, \quad \phi^{\text{stretch}}_{\text{sub-reg-2}}\Big|_{x_2=l_{\text{adj-half}}} = 0, \quad w^{\text{stretch}}_{\text{sub-reg-2}}\Big|_{x_2=l_{\text{adj-half}}} = 0 \tag{4.78}$$

4.3.3 *Solution for Internal Forces and Displacements*

The expressions for internal forces and displacements in Region 1 can be derived following the steps in Chap. 2. Here, 12 new integration constants namely F_8, F_9, F_{10}, F_{11}, F_{12}, F_{13}, F_{14}, F_{15}, F_{16}, F_{17}, F_{18}, and F_{19} are introduced. For Region 2, the expressions for the displacements of the substrate layer can be directly obtained by solving Eq. (4.72). The expressions for the internal forces can be deduced by combining the constitutive laws of Timoshenko beam, as given in Eqs. (4.79)–(4.84). The obtained expressions involve seven more integration constants namely F_{20}, F_{21}, F_{22}, F_{23}, F_{24}, F_{25}, and F_{26}.

For Region 2, the expressions for the mid-plane displacements of the substrate layer are as follows:

$$u^{\text{stretch}}_{\text{sub-reg-2}}(x_2) = F_{20}x_2 + F_{21} \tag{4.79}$$

$$\phi^{\text{stretch}}_{\text{sub-reg-2}}(x_2) = F_{22}x_2^2/2 + F_{23}x_2 + F_{24} \tag{4.80}$$

$$w^{\text{stretch}}_{\text{sub-reg-2}}(x_2) = -F_{22}x_2^3/6 - F_{23}x_2^2/2 + F_{25}x_2 + F_{26} \tag{4.81}$$

The expressions for the internal forces are as follows:

$$N^{\text{stretch}}_{\text{sub-reg-2}}(x_2) = A_{\text{sub}} F_{20} \tag{4.82}$$

$$M^{\text{stretch}}_{\text{sub-reg-2}}(x_2) = D_{\text{sub}}(F_{22}x_2 + F_{23}) \tag{4.83}$$

$$Q^{\text{stretch}}_{\text{sub-reg-2}}(x_2) = C_{\text{sub}}(F_{24} + F_{25}) \tag{4.84}$$

4.3.4 Determination of Integration Constants

A total of 26 integration constants should be determined. However, not all of them are concerned here, except for the first seven integration constants used in the expressions for the adhesive peel stress and shear stress. Furthermore, these integration constants are not all independent of each other. When the expressions for adhesive stresses and displacements in Region 1 are substituted into Eqs. (2.23)–(2.24), one can find seven relationships among the integration constants, as discussed in Chap. 2. For Region 2, an additional relationship is deduced by substituting the expressions for Q_3 and M_3 into Eq. (4.71). Namely

$$D_{\text{sub}}F_{22} - C_{\text{sub}}(F_{24} + F_{25}) = 0 \tag{4.85}$$

Hence, all the 26 integration constants can be determined by using the eight relationships among them and the 18 boundary and continuity conditions as given in Eqs. (4.73)–(4.78). Via the mathematical operation, one can find that eight integration constants of F_7, F_9, F_{12}, F_{14}, F_{15}, F_{17}, F_{18}, and F_{22} are equal to zero, while the first six integration constants of F_1, F_2, F_3, F_4, F_5, and F_6 can be obtained by the following linear equation set:

$$\sum_{n=1}^{6} F_n \frac{1}{\lambda_n} = 0 \tag{4.86}$$

$$\sum_{n=1}^{6} F_n \frac{1}{\lambda_n^3} = 0 \tag{4.87}$$

$$\sum_{n=1}^{6} F_n \lambda_n = 0 \tag{4.88}$$

$$\sum_{n=1}^{6} F_n \frac{\exp(\lambda_n l_{\text{chip-half}})}{\lambda_n} = 0 \tag{4.89}$$

$$\sum_{n=1}^{6} F_n \left\{ \alpha_1 \left[\frac{\lambda_n^2}{k_\sigma} - (c_{\text{chip}} + c_{\text{sub}}) \right] + \frac{\lambda_n}{k_\sigma} + \alpha_2 \left(\frac{1}{\lambda_n^3} - \frac{l_{\text{chip-half}}}{\lambda_n^2} \right) \right\} \exp(\lambda_n l_{\text{chip-half}}) = 0 \tag{4.90}$$

$$\sum_{n=1}^{6} F_n \left\{ \beta_1 \left[\frac{\lambda_n^2}{k_\sigma} - (c_{\text{chip}} + c_{\text{sub}}) \right] + \frac{\beta_2}{\lambda_n^2} \right\} \exp(\lambda_n l_{\text{chip-half}}) = 2a_{\text{sub}} h_{\text{sub-half}} (d_{\text{sub}} h_{\text{sub-half}} - d_{\text{chip}} h_{\text{chip-half}}) P_{\text{tension}} \tag{4.91}$$

where $\alpha_1 = l_{\text{adja-half}} - d_{\text{chip}} h_{\text{chip-half}} (l_{\text{chip-half}} + l_{\text{adja-half}}) / d_{\text{sub}} h_{\text{sub-half}}$, $\alpha_2 = d_{\text{chip}} (h_{\text{chip-half}} + h_{\text{sub-half}}) / h_{\text{sub-half}}$, $\beta_1 = (a_{\text{chip}} + a_{\text{sub}}) + d_{\text{chip}} h_{\text{chip-half}} (h_{\text{chip-half}} + h_{\text{sub-half}})$, and $\beta_2 = d_{\text{chip}} d_{\text{sub}} (h_{\text{chip-half}} + h_{\text{sub-half}})^2 + (a_{\text{chip}} + a_{\text{sub}})(d_{\text{chip}} + d_{\text{sub}})$

Here, MATLAB software is used to solve above linear equation set. Then, the adhesive stresses can be obtained by substituting the first seven integration constants into corresponding analytical expressions.

4.4 Adhesive Stresses for Adhesively Bonded Joints

4.4.1 Analysis of Balanced Adhesively Bonded Joints

To verify the analytical model proposed here, the adhesive peel and shear stresses for typical adhesively bonded joints subjected to uniform tensile force and bending moment are computed. Both the balanced and unbalanced conditions are considered. The theoretical results are compared with those obtained from other models [15, 17] and finite element simulations. All the results of stress are given under the plane strain condition along the mid-plane of the adhesive layer. The finite element analysis is conducted using Abaqus software. In the numerical model, all layers are assumed to be linear elastic, and the plane strain element CPE4 is adopted. In order to track the potential singular stress in the vicinity of overlapped edges, a finer mesh is applied to such areas. A high aspect ratio of the finite element mesh is utilized near the central region considering the almost constant stress distribution.

Several types of balanced joints are considered here, which consist of two identical aluminum adherends and one adhesive layer. The analytical predictions for the adhesive stresses can be obtained by introducing the integration constants, as given by Eqs. (4.5)–(4.10), (4.29)–(4.34), and (4.49)–(4.54), into the stress expressions. The material properties are assumed as follows: $E_{\text{chip}} = E_{\text{sub}} = 70$ GPa, $\nu_{\text{chip}} = \nu_{\text{sub}} = 0.34$, $E_{\text{adh}} = 2.5$ GPa, $\nu_{\text{adh}} = 0.25$. The geometric dimensions are as follows: $h_{\text{chip}} = h_{\text{sub}} = 5$ mm, $h_{\text{adh}} = 0.25$ mm, and $2l_{\text{chip-half}} = 50$ mm.

4.4.1.1 Stiffened Plate Joint

For a balanced stiffened plate joint, the normalized adhesive stresses versus the normalized distance from the symmetry axis is shown in Fig. 4.6. A very good agreement between analytical and numerical results can be found for the adhesive peel stress, while the corresponding results extracted from the literatures by Zou et al. [17] and Wang et al. [15] show some discrepancy. The prediction of the adhesive shear stress shows nearly the same trend through all approaches. The maximum values of both the adhesive peel stress and shear stress can be detected at the right-side end section of the joint, which indicates that a potential fracture of the adhesive layer should initiate from such locations.

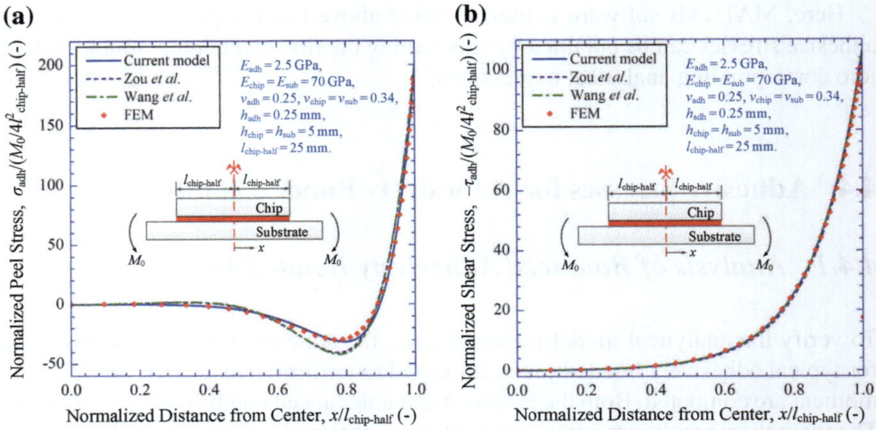

Fig. 4.6 Normalized adhesive stresses for a balanced stiffened plate joint under bending condition: **a** peel stress, **b** shear stress. Reprinted from Ref. [16], Copyright 2014, with permission from Elsevier

4.4.1.2 Single-Strap Joint

For a balanced single-strap joint, the normalized adhesive stresses versus the normalized distance from the symmetry axis is shown in Fig. 4.7. The result of comparison indicates that the proposed model is capable of predicting both the adhesive peel stress and shear stress along the mid-plane of the adhesive layer. The predictions extracted from the literature by Zou et al. [17] agree qualitatively with those of the finite element simulation, although the adhesive peel stress is somewhat apart from that of the present model. Very high adhesive stresses exist close to the interior edge of the bondline, indicating that the fracture is easy to occur at this section.

4.4.1.3 Single-Lap Joint

A single-lap joint with the adherends subjected to a tensile load P_0 is considered. To satisfy the static equilibrium conditions, additional bending moment and transverse shear load equivalent to a couple $P_0(h_{\text{chip-half}} + h_{\text{sub-half}})$ should be applied to the structure [19, 20, 26–28]. For example, if the structure is loaded through pin connection, then non-zero shear force and zero bending moment at the pins should be introduced [18]. The equilibrium condition is satisfied by applying both bending moments and transverse shear loads at the end sections of the structure. The bending moment factors $k_1 = \zeta_2/(\zeta_1 + 2\zeta_1\zeta_2 l_{\text{chip-half}} + \zeta_2)$ and $k_2 = \zeta_1/(\zeta_1 + 2\zeta_1\zeta_2 l_{\text{chip-half}} + \zeta_2)$, where $\zeta_1 = (P_0/D_{\text{chip}})^{0.5}$ and $\zeta_2 = (P_0/D_{\text{sub}})^{0.5}$ [20]. Obviously, the results will be dependent on the secondary loads applied to the structure to maintain the static equilibrium. The normalized adhesive stresses versus the normalized distance from the left-side end section is shown in Fig. 4.8. Very high adhesive stresses are formed at

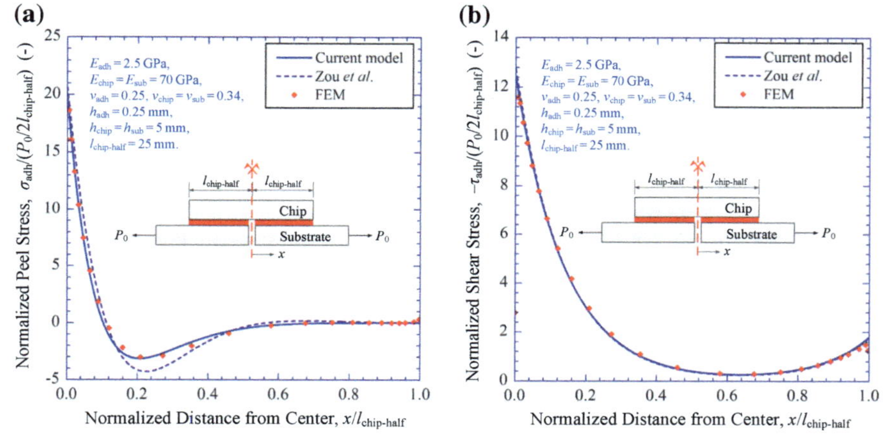

Fig. 4.7 Normalized adhesive stresses for a balanced single-strap joint under tension condition: **a** peel stress, **b** shear stress. Reprinted from Ref. [16], Copyright 2014, with permission from Elsevier

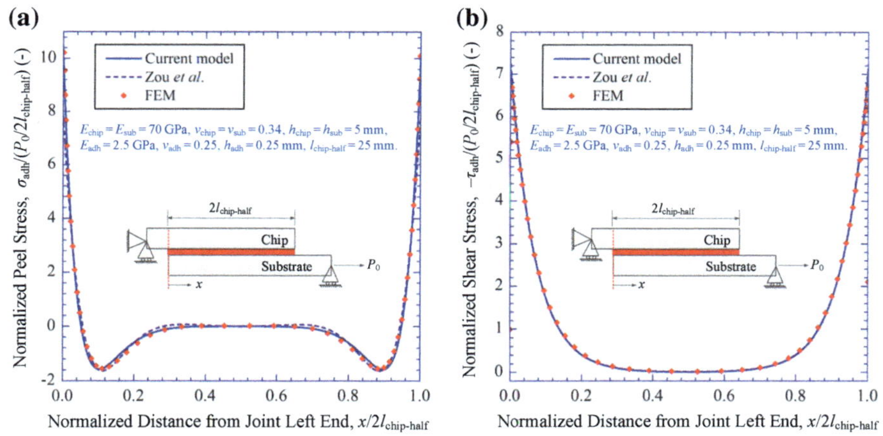

Fig. 4.8 Normalized adhesive stresses for a balanced single-lap joint under tension condition: **a** peel stress, **b** shear stress. Reprinted from Ref. [16], Copyright 2014, with permission from Elsevier

both ends of the overlapped area, hence the adhesive fracture is expected to initiate from one of such regions. Results of comparison show that the present solutions are in good agreement with the finite element simulation for both the adhesive peel stress and shear stress. In addition, the adhesive shear stress predicted by Zou et al. [17] agrees well with the finite element simulation, but there is a slight deviation in the adhesive peel stress compared with the current model.

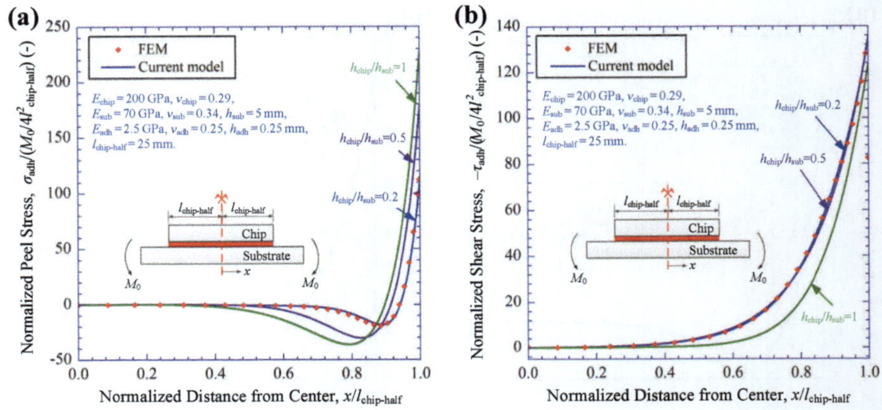

Fig. 4.9 Normalized adhesive stresses for an unbalanced stiffened plate joint under bending condition: **a** peel stress, **b** shear stress. Reprinted from Ref. [16], Copyright 2014, with permission from Elsevier

4.4.2 Analysis of Unbalanced Adhesively Bonded Joints

For the unbalanced adhesively bonded joints, the analytical solution relies on the integration constants obtained from Eqs. (4.15)–(4.20), (4.39)–(4.44), and (4.62)–(4.67). The steel-aluminum joints are adopted with following material properties: $E_{chip} =$ 200 GPa (steel), $\nu_{chip} = 0.29$ (steel), $E_{sub} = 70$ GPa (aluminum), $\nu_{sub} = 0.34$ (aluminum), and $E_{adh} = 2.5$ GPa (adhesive), $\nu_{adh} = 0.25$ (adhesive). The geometric dimensions are: $h_{sub} = 5$ mm, $h_{adh} = 0.25$ mm, and $2l_{chip\text{-}half} = 50$ mm. Three thickness ratios of the adherend are considered here for the analytical model, namely, $h_{chip}/h_{sub} = 0.2$, 0.5, and 1.0. The finite element simulation is only carried out under the condition of $h_{chip}/h_{sub} = 0.2$ to reduce the amount of computation.

4.4.2.1 Stiffened Plate Joint

The normalized adhesive stresses versus the normalized distance from the symmetry axis for an unbalanced stiffened plate joint under bending moment is shown in Fig. 4.9. Analytical predictions and numerical results match fairly well from both qualitative and quantitative point of view. It is observed that the concentration of the adhesive peel stress at the edge of overlapped area becomes more severe with the increase of the adherend thickness ratio. The corresponding maximum values of the adhesive shear stress, however, experiences a smaller variation. It suggests that the designer should pay attention not only to increasing the thickness of the cover layer to enhance the overall bending stiffness, but also preventing the adhesive cracking at the edge of overlapped area.

4.4.2.2 Single-Strap Joint

The normalized adhesive stresses versus the normalized distance from the symmetry axis for an unbalanced single-strap joint under tension condition is shown in Fig. 4.10. Like the balanced case, a very high value of both the adhesive peel stress and shear stress can be observed at the interior edges of overlapped area. The maximum values are much higher than those at exterior edges due to the existence of bending moment at the mid-span. Moreover, the maximum values of the adhesive peel stress are much severe than those of the adhesive shear stress for all the thickness ratios. Adhesive cracking at the section of mid-span is therefore expected to be the main failure mode, which can indeed be prevented by using double-sided strap joint to eliminate the parasite bending moment. It is observed that a larger value of h_{chip}/h_{sub} leads to a lower stress concentration level, which is related to the bending stiffness of the cover layer. In conclusion, stiffer and thicker cover layers are more preferred to enhance the mechanical durability of the single-strap joint.

4.4.2.3 Single-Lap Joint

The normalized adhesive stresses versus the normalized distance from the left-side end section for an unbalanced single-lap joint under tension condition is shown in Fig. 4.11. Predictions of the analytical model are in very good agreement with the finite element results. Maximum values of both the adhesive peel stress and shear stress are detected at the end sections of the adhesive layer. As to a smaller thickness ratio of h_{chip}/h_{sub}, because the bending stiffness of the chip layer is smaller than that of the substrate layer, the maximum values of the adhesive peel stress at $x = 0$ are

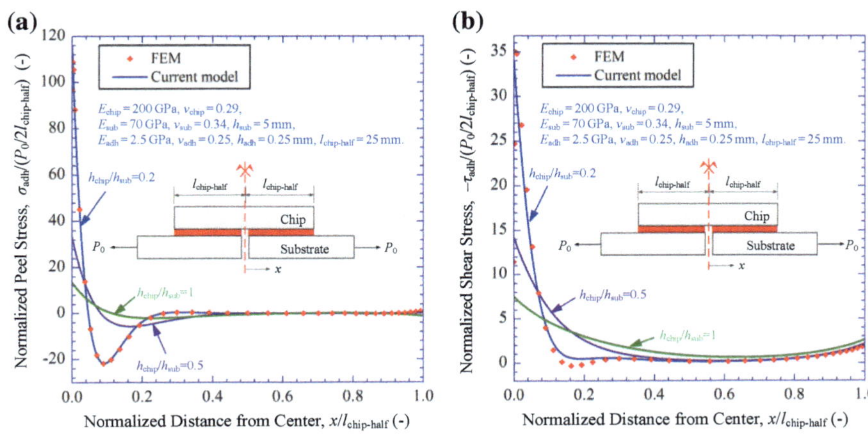

Fig. 4.10 Normalized adhesive stresses for an unbalanced single-strap joint under tension condition: **a** peel stress, **b** shear stress. Reprinted from Ref. [16], Copyright 2014, with permission from Elsevier

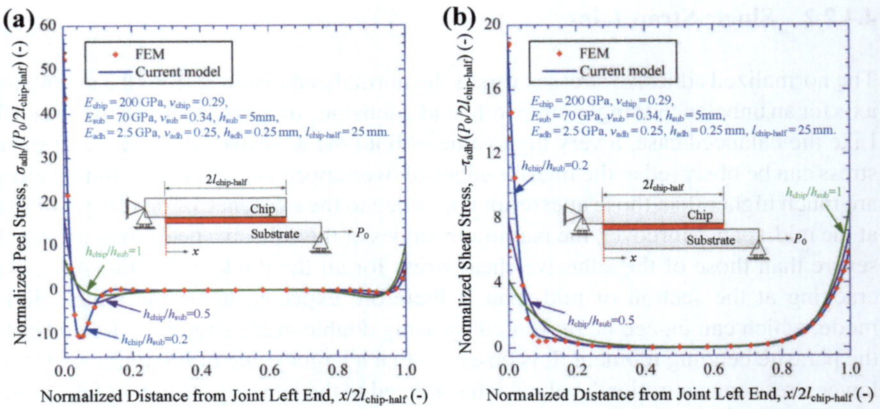

Fig. 4.11 Normalized adhesive stresses for an unbalanced single-lap joint under tension condition: **a** peel stress, **b** shear stress

greater than those at $x = 2l_{\text{chip-half}}$. This situation may be prevented with the increase of the thickness ratio of $h_{\text{chip}}/h_{\text{sub}}$.

4.4.3 Discussion of Results

In almost all above cases, results of comparison show that maximum values of the adhesive peel stress and shear stress are slightly overestimated by the analytical model. The maximum values are always detected at the end sections of overlapped areas in both the balanced and unbalanced joints. This is inherent in the assumptions of the mechanical model, where the adherends are treated as elastic beams and the adhesive layer is assumed to be independent and continuously distributed tensile-shear springs. Consequently, the kernels of the related integral equations only have logarithmic singularities and the adhesive stresses are bounded everywhere, including the end sections. In this sense, the boundary condition imposing that the adhesive shear stress be zero at the end sections of the overlapped area, applied in some research literature, is not consistent with the tensile-shear spring based adhesive model adopted here. This also explains why the overestimation of the analytical model is higher for the adhesive shear stress than for the adhesive peel stress. Like most analytical models available in literature, the current solution strategy may be unable to accurately predict the stress singularity at free edges of the overlapped area. However, the very good agreement of the analytical predictions with the numerical results, both in qualitative and quantitative aspects, confirms the validity and accuracy of the present model. In addition, the process for solving the linear equation sets to obtain the integration constants involved in the stress expressions can be performed conveniently by using the mathematical software. Obviously, the analytical solution strategy is more efficient compared with the finite element method in aspects of parametric studies, scaling analysis, design, and optimization of the adhesively bonded joints, since it only requires the input of material properties, basic geometric dimensions, and external loads.

4.5 Peeling Behavior for Chip Array on Stretched Substrate

4.5.1 Adhesive Stresses

The chip-adhesive-substrate structure with geometric dimensions and material properties as listed in Table 4.1 is considered here. The substrate layer is subjected to a uniformly distributed tensile stress of 5 MPa on the right-side end section of Region 2. The key parameters of periodic array of chips are the distance between adjacent chips and the chip length. Effects of these two geometric parameters on the debonding behavior will be investigated in following analysis.

In order to verify the analytical results, a finite element model of the chip-adhesive-substrate structure is established by using Abaqus software, as shown in Fig. 4.12. All layers are assumed to be linear elastic and isotropic materials. The whole structure is considered to deform under the condition of plane strain. The plane strain element CPE4 is employed. To impose the periodicity conditions, a reference point is defined and coupled with the right-side vertical surface in Region 2. The concentrated force applied to the reference point is taken as $P_{\text{concentrated}} = 0.5$ N, which corresponds to the uniform tensile stress, P_{stretch}, applied in the analytical model. At the same time, the vertical and rotational degrees of freedom of the reference point are restricted. Other boundary condition includes the symmetric about z axis.

Distribution of the adhesive shear stress versus the x_1-coordinate obtained by the finite element approach is shown in Fig. 4.13. $l_{\text{chip-half}} = 0.5$ mm and $l_{\text{adja-half}} = 0.02$ mm are adopted. Each of the seven curves corresponds to a specific value of the z-coordinate ranging from the bottom to the top of the adhesive layer. All the curves appear very close to each other, except for the maximum values in the vicinity of the joint right-side edge, which shows that the variation of the stresses in the adhesive layer along the z-direction is very small when the adhesive layer is much thinner than

Table 4.1 Geometric dimensions and material properties of the chip-adhesive-substrate structure [29]

Layer (–)	Thickness (μm)	Young's modulus (MPa)	Poisson's ratio (–)	Material (–)
Chip	100	129,000	0.28	Silicon
Adhesive	5	20	0.40	Acryl/epoxy resins
Substrate	100	160	0.45	Polyolefin

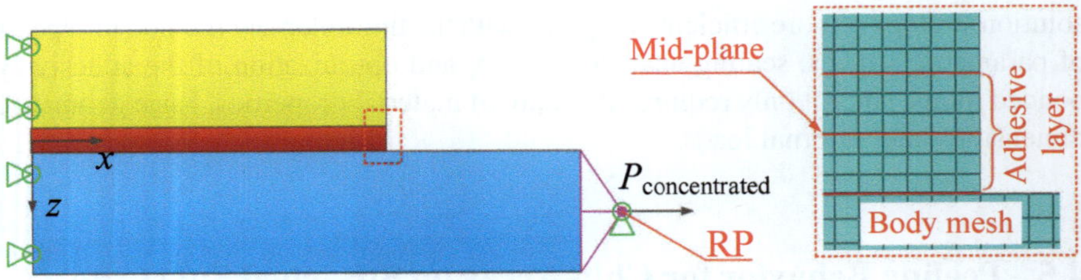

Fig. 4.12 Schematic description of the finite element model and local mesh of the adhesive layer (RP, reference point). Reprinted from Ref. [25], Copyright 2013, with permission from Elsevier

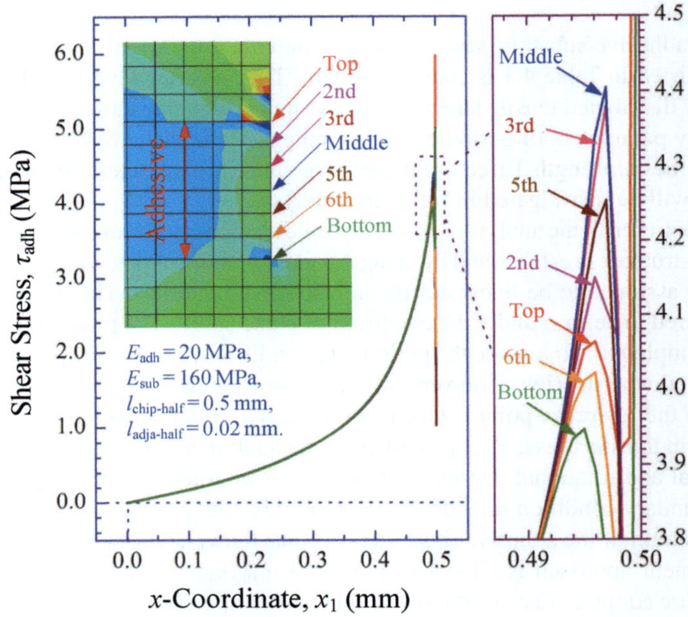

Fig. 4.13 Adhesive shear stress estimated by the finite element model in the adhesive layer versus the x_1-coordinate at different levels between the top and bottom adhesive interfaces. Reprinted from Ref. [25], Copyright 2013, with permission from Elsevier

the adherends. Moreover, the maximum values of the adhesive shear stress at the mid-plane level are higher than those obtained at other levels, except the singular stresses at the ends of both the top and bottom adhesive interfaces [30]. Therefore, attentions are mainly focused on the stresses at the mid-plane in the following analysis.

Distributions of the adhesive shear stress and peel stress are depicted in Fig. 4.14. The chip length is fixed at $2l_{\text{chip-half}} = 1.0$ mm. Several values of $l_{\text{adja-half}} = 20\ \mu\text{m}$, $100\ \mu\text{m}$, $500\ \mu\text{m}$, and $\infty\ \mu\text{m}$ are adopted to investigate the effects of the distance between adjacent chips. Here, ∞ represents the traction-free boundary condition, namely the case of a single isolated chip on an infinite substrate layer, or the case of many chips that are spaced far away from each other so that the interactions can be neglected.

The analytical predictions and finite element results agree very well except in the vicinity of the right-side end of the curves. The absolute differences of the two models versus x_1 indicate that the maximum stresses can be well estimated. The relative differences of both the adhesive shear stress and peel stress are 2.9%, 1.8%, 0.8%, 0.5% and 8.6%, 2.3%, 3.5%, 4.6% under the condition of $l_{\text{adja-half}} = 20$ μm, 100 μm, 500 μm, ∞ μm, respectively. The differences in the vicinity of the right-side occur due to that the adhesive behaves as an elastic material and the adhesive shear stress at the free edge must be null in the finite element model because of the boundary conditions (the condition is not even fulfilled exactly, because the stresses are evaluated at internal integration points in the finite element analysis). Instead, according to the analytical solution, the adhesive stresses attain maximum values at the free edge. In any case, it is worth noting that the adhesive layer will undergo plastic deformation and the stresses at the ends will be reduced for a real joint.

It can be seen from Fig. 4.14a that the adhesive shear stress decreases rapidly when moving away from the edge at $x_1 = l_{\text{chip-half}}$. As the distance between adjacent chips decreases (namely, as the chips are arrayed closer and therefore have stronger interaction), the adhesive shear stress increases rapidly. An opposite effect is observed from Fig. 4.14b for the adhesive peel stress which decreases rapidly as the distance between adjacent chips gets smaller. The value of the adhesive peel stress under the condition of $l_{\text{adja-half}} = \infty$ is not just two times as much as that for $l_{\text{adja-half}} = 20$ μm. It can also be observed that the adhesive peel stress has a self-equilibrated distribution, since the total force resulting from the adhesive peel stress must be vanished. Besides, the adhesive peel stress is negative near the right-side end section of the adhesive layer, which means that the adhesive layer is in compressive state at the edge when the chip layer is much stiffer than the substrate layer. Therefore, despite the presence of the adhesive peel stress, the crack propagation is expected to occur at pure mode II. Thus, only the mode II ERR is considered in the following analysis.

Based on the above discussion, it can be concluded that the adhesive stresses obtained by the analytical model, albeit built after some simplifying assumptions, provide quite accurate estimation on the adhesive shear and peel stresses in the mid-plane of the adhesive layer, in particular for the maximum values. The interactions among chips can be interpreted using the proposed approach. The theoretical predictions are quite accurate when the thickness of the adhesive layer is much thinner than the adherends.

4.5.2 *Effect of Geometric Dimensions*

The mode II ERR of interfacial peeling for a crack embedded in the mid-plane of the adhesive layer versus the distance between adjacent chips is depicted in Fig. 4.15. Results of the analytical model are represented by a solid line. The finite element predictions are represented by scatter plots. The chip length is fixed at 1.0 mm.

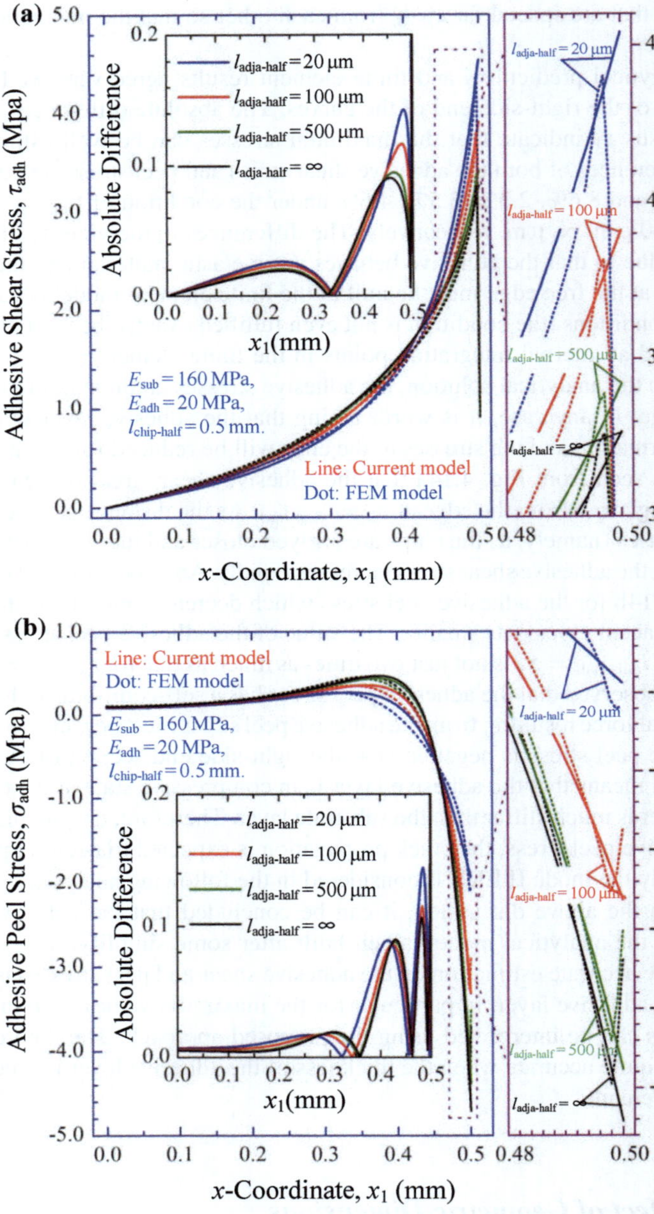

Fig. 4.14 Adhesive stresses estimated by both the analytical and numerical approaches with the absolute errors versus the x_1-coordinate: **a** shear stress, **b** peel stress. Reprinted from Ref. [25], Copyright 2013, with permission from Elsevier

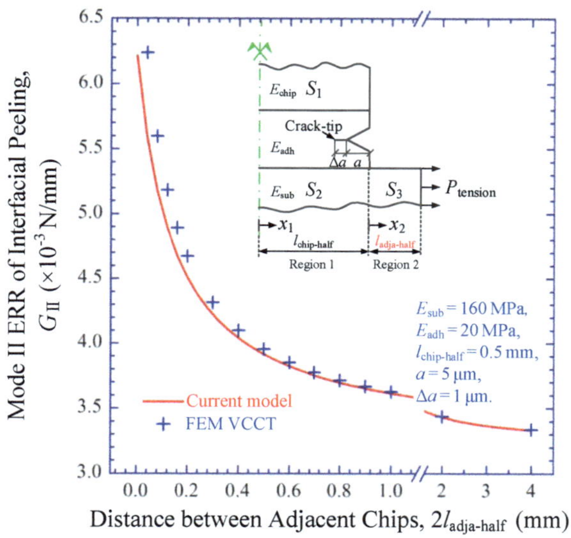

Fig. 4.15 Mode II ERR of interfacial peeling versus the distance between adjacent chips. The analytical results and finite element predictions are represented by a solid line and scatter plots, respectively. Reprinted from Ref. [25], Copyright 2013, with permission from Elsevier

The distance between adjacent chips is varied from 0.04 mm to 4.0 mm. Analytical and numerical results match fairly well in the entire variation range of the distance between adjacent chips. The mode II ERR of interfacial peeling decreases as the distance between adjacent chips gets larger. Greater variations in the mode II ERR of interfacial peeling can be observed when the distance between adjacent chips is smaller than 0.6 mm, which indicates that interaction is stronger when the chips are arrayed closer to each other. Instead, if the interval between adjacent chips is quite large, the strain resulting from the substrate tension is mainly accommodated by the substrate between the chips. In this case, the strain in the chips is negligible and the effect on the mode II ERR of interfacial peeling is small. In conclusion, one may say that the chip density has significant effect on the mode II ERR of interfacial peeling.

The mode II ERR of interfacial peeling versus the chip length is given in Fig. 4.16. Several values $l_{adja\text{-}half} = 0.04$ mm, 0.2 mm, 1.0 mm, and ∞ are considered in order to show the effect of the distance between adjacent chips. The chip length is varied from 0.2 mm to 2.0 mm, which covers both small and large chips used in industry. Analytical results are plotted as continuous curves, and finite element results are represented by scatter plots. As can be seen, the predictions agree well with each other, except for some slight deviations under conditions of small chip length. The mode II ERR of interfacial peeling increases as the chip length increases, because the strain level in the adhesive layer get larger as the chip size enlarges. However, the mode II ERR of interfacial peeling stays nearly constant when the chip length is larger than 0.8 mm. Concerning effects of the distance between adjacent chips, it can be found that smaller intervals correspond to higher values of the mode II ERR of interfacial peeling. However, slight variation in the mode II ERR of interfacial peeling can be observed when the distance between adjacent chips exceeds a certain

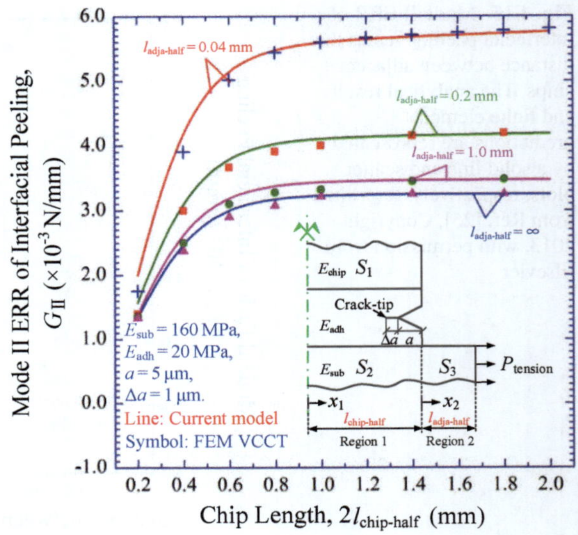

Fig. 4.16 Mode II ERR of interfacial peeling versus the chip length. Analytical results and finite element predictions are represented by solid lines and scatter plots, respectively. Reprinted from Ref. [25], Copyright 2013, with permission from Elsevier

value, which means that the effect of the distance between adjacent chips becomes weaker when the chips are spaced far away from each other. It may be concluded that the chip length has a strong effect on the mode II ERR of interfacial peeling when the chip length is relatively small. No obvious effect can be detected when the chip length is larger than a certain value, i.e., 0.8 mm. Additionally, the distance between adjacent chips aggravates the effect of the chip length on the mode II ERR of interfacial peeling.

4.5.3 Effect of Material Properties

The mode II ERR of interfacial peeling versus the Young's modulus of the substrate layer under different conditions of the distance between adjacent chips is depicted in Fig. 4.17. The chip length is fixed at 1.0 mm, other parameters are given in Table 4.1. In particular, a common adhesive (E_{adh} = 20 MPa) and substrate (E_{sub} ranges from 100 MPa to 1000 MPa) is adopted in Fig. 4.17a. Instead, a very compliant adhesive (E_{adh} = 0.5 MPa) and substrate (E_{sub} ranges from 5 MPa to 100 MPa) is adopted in Fig. 4.17b, which corresponds typically to the rubber substrate used in stretchable electronics. The analytical predictions and finite element estimations are respectively represented by solid lines and scatter plots. As can be seen, the results obtained from the finite element model and analytical model match very well. The mode II ERR of interfacial peeling increases as the substrate layer becomes more compliant, which means that the premature debonding of chips is easier to occur for the chip-adhesive-substrate structure with more compliant substrate. The trend can be explained as

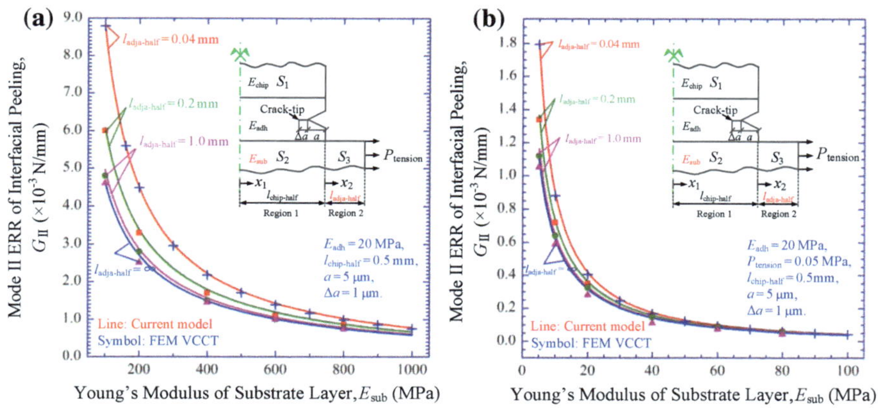

Fig. 4.17 Mode II ERR of interfacial peeling versus the Young's modulus of the substrate layer: **a** common adhesive and substrates, **b** very compliant adhesive and substrates. Analytical results and finite element predictions are represented by solid lines and scatter plots, respectively. Reprinted from Ref. [25], Copyright 2013, with permission from Elsevier

follows: if the substrate layer is more compliant, the strain of the adhesive layer strain gets larger to accommodate the deformation of the substrate layer, the resulting strain energy is thus increased. Based on the above analysis, it can be concluded that mechanical properties of the substrate layer have important effects on the mode II ERR of interfacial peeling. Additionally, the analytical model presented here can accurately predict the debonding behavior of the chips from different substrates, ranging from the rubber to the polymer.

The mode II ERR of interfacial peeling versus the Young's modulus of the adhesive layer is shown in Fig. 4.18. The chip length is fixed at 1.0 mm. In practical application, the adhesive layer should be more compliant than the substrate layer. Here the Young's modulus of the adhesive layer is varied from 20 MPa to 120 MPa. The analytical predictions and finite element estimations are respectively represented by solid lines and scatter plots. As can be seen the analytical predictions and numerical results agree very well. A monotonous decreasing trend of the mode II ERR of interfacial peeling with the increase of the Young's modulus of the adhesive layer can be observed in both approaches. However, the trend is not that sensitive to the variation of the Young's modulus of the adhesive layer. The above results can be explained as follows: although the relative axial displacement increases as the adhesive layer gets more compliant, the decrease in the axial force results in the small variation of the mode II ERR of interfacial peeling. The discrepancy between analytical predictions and numerical results are increased with the increase of the Young's modulus of the adhesive layer. Finally, by observing the curves plotted under different conditions of the distance between adjacent chips, one can find that the effect of the variation of the Young's modulus of the adhesive layer gets more significant when the distance between adjacent chips get larger [31].

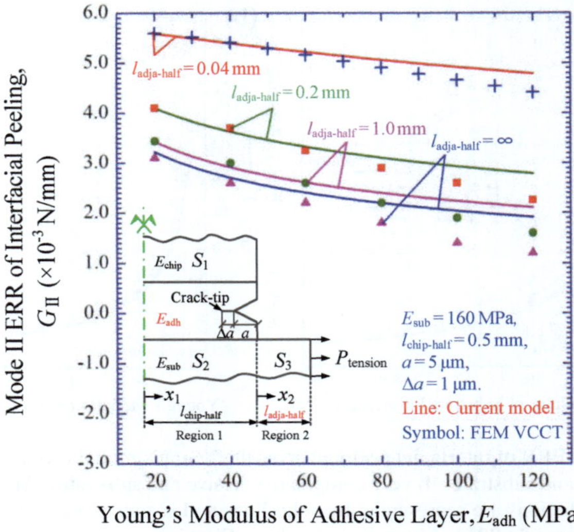

Fig. 4.18 Mode II ERR of interfacial peeling versus the Young's modulus of the adhesive layer. Analytical results and finite element predictions are represented by solid lines and scatter plots, respectively. Reprinted from Ref. [25], Copyright 2013, with permission from Elsevier

4.5.4 Process Optimization

In practice, for a given chip-adhesive-substrate structure, the geometric dimensions and material properties are generally predetermined, depending on the service and manufacturing conditions. The parameter that can be easily changed is the value of applied tension. Based on the above analysis, the process can be optimized by calculating in advance the most suitable tension force. The premature debonding of the chip from the substrate layer can thus be prevented. The mode II ERR of interfacial peeling versus the chip length under different conditions of the Young's modulus of the substrate layer, ranging from 150 MPa to 750 MPa is shown in Fig. 4.19. The distance between adjacent chips is fixed at 0.08 mm. If the adhesive fracture energy of the adhesive layer is known, the critical Young's modulus of the substrate layer or the critical chip length for adhesive debonding under prescribed uniaxial tension can be determined from Fig. 4.19. For example, assuming that $G^c_{adh} = 0.002N/mm$, the critical chip length triples as the Young's modulus of the substrate layer increases from 300 MPa to 450 MPa. If $G^c_{adh} = 0.0035$ N/mm, the chips will not delaminate from the substrate layer in the entire variation range of the chip length as long as the Young's modulus of the substrate layer is not less than 300 MPa.

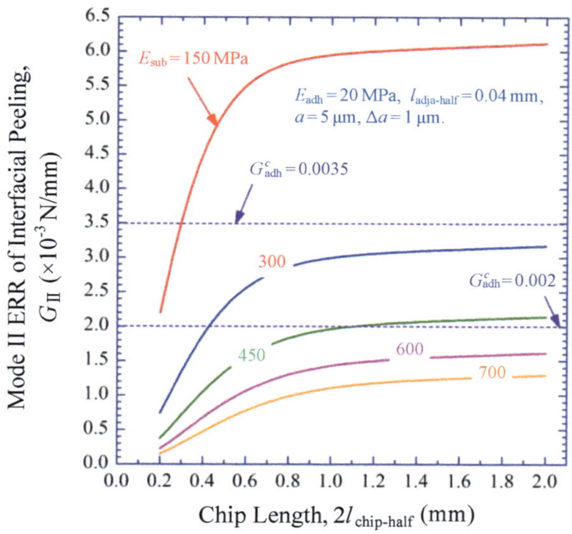

Fig. 4.19 Mode II ERR of interfacial peeling versus the chip length. Reprinted from Ref. [25], Copyright 2013, with permission from Elsevier

4.6 Summary

This chapter presents mechanical approaches to predict the peeling behavior of the adhesively bonded stiffened plate joint, single-strap joint, single-lap joint, as well as chip arrays on an elastomer substrate under stretching condition.

An analytical solution strategy was first developed to investigate the adhesive stresses in both balanced and unbalanced adhesively bonded joints. A generalized analytical solution was determined for adhesive joints with identical or different materials under mixed force and displacement boundary conditions. The analytical predictions agreed well with finite element results, not only with regard to the trends but also from the quantitative point of view. It can be used in particular for the analysis of the adhesively bonded joints with relatively small adhesive thickness.

An analytical model was also presented to investigate the adhesive stresses and the ERR of interfacial peeling of layered structure consisting of a periodic array of chips adhesively bonded to a stretched substrate. Although it is necessary to use a computer-based implementation, the proposed approach is still advantageous compared with other closed-form solutions considering the requirement on computing capability. Good agreements could be found between the theoretical and finite element predictions considering effects of key factors including the distance between adjacent chips, chip length, and material properties. Results showed that the adhesive peel stress at the crack tip was always compressive. The crack propagation was expected to occur at pure mode II. The interaction between adjacent chips showed a remarkable effect on the adhesive stresses, which became stronger as the chips got closer to each other. Most of the deformation was accommodated by the substrate layer under the condition of high ratio of the distance between adjacent chips to the chip length. The probability of the chip debonding from the substrate increased as the substrate and adhesive layers became more compliant.

References

1. Deng J, Lee MM, Moy SS (2004) Stress analysis of steel beams reinforced with a bonded CFRP plate. Compos Struct 65(2):205–215
2. Campilho RD, De Moura M, Domingues J (2008) Using a cohesive damage model to predict the tensile behaviour of CFRP single-strap repairs. Int J Solids Struct 45(5):1497–1512
3. Cheng J, Taheri F (2006) A smart single-lap adhesive joint integrated with partially distributed piezoelectric patches. Int J Solids Struct 43(5):1079–1092
4. Sun J-Y, Lu N, Yoon J, Oh K-H, Suo Z, Vlassak JJ (2012) Debonding and fracture of ceramic islands on polymer substrates. J Appl Phys 111(1):013517
5. Jiang Z, Huang Y, Chandra A (1997) Thermal stresses in layered electronic assemblies. J Electron Packag 119(2):127–132
6. Cheng H, Wu J, Li M, Kim D-H, Kim Y-S, Huang Y, Kang Z, Hwang K, Rogers J (2011) An analytical model of strain isolation for stretchable and flexible electronics. Appl Phys Lett 98(6):061902
7. Huang Y, Liu H, Xu Z, Chen J, Yin Z (2018) Conformal peeling of device-on-substrate system in flexible electronic assembly. IEEE Trans Compon Packag Manuf Technol 8(8):1496–1506
8. Liu Z, Huang Y, Chen J, Yin Z (2014) Tunable peeling technique and mechanism of thin chip from compliant adhesive tapes. IEEE Trans Compon Packag Manuf Technol 4(4):560–568
9. Liu Z, Huang YA, Liu H, Chen J, Yin Z (2014) Reliable peeling of ultrathin die with multineedle ejector. IEEE Trans Compon Packag Manuf Technol 4(9):1545–1554
10. Huang Y, Chen J, Yin Z, Xiong Y (2011) Roll-to-roll processing of flexible heterogeneous electronics with low interfacial residual stress. IEEE Trans Compon Packag Manuf Technol 1(9):1368–1377
11. Peng B, Huang Y, Yin Z, Xiong Y (2011) Analysis of interfacial peeling in IC chip pick-up process. J Appl Phys 110(7):073508
12. Feng Y, Wu L (2001) Analysis of interfacial thermal stresses of chip-substrate structure. Int J Solids Struct 38(9):1551–1562
13. Park SI, Ahn JH, Feng X, Wang S, Huang Y, Rogers JA (2008) Theoretical and experimental studies of bending of inorganic electronic materials on plastic substrates. Adv Func Mater 18(18):2673–2684
14. Liu Z, Wan X, Huang Y, Chen J, Yin Z (2018) Theoretical and experimental studies of competing fracture for flexible chip-adhesive-substrate composite structure. IEEE Trans Compon Packag Manuf Technol 8(1):57–64
15. Wang K, Huang Y, Chandra A, Hu KX (2000) Interfacial shear stress, peeling stress, and die cracking stress in trilayer electronic assemblies. In: ITHERM 2000. The seventh inter-society conference on thermal and thermomechanical phenomena in electronic systems (Cat. No.00CH37069), 23–26 May 2000. IEEE, pp 56–64
16. Liu Z, Huang Y, Yin Z, Bennati S, Valvo PS (2014) A general solution for the two-dimensional stress analysis of balanced and unbalanced adhesively bonded joints. Int J Adhes Adhes 54:112–123
17. Zou G, Shahin K, Taheri F (2004) An analytical solution for the analysis of symmetric composite adhesively bonded joints. Compos Struct 65(3):499–510
18. Delale F, Erdogan F, Aydinoglu M (1981) Stresses in adhesively bonded joints: a closed-form solution. J Compos Mater 15(3):249–271
19. Cheng S, Chen D, Shi Y (1991) Analysis of adhesive-bonded joints with nonidentical adherends. J Eng Mech 117(3):605–623

20. Zhao X, Adams R, da Silva LF (2010) A new method for the determination of bending moments in single lap joints. Int J Adhes Adhes 30(2):63–71
21. Yang C, Pang S-S (1996) Stress-strain analysis of single-lap composite joints under tension. J Eng Mater Technol 118(2):247–255
22. Yang C, Chadegani A, Tomblin JS (2008) Strain energy release rate determination of prescribed cracks in adhesively-bonded single-lap composite joints with thick bondlines. Compos B Eng 39(5):863–873
23. Chadegani A, Batra RC (2011) Analysis of adhesive-bonded single-lap joint with an interfacial crack and a void. Int J Adhes Adhes 31(6):455–465
24. Chadegani A, Yang C, Smeltzer SS (2012) Adhesive-bonded composite joints analysis with delaminated surface ply using strain-energy release rate. J Aircraft 49(2):503–520
25. Liu Z, Valvo PS, Huang Y, Yin Z (2013) Cohesive failure analysis of an array of IC chips bonded to a stretched substrate. Int J Solids Struct 50(22–23):3528–3538
26. Goland M, Reissner E (1944) The stresses in cemented joints. J Appl Phys 11(1):A17–A27
27. Hart-Smith LJ (1973) Adhesive-bonded single-lap joints. National Aeronautics and Space Administration (NASA)
28. Adams R, Mallick V (1992) A method for the stress analysis of lap joints. J Adhes 38(3–4):199–217
29. Saiki N, Inaba K, Kishimoto K, Seno H, Ebe K (2010) Study on peeling behavior in pick-up process of IC chip with adhesive tapes. J Solid Mech Mater Eng 4(7):1051–1060
30. Gleich D, Van Tooren M, Beukers A (2001) Analysis and evaluation of bondline thickness effects on failure load in adhesively bonded structures. J Adhes Sci Technol 15(9):1091–1101
31. da Silva LF, das Neves PJ, Adams R, Spelt J (2009) Analytical models of adhesively bonded joints—part I: literature survey. Int J Adhes Adhes 29(3):319–330

Chapter 5
Single-needle Peeling

5.1 Introduction

By now, the development of thinner/lighter consumer electronics and flexible/stretchable electronic devices has demanded the chips to be thinner and/or smaller [1], which is challenging the current electronic assembly technique [2–5]. Thinner semiconductor chips are increasingly difficult to be handled [6, 7] and easy to be broken during the packaging process, especially for the chip peeling-off process by ejector needle [8–10]. The chip cracking tends to occur more easily for thinner chips resulting from inappropriate handling technique, which significantly reduces the chip peeling-off success rate. The chips with 0.1 mm thickness are more likely to fail during the needle ejecting process, while the peeling-off success rate of 0.34-mm-thick chip is nearly 100% [11]. Therefore, it is eager to understand fundamentally the mechanism of the chip peeling-off from the adhesive tape.

A schematic illustration of the single-needle peeling-off process is shown in Fig. 5.1 [12]. As depicted in Fig. 5.1a, an ejector needle rises to push the substrate and chip upward and make the chip peeled from the adhesive tape. When about 80% of the interface region is debonded, the pickup head draws the chip up. Generally, the investigation of the chip peeling-off is similar to the fracture analysis of the adhesive layer between the chip and substrate. Numerous analytical models of adhesively bonded/composite structures have been proposed over the past decades as described by da Silva et al. [13]. The classical and typical analytical solutions include the earlier work by Goland and Reissner [14], Adams and Peppiatt [15], Delale et al. [16], Shahin et al. [17], and Wang and Zeng [18], etc. However, these analytical models showed helplessness for mixed boundary conditions. Lately, Bennati et al. [19] and Liu et al. [20] developed an effective solution strategy to obtain the analytical expressions for the adhesive stresses, internal forces, and displacements in the adherends as well as ERR for mixed-mode cohesive failure.

Y. Huang et al., *Modeling and Application of Flexible Electronics Packaging*,
https://doi.org/10.1007/978-981-13-3627-0_5

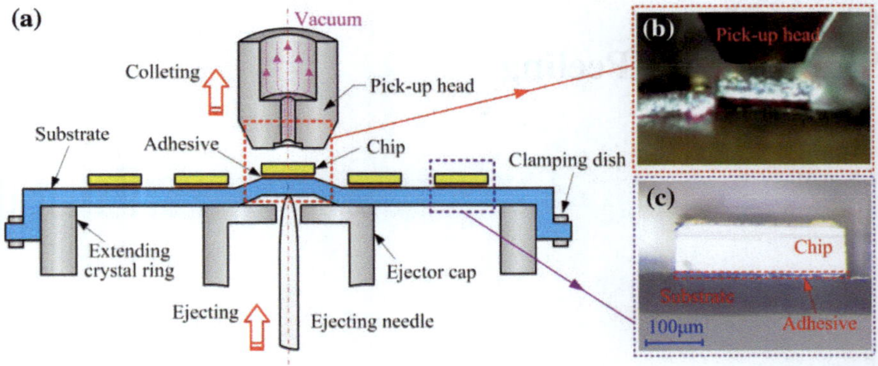

Fig. 5.1 Single-needle peeling-off process: **a** schematics of the needle ejecting process, **b** image of chip peeling from the adhesive substrate, and **c** image of the chip-adhesive-substrate structure. © 2014 IEEE. Reprinted, with permission, from Ref. [12]

The chip peeling-off from the substrate is demanded for the needle ejecting process. The cracking on the top surface center of chips, which arises from inappropriate handling processes, occurs frequently in the peeling-off process. It is more challenging for thin chips since the risk of cracking gets higher [11]. However, previous work investigated these two aspects separately. The chip peeling and chip cracking can be viewed as two kinds of fracture under the action of load. One is interfacial delamination, the other is brittle fracture at the top surface center of the chip. It is of significance to understand fundamentally the relationship between the chip peeling and the chip cracking, predict the fracture mode, and give guidelines to avoid the occurrence of the chip cracking.

This chapter is outlined as follows. First, an analytical model of the chip peeling-off process was proposed to obtain the equivalent loads and the ERR of interfacial peeling of the chip-on-substrate structure. Second, mixed boundary conditions were introduced into conventional fracture analysis of the chip-adhesive-substrate structure to address the single-needle ejecting process. The adhesive stresses and ERR of interfacial peeling were obtained analytically and compared with those obtained by the VCCT-based finite element simulation. Subsequently, a theoretical model to establish the competing fracture of the chip-on-substrate structure based on the competing index was proposed. The chip peelability, process window, and limitation of the needle ejecting process were explored. Lastly, the dynamic contact-impact effect in the needle ejecting process was investigated considering the factors of the impact speed, distance from the contact center, and substrate penetration.

5.2 A Case for Chip-on-Substrate Structure

To verify the theoretical analysis on the chip-on-substrate structure as proposed in Chap. 2, a common case for the peeling-off process is provided. As indicated in Chap. 2, the total ERR of interfacial peeling can be expressed combining Eqs. (2.13)–(2.15). The basic material properties and geometry dimensions are [11]: $E_{\text{chip}} = 1.29 \times 10^5$ MPa, $E_{\text{sub}} = 3.0 \times 10^3$ MPa, $\nu_{\text{chip}} = 0.28$, $\nu_{\text{sub}} = 0.38$, $h_{\text{chip}} = 0.1$ mm, $h_{\text{sub}} = 0.1$ mm, $l_{\text{chip}} = 1$ mm, $l_{\text{sub}} = 3$ mm, $P_{\text{needle}} = 1.5$ N. The intermediate parameters are defined as: $\overline{E}_{\text{chip}} = 1.40 \times 10^5$ MPa, $\overline{E}_{\text{sub}} = 3.51 \times 10^3$ MPa, $\sum = 39.9$, $H = 1$, $\Delta = 1.476$, $\tilde{I}_{(2)} = 4.386$, $C_2 = 0.222$, $C_3 = 0.759$, $\alpha = 0.951$, $\beta = 0$, $f_M = 15.669$, and $f_N = 16.124$. By substituting above values into Eqs. (2.13)–(2.15) and referring to the Table in the appendix of Ref. [21]. for the numerical values of phase angles, ψ_M, ψ_N, ψ_{Vd}, ψ_{Vs}, and nondimensional functions, f_{Vd}, f_{Vs}, the ERR of interfacial peeling for the chip-on-substrate structure can be obtained by

$$G_{\text{crack-tip}} = 0.0061 \times [(3.07 - 2l_{\text{chip-half}} + 2a)^2 + 0.262] \tag{5.1}$$

If the initiation of interfacial peeling is only considered here, the crack length is taken to be zero, i.e., $a = 0$. Then Eq. (5.1) can be simplified into

$$G_{\text{crack-tip}} = 0.0061 \times [(3.07 - 2l_{\text{crack-half}})^2 + 0.262] \tag{5.2}$$

If the chip length is taken to be constant at 1 mm and the crack propagation is considered. Then Eq. (5.1) can be simplified into

$$G_{\text{crack-tip}} = 0.0061 \times [(2.07 + 2a)^2 + 0.262] \tag{5.3}$$

Through observing the analytical results in Eqs. (2.14) and (2.14), dimensional considerations indicate that the ERR should take the following form [22]:

$$G_{\text{crack-tip}} = \frac{P_{\text{needle}}^2}{\overline{E}_{\text{chip}} h_{\text{chip}}} \times g\left(\frac{E_{\text{chip}}}{E_{\text{sub}}}, \frac{l_{\text{chip-half}}}{h_{\text{chip}}}, \frac{a}{h_{\text{chip}}}, \frac{h_{\text{chip}}}{h_{\text{sub}}}\right) \tag{5.4}$$

where g is a dimensionless function determined by finite element method. A normalizing factor of $P_{\text{needle}}^2/\overline{E}_{\text{chip}} h_{\text{chip}}$ can be chosen from Eq. (5.4) to examine effects of the chip length and the crack length. The normalized ERR of interfacial peeling is then given by

$$G_{\text{crack-tip}}/\left(\frac{P_{\text{needle}}^2}{\bar{E}_{\text{chip}} h_{\text{chip}}}\right) = g\left(\frac{E_{\text{chip}}}{E_{\text{sub}}}, \frac{l_{\text{chip-half}}}{h_{\text{chip}}}, \frac{a}{h_{\text{chip}}}, \frac{h_{\text{chip}}}{h_{\text{sub}}}\right) \tag{5.5}$$

5.2.1 *Effect of Crack Length*

The effect of the normalized crack length on the ERR of interfacial peeling and the mode-mixity angle is shown in Fig. 5.2. As can be seen, analytical results match the numerical simulations well throughout the entire range of the crack length and match even better when the crack length becomes smaller. Additionally, the ERR of interfacial peeling increases with the increase in the crack length. Once the interfacial peeling initiates, it can propagate by itself under the action of the needle ejecting, which is beneficial to the design of process parameters and selection of proper tools and materials for the chip peeling-off process. An initial chip peeling can be achieved by selecting suitable ejector needle and interfacial adhesive material. It also indicates that the mode-mixity angle is insensitive to the change of the crack length until the crack length is too long to attain the steady-state debonding.

5.2.2 *Effect of Geometric Dimensions*

The effect of the chip length on the ERR of interfacial peeling and the mode-mixity angle is shown in Fig. 5.3. The chip length varies from 0.4 to 2.0 mm. It is shown that analytical results match the numerical simulations well throughout the entire range of the chip length. When the chip length becomes larger, the results obtained by the two methods match even better. For the initiation of interfacial chip peeling, the mode-mixity angle is found insensitive to the change of the chip length. It can also be concluded that the larger chip length corresponds to a smaller ERR of interfacial

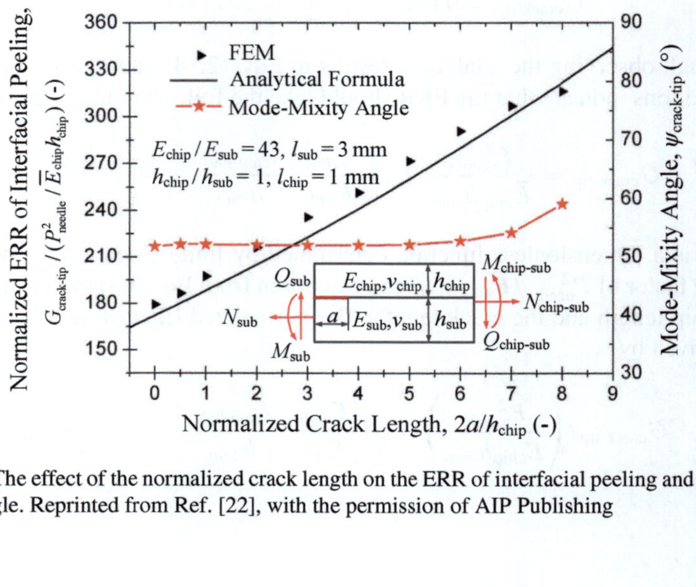

Fig. 5.2 The effect of the normalized crack length on the ERR of interfacial peeling and the mode-mixity angle. Reprinted from Ref. [22], with the permission of AIP Publishing

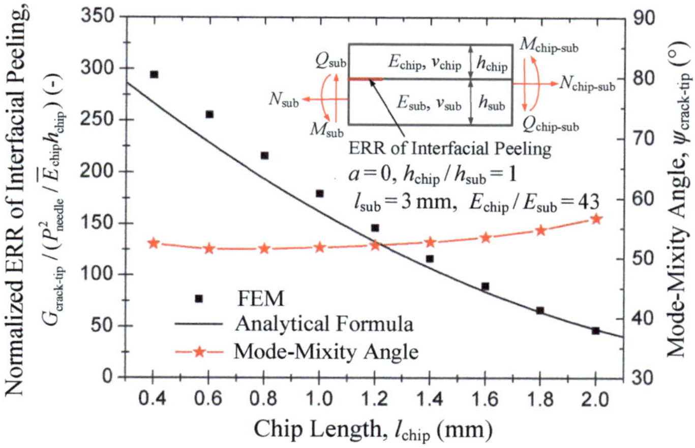

Fig. 5.3 The effect of the chip length on the ERR of interfacial peeling and the mode-mixity angle. Reprinted from Ref. [22], with the permission of AIP Publishing

peeling, which indicates that the larger chip is more difficult to be peeled off. In fact, as the chip becomes larger, the chip-on-substrate structure as a whole is strengthened, which leads to a smaller deformation and less strain in the structure. Thus, the ERR of interfacial peeling gets smaller. In order to achieve high success rate and suppress damage to the chip in the peeling-off process, it is suggested that the weaker adhesive strength should be chosen and multiple needles should be used for large chips.

The effect of the chip thickness on the ERR of interfacial peeling and the mode-mixity angle is shown in Fig. 5.4. The chip thickness varies from 0.005 mm to 0.3 mm. It is shown that the thinner chip corresponds to a smaller ERR of interfacial peeling. The ERR of interfacial peeling remains almost the same when the chip is thick enough, i.e., $h_{chip} > 0.20$ mm, which results from the huge mismatch of the bending stiffness between the chip and the substrate layers. When the chip becomes thinner, the mismatch of the flexural stiffness between the chip and the substrate gets much smaller since the flexural stiffness is related to the cubic of thickness. More compatible deformation and weaker interaction at the interface can then be found. The reduction in the ERR of interfacial peeling can thus be detected. It is also shown that the mode-mixity angle can attain steady-state values as the chip is thick enough, i.e., $h_{chip} > 0.20$ mm, and the interfacial peeling is gradually shear-dominated as the chip becomes thinner.

The above results also indicate that the interfacial peeling gets tougher for thinner chips during the peeling-off process, which is in accordance with the experimental testing results. Previous studies showed that the peeling-off process with chips of 5 mm × 5 mm × 0.1 mm failed easily, while the chips of 5 mm × 5 mm × 0.34 mm got 100% success rate of the peeling-off with an even wider process window [11]. It is suggested that a more compliant and thinner substrate and weaker adhesive strength can be adopted to achieve a high success rate of peeling-off. By comparing the result

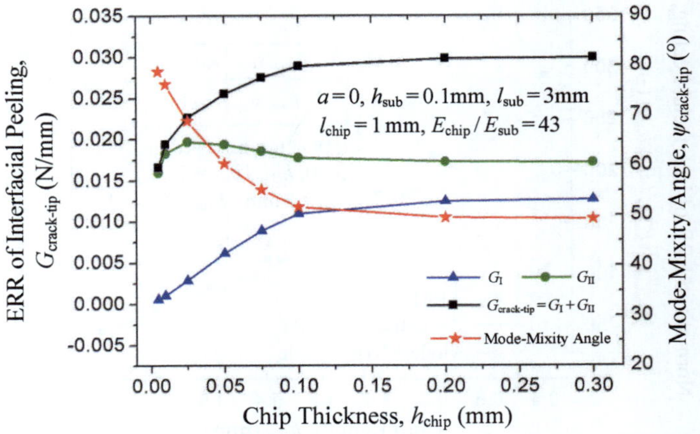

Fig. 5.4 The effect of the chip thickness on the ERR of interfacial peeling and the mode-mixity angle. Reprinted from Ref. [22], with the permission of AIP Publishing

in Fig. 5.3 with Fig. 5.4, it can be concluded that the effect of the chip length is greater than the effect of the chip thickness. Based on chips with the dimension of 1.0 mm × 0.1 mm, as the ratio of the chip length to the chip thickness varies from 10 (1.0 mm × 0.1 mm) to 20 (2.0 mm × 0.1 mm or 1.0 mm × 0.05 mm), the ERR of interfacial peeling decreases by 3.8 times for 2.0 mm × 0.1 mm chips while decreases merely 1.2 times for 1.0 mm × 0.05 mm chips.

5.2.3 *Effect of Elastic Mismatch*

The effect of the ratio of Young's modulus E_{chip}/E_{sub} on the ERR of interfacial peeling and the mode-mixity angle is shown in Fig. 5.5. It can be found that the ERR of interfacial peeling gets larger as the substrate becomes more compliant under the condition of the constant needle force. The mode-mixity angle shows little correlation with the elastic mismatch. It can be concluded that the successful peeling-off of the chip is easier to be achieved for a more compliant substrate, which can be explained by the much higher compliance of deformation. As the mismatch of the flexural stiffness between the chip and the substrate gets higher, more incompatible deformation can be detected and stronger interaction at the interface occurs, which leads to the increase in the ERR of interfacial peeling. It can thus be implied that the chip peeling gets easier for a softer substrate. More compliant substrate should be adopted to improve the success rate of the chip peeling-off.

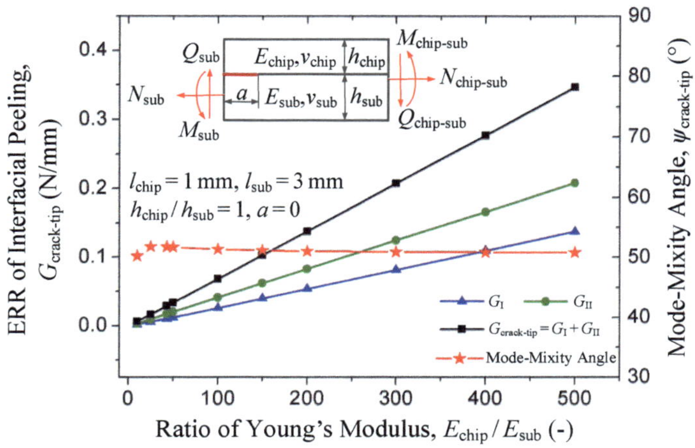

Fig. 5.5 The effect of the ratio of Young's modulus E_{chip}/E_{sub} on the ERR of interfacial peeling and the mode-mixity angle. Reprinted from Ref. [22], with the permission of AIP Publishing

5.3 A Case for Chip-Adhesive-Substrate Structure

5.3.1 Geometric Model Description

A geometric description of the chip-adhesive-substrate structure is shown in Fig. 5.6. In practice, the chip-adhesive-substrate structure usually undergoes prestress P_{pre} before the peeling-off process, as illustrated in Fig. 5.6a. Here, the chip-adhesive-substrate model in Fig. 5.6b is adopted. The substrate is under clamped-clamped boundary conditions, and subjected to a concentrated force P_{needle} resulting from the ejector needle. Then, the effect of the prestress is included to give insight into the mechanism of tunable mode peeling. The thicknesses of the chip and substrate layers are denoted by h_{chip} and h_{sub}, respectively. The corresponding lengths are represented by l_{chip} and l_{sub}, respectively. The thickness of the adhesive layer is represented by h_{adh}, where $h_{adh} \ll h_{chip}, h_{sub}$. The materials are assumed to be isotropic, and the corresponding Young's modulus and Poisson's ratio are E_{chip}, ν_{chip}, E_{sub}, ν_{sub}, and E_{adh}, ν_{adh} for the chip layer, substrate layer, and adhesive layer, respectively. The mechanical model can be limited to the right-side half portion considering the geometric symmetry, as seen in Fig. 5.6c. It is assumed that a crack will initiate from the free end of the adhesive layer and propagate along the midplane. The initial crack length and virtual crack propagation length are denoted by a and Δa, respectively.

As depicted in Fig. 5.6c, starting from the symmetry axis, the body is divided into Regions 1 and 2 with lengths $l_{reg\text{-}1} = l_{chip}/2\text{-}a$ and $l_{reg\text{-}2} = (l_{sub}\text{-}l_{chip})/2 + a$, respectively. For each segment $S_i (i = 1, 2, 3)$, the local rectangular coordinates (x_j, y_i, z_i) are utilized (here, $j = 1, 2$ represents Regions 1 and 2, respectively). The origin of the local coordinate system is defined at the midpoint of the left edge of each

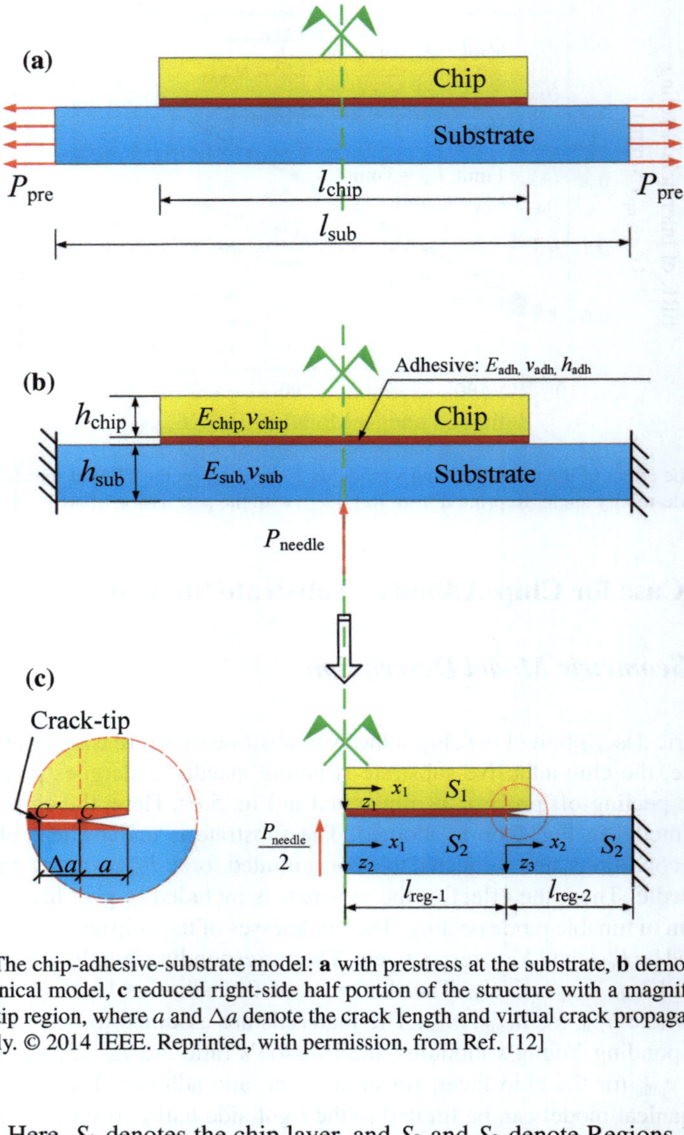

Fig. 5.6 The chip-adhesive-substrate model: **a** with prestress at the substrate, **b** demonstration of the mechanical model, **c** reduced right-side half portion of the structure with a magnified view of the crack-tip region, where a and Δa denote the crack length and virtual crack propagation length, respectively. © 2014 IEEE. Reprinted, with permission, from Ref. [12]

segment. Here, S_1 denotes the chip layer, and S_2 and S_3 denote Regions 1 and 2 of the substrate layer, respectively. Local coordinates x_1 and x_2 measure the distance in the axial direction from left-side end sections of Regions 1 and 2, respectively. Local coordinates z_1, z_2, and z_3 denote the distances in the transverse direction from the midplanes of segments S_1, S_2, and S_3, respectively. Here, each segment is modeled as an elastic beam according to Timoshenko theory. More details on the construction of the adhesive model for Regions 1 and 2 can be referred to Chap. 2.

5.3.1.1 Boundary and Continuity Conditions

(a) Symmetric conditions at the left-side end section of Region 1:

$$u_{\text{chip-reg-1}}^{\text{single}}\Big|_{x_1=0} = 0, \quad \phi_{\text{chip-reg-1}}^{\text{single}}\Big|_{x_1=0} = 0, \quad Q_{\text{chip-reg-1}}^{\text{single}}\Big|_{x_1=0} = 0 \tag{5.6}$$

$$u_{\text{sub-reg-1}}^{\text{single}}\Big|_{x_1=0} = 0, \quad \phi_{\text{sub-reg-1}}^{\text{single}}\Big|_{x_1=0} = 0, \quad Q_{\text{sub-reg-1}}^{\text{single}}\Big|_{x_1=0} = \frac{1}{2}P_{\text{needle}} \tag{5.7}$$

(b) Free end conditions at the right-side end of Region 1 in the chip layer:

$$N_{\text{chip-reg-1}}^{\text{single}}\Big|_{x_1=l_{\text{reg-1}}} = 0, \quad Q_{\text{chip-reg-1}}^{\text{single}}\Big|_{x_1=l_{\text{reg-1}}} = 0, \quad M_{\text{chip-reg-1}}^{\text{single}}\Big|_{x_1=l_{\text{reg-1}}} = 0 \tag{5.8}$$

(c) Continuity conditions at the cross sections connecting segments S_2 and S_3 of the substrate layer:

$$\begin{cases} u_{\text{sub-reg-1}}^{\text{single}}\Big|_{x_1=l_{\text{reg-1}}} = u_{\text{sub-reg-2}}^{\text{single}}\Big|_{x_2=0} \\ \phi_{\text{sub-reg-1}}^{\text{single}}\Big|_{x_1=l_{\text{reg-1}}} = \phi_{\text{sub-reg-2}}^{\text{single}}\Big|_{x_2=0} \\ w_{\text{sub-reg-1}}^{\text{single}}\Big|_{x_1=l_{\text{reg-1}}} = w_{\text{sub-reg-2}}^{\text{single}}\Big|_{x_2=0} \end{cases} \tag{5.9}$$

$$\begin{cases} N_{\text{sub-reg-1}}^{\text{single}}\Big|_{x_1=l_{\text{reg-1}}} = N_{\text{sub-reg-2}}^{\text{single}}\Big|_{x_2=0} \\ Q_{\text{sub-reg-1}}^{\text{single}}\Big|_{x_1=l_{\text{reg-1}}} = Q_{\text{sub-reg-2}}^{\text{single}}\Big|_{x_2=0} \\ M_{\text{sub-reg-1}}^{\text{single}}\Big|_{x_1=l_{\text{reg-1}}} = M_{\text{sub-reg-2}}^{\text{single}}\Big|_{x_2=0} \end{cases} \tag{5.10}$$

(d) Displacement conditions at the right-side end of Region 2 in the substrate layer:

$$u_{\text{sub-reg-2}}^{\text{single}}\Big|_{x_2=l_{\text{reg-2}}} = 0, \quad \phi_{\text{sub-reg-2}}^{\text{single}}\Big|_{x_2=l_{\text{reg-2}}} = 0, \quad w_{\text{sub-reg-2}}^{\text{single}}\Big|_{x_2=l_{\text{reg-2}}} = 0 \tag{5.11}$$

5.3.1.2 Determination of the Integration Constants

To sum up, there are a total of 26 integration constants to be determined. However, not all of them are concerned here, except for the first seven integration constants involved in the expressions for adhesive peel stress and shear stress. Typically, these integration constants are not all independent of each other. Hence, all the 26 integration constants can be determined by using the eight relationships among the integration constants and the 18 boundary and continuity conditions in Eqs. (5.6)–(5.11). The first six integration constants F_1, F_2, F_3, F_4, F_5, and F_6 can be given by the solution of the following linear equation set:

$$\sum_{n=1}^{6} f_n \frac{\lambda_n}{k_\sigma} = \frac{1}{2} c_2 \tag{5.12}$$

$$\sum_{n=1}^{6} f_n \frac{1}{\lambda_n} = -f_9 \tag{5.13}$$

$$\sum_{n=1}^{6} f_n \frac{1}{\lambda_n^3} = \alpha_1 \tag{5.14}$$

$$\sum_{n=1}^{6} f_n \frac{1}{\lambda_n} \exp\left(\lambda_n l_{\text{reg-1}}\right) = -f_9 \tag{5.15}$$

$$\sum_{n=1}^{6} f_n \left[\beta_1 \left(\frac{\lambda_n^2}{k_\sigma} - c_1 - c_2 \right) - \left(\frac{1}{\lambda_n^3} - \frac{\beta_2}{\lambda_n^2} + \frac{l_{\text{reg-1}}^2}{2\lambda_n} \right) \right] \exp\left(\lambda_n l_{\text{reg-1}}\right) = \beta_3 - \alpha_1 \tag{5.16}$$

$$\sum_{n=1}^{6} f_n \left[\gamma_1 \left(\frac{\lambda_n^2}{k_\sigma} - c_1 - c_2 \right) + \frac{\lambda_n}{k_\sigma} + \frac{\gamma_2}{\lambda_n^2} \right] \exp\left(\lambda_n l_{\text{reg-1}}\right) = \gamma_3 \tag{5.17}$$

where $f_n = F_n/P_{\text{needle}}$, $\alpha_1 = -[f_7/k_\tau + (a_{\text{chip}} + a_{\text{sub}} + d_{\text{chip}}h^2_{\text{chip-half}} + d_{\text{sub}}h^2_{\text{sub-half}})(c_{\text{chip}}f_9 + c_{\text{sub}}f_{12})]/[(a_{\text{chip}} + a_{\text{sub}})(d_{\text{chip}} + d_{\text{sub}}) + d_{\text{chip}}d_{\text{sub}}(h_{\text{chip-half}} + h_{\text{sub-half}})^2]$, $\beta_1 = (l_{\text{reg-1}} + l_{\text{reg-2}})[a_{\text{chip}}d_{\text{sub}}h_{\text{sub-half}} + a_{\text{sub}}d_{\text{chip}}h_{\text{chip-half}} + d_{\text{chip}}d_{\text{sub}}h_{\text{chip-half}}h_{\text{sub-half}}(h_{\text{chip-half}} + h_{\text{sub-half}})]/[a_{\text{sub}}d_{\text{sub}}(d_{\text{chip}}h_{\text{chip-half}} - d_{\text{sub}}h_{\text{sub-half}})]$, $\beta_2 = l_{\text{reg-1}} + h_{\text{sub-half}}(l_{\text{reg-1}} + l_{\text{reg-2}})[(a_{\text{chip}} + a_{\text{sub}})(d_{\text{chip}} + d_{\text{sub}}) + d_{\text{chip}}d_{\text{sub}}(h_{\text{chip-half}} + h_{\text{sub-half}})^2]/a_{\text{sub}}(d_{\text{chip}}h_{\text{chip-half}} - d_{\text{sub}}h_{\text{sub-half}})$, $\beta_3 = -0.25l^2_{\text{reg-1}} + 0.25l^2_{\text{reg-2}}$, $\gamma_1 = l_{\text{reg-2}} + d_{\text{chip}}(l_{\text{reg-1}} + l_{\text{reg-2}})[a_{\text{chip}}d_{\text{sub}}(h_{\text{chip-half}} + h_{\text{sub-half}}) + a_{\text{sub}}h_{\text{chip-half}}(d_{\text{chip}} + d_{\text{sub}}) + d_{\text{chip}}d_{\text{sub}}h_{\text{chip-half}}(h_{\text{chip-half}} + h_{\text{sub-half}})^2]/a_{\text{sub}}d_{\text{sub}}(d_{\text{chip}}h_{\text{chip-half}} - d_{\text{sub}}h_{\text{sub-half}})$, $\gamma_2 = d_{\text{chip}}(h_{\text{chip-half}} + h_{\text{sub-half}})(l_{\text{reg-1}} + l_{\text{reg-2}})[(a_{\text{chip}} + a_{\text{sub}})(d_{\text{chip}} + d_{\text{sub}}) + d_{\text{chip}}d_{\text{sub}}(h_{\text{chip-half}} + h_{\text{sub-half}})^2]/a_{\text{sub}}(d_{\text{chip}}h_{\text{chip-half}} - d_{\text{sub}}h_{\text{sub-half}})$, $\gamma_3 = 0.5c_{\text{sub}} - 0.25d_{\text{chip}}l^2_{\text{reg-1}} + 0.25(d_{\text{chip}} + d_{\text{sub}})l^2_{\text{reg-2}}$, and

$$f_7 = -\frac{1}{2} \frac{d_{\text{chip}}d_{\text{sub}}\left(h_{\text{chip-half}} + h_{\text{sub-half}}\right)\left(d_{\text{chip}}h_{\text{chip-half}}\text{-}d_{\text{sub}}h_{\text{sub-half}}\right)}{d_{\text{chip}}d_{\text{sub}}\left(h_{\text{chip-half}} + h_{\text{sub-half}}\right)^2 + \left(a_{\text{chip}} + a_{\text{sub}}\right)\left(d_{\text{chip}} + d_{\text{sub}}\right)} \tag{5.18}$$

$$f_9 = -\frac{1}{2} \frac{d_{\text{sub}}\left[d_{\text{chip}}h_{\text{chip-half}}\left(h_{\text{chip-half}} + h_{\text{sub-half}}\right) + \left(a_{\text{chip}} + a_{\text{sub}}\right)\right]}{d_{\text{chip}}d_{\text{sub}}\left(h_{\text{chip-half}} + h_{\text{sub-half}}\right)^2 + \left(a_{\text{chip}} + a_{\text{sub}}\right)\left(d_{\text{chip}} + d_{\text{sub}}\right)} \tag{5.19}$$

$$f_{12} = \frac{1}{2} \frac{d_{\text{chip}}\left[d_{\text{sub}}h_{\text{sub-half}}\left(h_{\text{chip-half}} + h_{\text{sub-half}}\right) + \left(a_{\text{chip}} + a_{\text{sub}}\right)\right]}{d_{\text{chip}}d_{\text{sub}}\left(h_{\text{chip-half}} + h_{\text{sub-half}}\right)^2 + \left(a_{\text{chip}} + a_{\text{sub}}\right)\left(d_{\text{chip}} + d_{\text{sub}}\right)} \tag{5.20}$$

MATLAB software is used here to solve above linear equation set. Then, the analytical adhesive stresses are obtained by substituting the first seven integration constants into the general expressions for the adhesive peel stress and shear stress.

5.3.2 *Theoretical Calculation of ERR*

Based on the analytical model, the total ERR of interfacial peeling can be approximated as [23, 24]

$$G_{\mathrm{adh}}^{\text{crack-tip}} = G_{\mathrm{I}} + G_{\mathrm{II}} \tag{5.21}$$

where $G_{\mathrm{I}} = \frac{\sigma_{\text{crack-tip}}^2}{2k_\sigma}$ and $G_{\mathrm{II}} = \frac{\tau_{\text{crack-tip}}^2}{2k_\tau}$, and there exist

$$\sigma_{\text{crack-tip}} = \sigma|_{x_1 = l_{\text{chip-half}}} = P_{\text{needle}} \sum_{n=1}^{6} f_n \exp(\lambda_n l_{\text{chip-half}}) \tag{5.22}$$

$$\begin{aligned} \tau_{\text{crack-tip}} = \tau|_{x_1 = l_{\text{chip-half}}} = &\frac{P_{\text{needle}}}{2(d_{\text{chip}} h_{\text{chip-half}} \text{-} d_{\text{sub}} h_{\text{sub-half}})} \\ &\times \left[d_{\text{sub}} + 2\sum_{n=1}^{6} f_n \lambda_n \left(c_{\text{chip}} + c_{\text{sub}} \text{-} \frac{\lambda_n^2}{k_\sigma} \right) \exp(\lambda_n l_{\text{chip-half}}) \right] \end{aligned} \tag{5.23}$$

The adhesive peel stress and shear stress at the crack tip are represented by Eqs. (5.22) and (5.23), respectively. It is worth noting that Eq. (5.21) results in finite values even when no crack is presented. The mode-mixity angle $\psi_{\mathrm{adh}}^{\text{crack-tip}}$ characterizing the relative portion of these two modes is defined as $\psi_{\mathrm{adh}}^{\text{crack-tip}} = \arctan\left[(G_{\mathrm{II}}/G_{\mathrm{I}})^{0.5}\right]$, and can be expressed as

$$\psi_{\mathrm{adh}}^{\text{crack-tip}} = \arctan\left[\sqrt{\frac{k_\sigma}{k_\tau}} \frac{d_{\text{sub}} + 2\sum_{n=1}^{6} f_n \lambda_n \left(c_{\text{chip}} + c_{\text{sub}} \text{-} \lambda_n^2 / k_\sigma\right) \exp(\lambda_n l_{\text{chip-half}})}{2(d_{\text{chip}} h_{\text{chip-half}} \text{-} d_{\text{sub}} h_{\text{sub-half}}) \sum_{n=1}^{6} f_n \exp(\lambda_n l_{\text{chip-half}})} \right] \tag{5.24}$$

In general, $\psi_{\mathrm{adh}}^{\text{crack-tip}} = 0°$ and $90°$ correspond to the pure mode I and mode II cracking, respectively. To verify the theoretical predictions by Eqs. (5.22)–(5.24), the analytical results are compared with those obtained through the VCCT method [8, 20, 22].

5.3.3 Effect of Crack Length

The effect of the crack length on the ERR of interfacial peeling and the mode-mixity angle is shown in Fig. 5.7. Here, the lengths of the chip and substrate layers are fixed at 1 mm and 3 mm, respectively. It is shown that the analytical predictions and numerical results match well, especially when the crack length is relatively small. The chips are peeled off from adhesive tapes mainly by mode I fracture and the tendency will be maintained unless the crack is too large to attain the steady-state debonding [25]. It can also be depicted that the mode-mixity angle is below 45° and insensitive to the change of the crack length when $2a/l_{\mathrm{chip}}$ is smaller than 0.6. An obvious increasing trend of the mode-mixity angle can be found when $2a/l_{\mathrm{chip}}$ becomes larger than 0.6, which implies that the dominant fracture of the initial chip peeling changes from the mode I to the mode II. The total ERR of interfacial peeling increases with the crack propagation until the crack extends to convergent debonding after $2a/l_{\mathrm{chip}} = 0.8$. In fact, it is the reason why the pickup head needs to draw the chip up when about 80% of the bonded region has been debonded [22, 26]. Thus, it is of significance to optimize process parameters for the chip peeling-off process, aiming at the initial peeling, e.g., choosing proper ejector needle force and adhesive material.

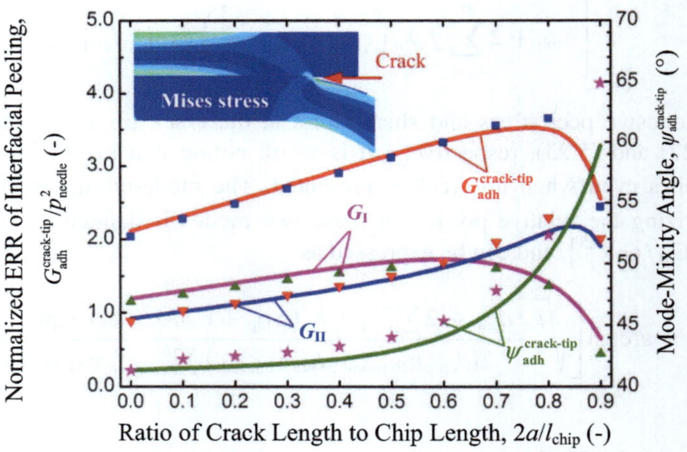

Fig. 5.7 The effect of the crack length on the ERR of interfacial peeling and the mode-mixity angle, where analytical predictions and numerical results are represented by solid lines and scatter plots, respectively. © 2014 IEEE. Reprinted, with permission, from Ref. [12]

5.3.4 *Effect of Geometric Dimensions*

The key geometric dimensions determining whether the chips can be successfully peeled off include the chip thickness and the chip length. The effect of the chip length on the ERR of interfacial peeling and the mode-mixity angle is shown in Fig. 5.8. The substrate length is fixed at 3 mm. The chip length varies from 0.4 mm to 2.0 mm. As can be seen, the analytical results match well with those from the finite element analysis within the entire range of the chip length. The longer chip tends to promote a smaller total ERR of interfacial peeling, which means that the longer chip is more difficult to be peeled off. Additionally, it can also be found that the mode I contribution to the ERR of interfacial peeling is larger than that of the mode II within the range of the chip length. Besides, a very limited variation of the mode-mixity angle, approximately from 39° to 44°, can be observed.

The effect of the chip thickness on the ERR of interfacial peeling and the mode-mixity angle is shown in Fig. 5.9. The lengths of the chip and substrate layers are fixed at 1 mm and 3 mm, respectively. The chip thickness varies from 0.02 mm to 0.3 mm. It can be observed that the thinner chip corresponds to a lower total ERR of interfacial peeling. The ERR of interfacial peeling as well as the mode-mixity angle are almost constant when the chip is thick enough, i.e., $h_{\text{chip-half}} > 0.1$ mm. The conclusion is in accordance with that obtained by Peng et al. [22]. In addition, the peeling mode turns to be shear-dominated when the chip becomes very thin.

Therefore, the chip peeling-off from the adhesive tape, normally starting at the chip ends, would get tougher for thinner and/or larger chips, which is consistent with the experimental testing results by Cheng et al. [11]. In practice, the chip peeling-off

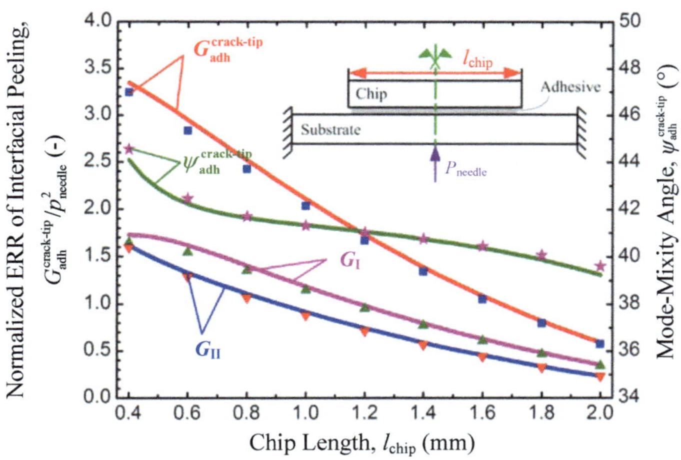

Fig. 5.8 The effect of the chip length on the ERR of interfacial peeling and the mode-mixity angle, where analytical predictions and numerical results are represented by solid lines and scatter plots, respectively. © 2014 IEEE. Reprinted, with permission, from Ref. [12]

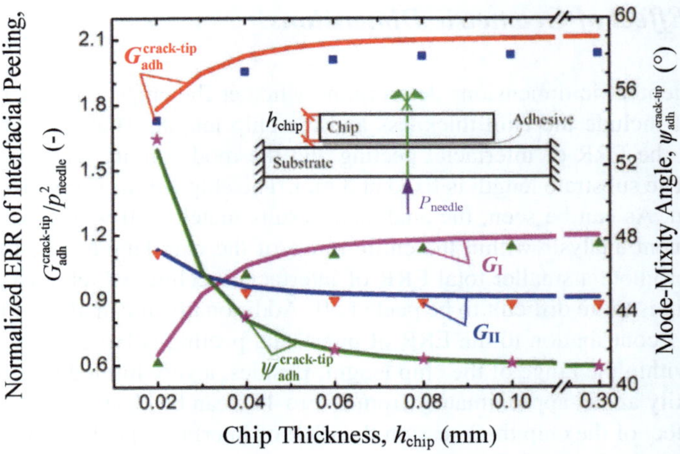

Fig. 5.9 The effect of the chip thickness on the ERR of interfacial peeling and the mode-mixity angle, where analytical predictions and numerical results are represented by solid lines and scatter plots, respectively. © 2014 IEEE. Reprinted, with permission, from Ref. [12]

by single ejector needle has already reached the process limit for the chip thinner than 50 μm [2]. For thin and large chips, it is suggested that the adhesive tape need to undergo appropriate prestress, and forced by more needles. Then, the chips can be lifted and picked up by a vacuum tool. Even so, it is still insufficient to handle chips thinner than 20 μm.

5.3.5 *Effect of Material Properties*

The effect of Young's modulus of the substrate layer on the ERR of interfacial peeling and the mode-mixity angle is shown in Fig. 5.10. The geometric dimensions are maintained the same as previous subsections. Young's modulus of the substrate is selected varied from 160 MPa to 2.0 GPa, which covers most of the substrate materials used in the industry. It seems that not only the total ERR of interfacial peeling but also the modes I and II ERR of interfacial peeling increase exponentially as the substrate becomes more compliant, indicating that material properties of the substrate layer have important effects on the ERR. In addition, the mode-mixity angle is always smaller than 45°, which implies that the mode I fracture remains the dominant peeling mode.

The effect of Young's modulus of the adhesive layer on the ERR of interfacial peeling and the mode-mixity angle is shown in Fig. 5.11. Young's modulus of the adhesive layer varies from 20 MPa to 140 MPa. The agreement between analytical model and finite element method is not very good from the quantitative aspect, but the trend is identical. The deviations mainly result from the model itself which is more

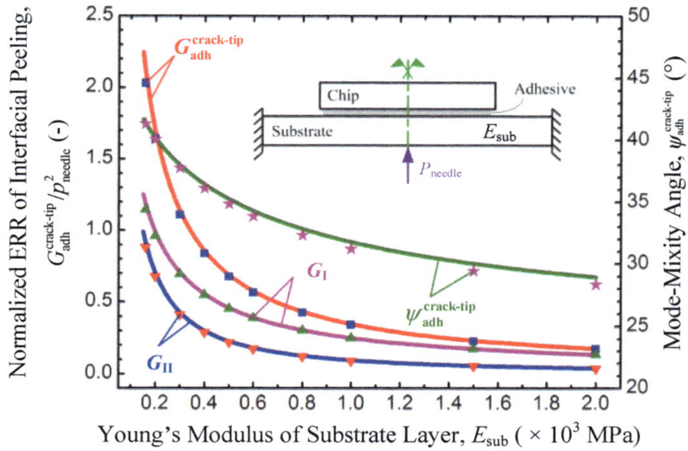

Fig. 5.10 The effect of Young's modulus of the substrate layer on the ERR of interfacial peeling and the mode-mixity angle, where analytical predictions and numerical results are represented by solid lines and scatter plots, respectively. © 2014 IEEE. Reprinted, with permission, from Ref. [12]

accurate on the condition that the chip and substrate layers are much thicker and stiffer than the adhesive layer [27]. Therefore, the loss of accuracy will be introduced as Young's modulus of the adhesive layer gets closer to that of the substrate layer. Even though the ERR shows decreasing trends with the increase of Young's modulus of the adhesive layer, the dependency appears quite unobvious, suggesting the insensitivity of the ERR to the narrow range of Young's modulus of the adhesive layer. Obviously, a more compliant adhesive layer should be used to improve the success rate of the chip peeling-off process. Thus, some aided processes like the ultraviolet (UV) curing can be adopted, whose purpose is to reduce the strength of both the adhesive and the substrate layers [28].

5.3.6 Analysis on Tunability of Peeling Mode

It can be concluded from the above analysis that the chip thickness and the adhesive material property play important roles in the ERR of interfacial peeling and mode transformation. As discussed in Chap. 4 [20], when the substrate of the chip-adhesive-substrate model is subjected to a prestress as shown in Fig. 5.6a, the mode II peeling is promoted. Meanwhile, the adhesive layer is subjected to compression at the edge, namely, a negative contribution to the mode I type of peeling can be detected. It may be an effective way to achieve the tunability of the peeling mode if the prestress is superimposed onto the ejector needle force, as shown in Fig. 5.6b.

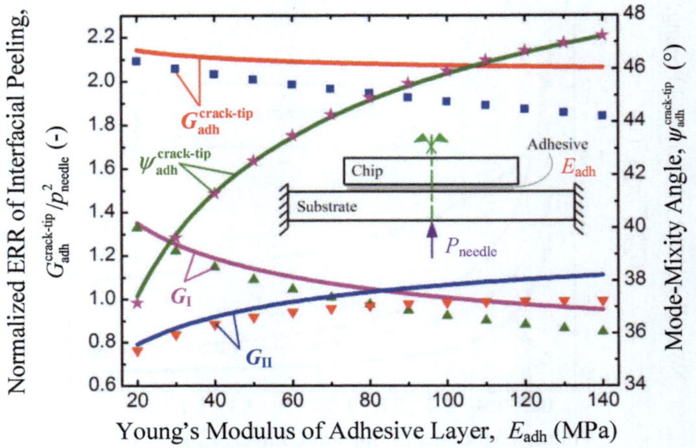

Fig. 5.11 The effect of Young's modulus of the adhesive layer on the ERR of interfacial peeling and the mode-mixity angle, where analytical predictions and numerical results are represented by solid lines and scatter plots, respectively. © 2014 IEEE. Reprinted, with permission, from Ref. [12]

Comparisons of the contributions of modes I and II peeling to the mixed-mode peeling without and with the prestress are shown in Fig. 5.12. Here, the prestress is fixed at $P_{pre} = 2.5\,\mathrm{P}_{needle}$. In Fig. 5.12a, the critical chip thickness for the transformation of the peeling mode under pure action of the ejector needle is represented by h_{chip}^c. The critical chip thickness gets larger as the prestress is superimposed, which suggests that the process window of the mode-II-dominated peeling can be widened to thicker chip by increasing the prestress. Similarly, the critical Young's modulus of the adhesive layer, E_{adh}^c, can also be changed by the prestress, as shown in Fig. 5.12b. The critical Young's modulus moves towards more compliant adhesive materials as the prestress increases.

Tunability analysis of the peeling mode by the implementation of the prestress is schematically illustrated in Fig. 5.13, from which it can be concluded that the prestress contributes to promoting the mode-II-dominated peeling for thicker chips and more compliant adhesive tapes. In practical chip peeling-off process with a single ejector needle, deformation of the thin and large chips remains consistent with the adhesive tape even under high degree of deformation, which implies that the chip peeling is difficult to be initiated in the mode I. Therefore, an effective approach to promote the initial peeling can be achieved by implementing the prestress, through which the contribution of the mode II peeling can be increased [2].

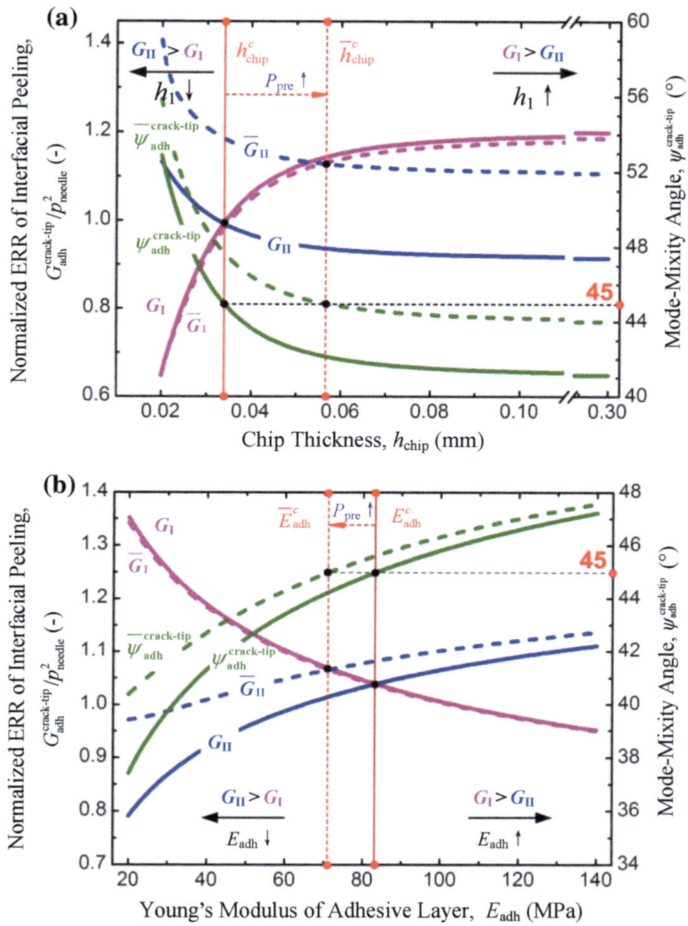

Fig. 5.12 Comparisons of the contributions of modes I and II peeling to the mixed-mode peeling without and with the prestress: **a** the effect of the chip thickness and **b** the effect of Young's modulus of the adhesive layer. The results obtained without and with the prestress are represented by solid and dash lines, respectively. © 2014 IEEE. Reprinted, with permission, from Ref. [12]

5.4 Analysis of Fracture Mode Based on Competing Index

5.4.1 Typical Fracture Modes

The chip-on-substrate model ignoring the adhesive layer is illustrated in Fig. 5.14a. In the model, the silicon chip is bonded to the top surface of the substrate with the center coinciding with the origin of the coordinate system. The substrate is subjected to a transversely concentrated force resulting from the ejector needle under clamped—

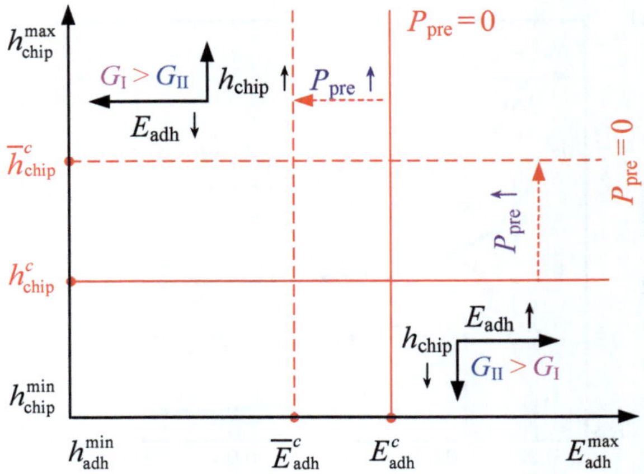

Fig. 5.13 Tunability analysis of the peeling mode, where the region described by the solid and dashed lines respectively represent the condition without and with the prestress. In both of the crisscross regions, the top left corner represents the mode-I-dominated condition, i.e., $G_I > G_{II}$, and the bottom right corner represents the mode-II-dominated condition, i.e., $G_{II} > G_I$. © 2014 IEEE. Reprinted, with permission, from Ref. [12]

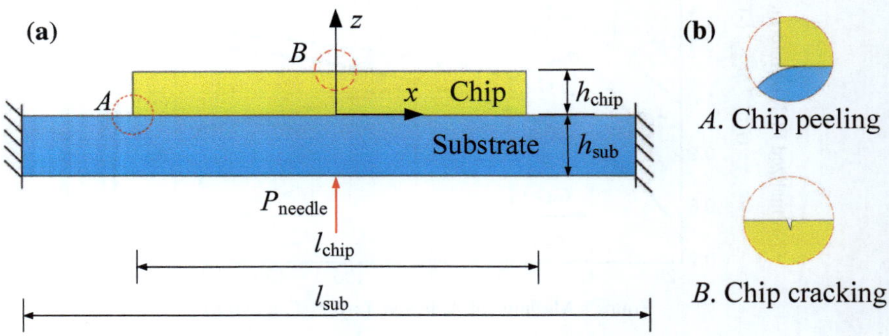

Fig. 5.14 Chip-on-substrate structure and typical fracture modes: **a** mechanical model, **b** two competing fracture modes

clamped boundary conditions. The typical fracture modes of the chip peeling and chip cracking are described in Fig. 5.14b. In general, the chip peeling occurs at the chip corner and propagates along the interface between the chip and the substrate, while the chip cracking occurs at the top surface center of the chip.

5.4.2 Competing Fracture Model

As to the analysis of the competing fracture behavior based on the chip-on-substrate structure, four mechanical parameters should be considered, i.e., $G_{\mathrm{adh}}^{\text{crack-tip}}$, G_{adh}^{c}, $\sigma_{\mathrm{chip}}^{\max}$, and $\sigma_{\mathrm{chip}}^{c}$. To be specific, $G_{\mathrm{adh}}^{\text{crack-tip}}$ is the ERR of interfacial peeling contributing to the chip peeling, G_{adh}^{c} is the adhesive fracture energy of the adhesive layer, $\sigma_{\mathrm{chip}}^{\max}$ denotes the maximum tensile stress on the top surface of the chip contributing to the chip cracking, and $\sigma_{\mathrm{chip}}^{c}$ represents the ultimate tensile strength of the silicon chip.

On one hand, as to the chip peeling, a wedge with 90° can be viewed as an initial crack with zero length [29, 30]. The interfacial crack initiates once the ERR of interfacial peeling reaches the adhesive fracture energy based on Griffith fracture criterion. The chip peeling is similar to the steady-state propagation of an interfacial crack from the substrate. On the other hand, the chip cracking can be predicted by comparing the maximum tensile stress of the layer with the corresponding ultimate tensile strength. Thus, the criteria for these two fracture modes are as follows: $G_{\mathrm{adh}}^{\text{crack-tip}} > G_{\mathrm{adh}}^{c}$ for chip peeling, and $\sigma_{\mathrm{chip}}^{\max} > \sigma_{\mathrm{chip}}^{c}$ for chip cracking.

Furthermore, there exists a competing relationship between these two kinds of fracture modes. If the chip peeling occurs first, the strain energy of the chip-on-substrate structure is released, the maximum tensile stress of the chip layer stops increasing, and the chip will be intact. Conversely, if the chip cracking occurs first, the chip peeling will be suppressed. In practice, only one of these two modes will occur finally. For the peeling-off process, the chip peeling is desired and the chip cracking should be avoided. It is thus essential to examine the competing fracture mechanism.

Based on the above discussion, the criterion of the competing fracture can be obtained by

$$\frac{G_{\mathrm{adh}}^{\text{crack-tip}}}{G_{\mathrm{adh}}^{c}} > \frac{\sigma_{\mathrm{chip}}^{\max}}{\sigma_{\mathrm{chip}}^{c}}, \text{ for chip peeling} \tag{5.25}$$

$$\frac{G_{\mathrm{adh}}^{\text{crack-tip}}}{G_{\mathrm{adh}}^{c}} < \frac{\sigma_{\mathrm{chip}}^{\max}}{\sigma_{\mathrm{chip}}^{c}}, \text{ for chip cracking} \tag{5.26}$$

The fracture mode is uncertain under the critical condition of $G_{\mathrm{adh}}^{\text{crack-tip}}/G_{\mathrm{adh}}^{c} = \sigma_{\mathrm{chip}}^{\max}/\sigma_{\mathrm{chip}}^{c}$. Combining Eqs. (5.25) and (5.26), the competing fracture criterion can be rewritten as

$$\frac{G_{\mathrm{adh}}^{\text{crack-tip}}}{\sigma_{\mathrm{chip}}^{\max}} > \frac{G_{\mathrm{adh}}^{c}}{\sigma_{\mathrm{chip}}^{c}}, \text{ for chip peeling} \tag{5.27}$$

$$\frac{G_{\mathrm{adh}}^{\text{crack-tip}}}{\sigma_{\mathrm{chip}}^{\max}} < \frac{G_{\mathrm{adh}}^{c}}{\sigma_{\mathrm{chip}}^{c}}, \text{ for chip cracking} \tag{5.28}$$

A competing index can thus be defined here to characterize the competing fracture mode, as given by

$$\Gamma_{\text{fracture}} = \frac{G_{\text{adh}}^{\text{crack-tip}}}{\sigma_{\text{chip}}^{\max}} \tag{5.29}$$

Particularly, $\Gamma_c = G_{\text{adh}}^c/\sigma_{\text{chip}}^c$ denotes the critical competing index determined by the material properties of the adhesive and chip layers. Therefore, the result of competing fracture between chip peeling and chip cracking depends on the competing index and the critical competing index. The criterion of competing fracture can be given by

$$\Gamma_{\text{fracture}} > \Gamma_c, \text{ for chip peeling} \tag{5.30}$$

$$\Gamma_{\text{fracture}} < \Gamma_c, \text{ for chip cracking} \tag{5.31}$$

The geometric dimensions affecting the competing index including the chip thickness and chip length are examined. Take the chip thickness for example, thick chips are easy to be peeled off and widely used in the semiconductor industry, while the chip cracking tends to occur easily for thin ones, leading to a high damage risk for integrated circuit (IC) packaging [11, 26, 31]. The chip peeling occurs at relative large thickness and the chip cracking occurs at relative small thickness, which suggests that there exists a critical value of the chip thickness h_{chip}^c separating the chip peeling and chip cracking. The critical chip thickness can be obtained on condition that $\Gamma_{\text{fracture}}(h_{\text{chip}}^c) = \Gamma_c$. For a chip thicker than the critical chip thickness, the chip peeling occurs, otherwise, the chip cracking can be detected. Similarly, there exists a critical value of the chip length separating the chip peeling and chip cracking.

5.4.3 *Basic Fracture Parameters*

Abaqus software is employed to calculate the stress distribution and the ERR of interfacial peeling. A common case of the chip-on-substrate structure with material properties and geometric dimensions the same as those in Chap. 2 is adopted. A detailed description of the FEM model has already been given in Chap. 2. Besides, the critical competing index is determined here as $\Gamma_c = 0.23 - 1.08 \times 10^{-4}$ mm.

To validate the proposed competing fracture mechanism, the Von Mises stress contour of one half of the chip-on-substrate structure is shown in Fig. 5.15, the upper part is the silicon chip and the lower part is the substrate. The stress distribution is discontinuous around the bonded interface due to the discontinuity of the interfacial displacement and the mismatch of Young's modulus. The stress at the edge of the interface is relatively high. High stress distribution can also be found around the center of the top surface of the chip. Thus, it shows clearly that there are two critical regions

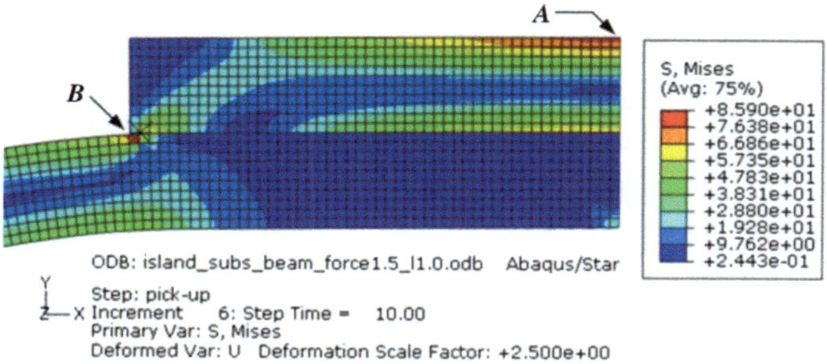

Fig. 5.15 The Von Mises stress contour of one half of the chip-on-substrate structure. © 2012 IEEE. Reprinted, with permission, from Ref. [8]

where the fracture tends to occur. The corresponding locations are represented by regions A and B, as illustrated in Fig. 5.15. The chip cracking and chip peeling tend to respectively occur at regions A and B.

5.4.4 *Effect of Geometric Dimensions*

The effect of the chip thickness on the ERR of interfacial peeling and the maximum tensile stress of the chip layer is shown in Fig. 5.16, in which the chip length and substrate length are, respectively, fixed at 1.0 mm and 3.0 mm. The adhesive thickness is selected as 0.1 mm. The ejector needle force is selected as 1.5 N. The ratio of Young's modulus $E_{\text{chip}}/E_{\text{sub}} = 43$. The chip thickness is designed to be varied from 0.005 mm to 0.3 mm. As can be seen, the thinner chip corresponds to a smaller total ERR, which indicates that the chip peeling is tougher for thinner chips. On the other hand, the maximum tensile stress of the chip layer gets higher as the chip becomes thinner. Therefore, the conclusion can be used to interpret the experimental results obtained by Cheng et al. [11, 32] concerning the effect of the chip thickness, that is, why the chips of 5 mm × 5 mm × 0.1 mm tended to fail while the chips of 5 mm × 5 mm × 0.34 mm got 100% success rate of the chip peeling-off under the same condition. In addition, there must be a critical thickness which separates the mode of fracture from chip peeling to chip cracking. Moreover, when the chip is thin enough, both the ERR of interfacial peeling and the maximum tensile stress of the chip layer vary remarkably, showing high sensitivity to the chip thickness.

The effect of the chip thickness on the competing index is shown in Fig. 5.17, in which the semilogarithmic coordinates are used. The horizontal line represents $\Gamma = \Gamma_c$. As can be found, the competing index increases monotonically with increase in the chip thickness. The point of intersection indicates the critical chip thickness. The chip-on-substrate structure depicts the chip cracking and chip peeling behavior when the chip thickness is smaller and larger than the critical chip thickness, respectively.

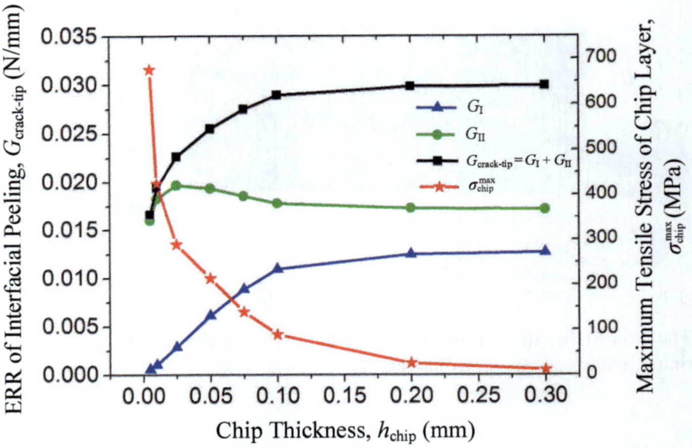

Fig. 5.16 The effect of the chip thickness on the ERR of interfacial peeling and the maximum tensile stress of the chip layer. © 2012 IEEE. Reprinted, with permission, from Ref. [8]

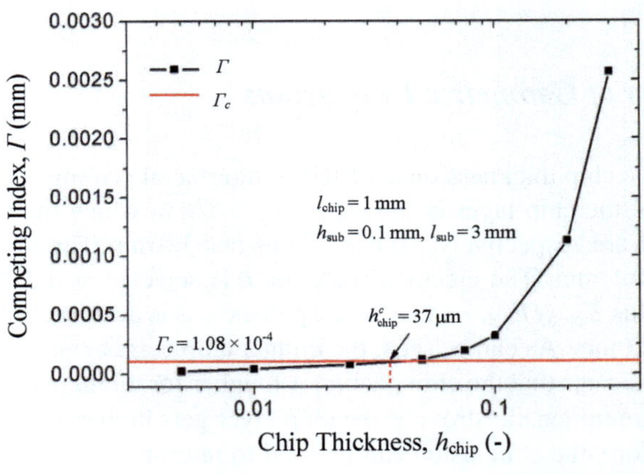

Fig. 5.17 The effect of the chip thickness on the competing index. © 2012 IEEE. Reprinted, with permission, from Ref. [8]

For the chip peeling-off process, the desired chip peeling will not occur when $\Gamma_{max} < \Gamma_c$. For relatively small chips, i.e., smaller than 2 mm × 2 mm, single ejector needle is generally used for the peeling-off process. The process limit can be generally indicated by the critical chip thickness. Let $G_{adh}^c = 14\,\mathrm{J/m^2}$ and $\sigma_{chip}^c = 130\,\mathrm{MPa}$, then $\Gamma_c = G_{adh}^c/\sigma_{chip}^c = 1.08 \times 10^{-4}\mathrm{mm}$ and $h_{chip}^c = 0.037\,\mathrm{mm}$ can be obtained. It can thus be deduced that the chip can be peeled off from the substrate by the single ejector needle only when the chip thickness is larger than 37 μm.

The effect of the chip length on the ERR of interfacial peeling and the maximum tensile stress of the chip layer is shown in Fig. 5.18, in which $h_{chip} = h_{sub} =$

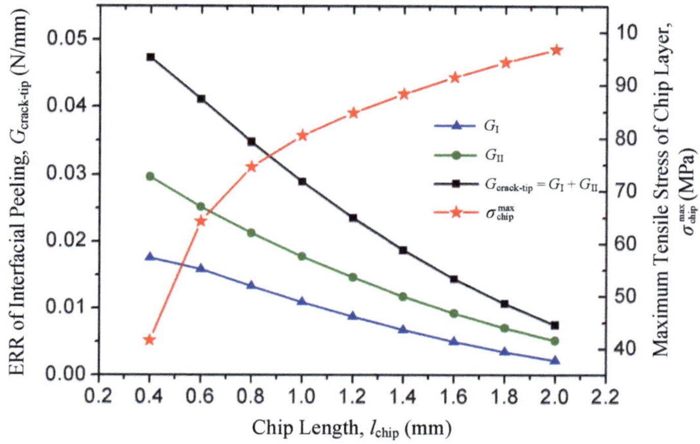

Fig. 5.18 The effect of the chip length on the ERR of interfacial peeling and the maximum tensile stress of the chip layer. © 2012 IEEE. Reprinted, with permission, from Ref. [8]

0.1 mm, $E_{\text{sub}} = 43$, and $P_{\text{needle}} = 1.5\,\text{N}$. The chip length varies from 0.4 mm to 2.0 mm. It can be concluded that the total ERR of interfacial peeling decreases with the increase of the chip length, on the contrary, the maximum tensile stress of the chip layer increases as the chip length enlarges, which indicates that larger chips are more difficult to be peeled off under the same process condition. It can also be inferred that there should exist a critical chip length that separates the chip peeling from chip cracking. The chip peeling occurs when the chip length is smaller than the critical value, and the chip cracking should occur when the chip length is larger than the critical value.

The effect of the chip length on the competing index is shown in Fig. 5.19, which clearly reveals the competing relationship. The horizontal line represents $\Gamma = \Gamma_c$. As can be seen, the competing index decreases monotonically with the increase in the chip thickness. The point of intersection denotes the critical chip length l_{chip}^c. Here, the critical chip length can be derived as 1.84 mm. It can thus be concluded that for silicon chips with the thickness of 100 μm, chips with the length smaller than 1.84 mm can be peeled off successfully under the condition of the single-needle ejecting. It can also explain the easy occurrence of the cracking of the chips with dimensions of 5 mm × 5 mm × 0.1 mm during the peeling-off process [11]. At present, it cannot interpret the nearly 50% success rate of the pickup of the chips with dimensions of 5 mm × 5 mm × 0.1 mm.

5.4.5 Discussion on Results

From the perspective of engineering applications, silicon chips with the thickness over 100 μm have been widely used in the semiconductor industry. Meanwhile,

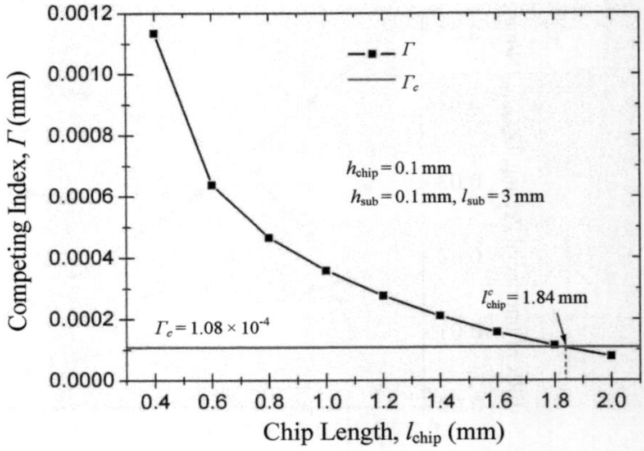

Fig. 5.19 The effect of the chip length on the competing index. © 2012 IEEE. Reprinted, with permission, from Ref. [8]

Table 5.1 Chip series from companies and research institutions

IC Chip Series	Material	Thickness (μm)	Corp.
Monaco/64: IPJ-W1101-C00	Si	150	Impinj, Inc.
Monza 4E: IPJ-W1510-C00	Si	80	Impinj, Inc.
Monza 5: IPJ-W1600-E00	Si	100	Impinj, Inc.
ICODE: SL2 ICS5301EW/V7	Si	150	NXP Semiconductors
UCODE G2iM: SL3S1003	Si	120	NXP Semiconductors
Ultra-thin chips from labs [2, 33]	Si	6–30	Laboratories

thinner chips with thickness down to several microns with corresponding packaging processes have been studied extensively in laboratory [33–37]. Some chip products from companies and research institutions are listed in Table 5.1.

It can be seen from Table 5.1 that the chip thickness varies remarkably from several microns to several hundreds of microns. The peeling-off process with single ejector needle would reach the limit. The determination of the critical chip size is meaningful for the selection of the chip, optimization of the chip peeling-off process, and development of new peeling-off processes. To be specific, some comments on the chip peeling-off process based on the competing index are as follows:

(a) The choice of the suitable strength of the interfacial adhesive for a successful chip peeling-off.

The concept of process margin is first defined combining the competing index. For a given silicon chip, the corresponding competing index is also the upper limit for choosing suitable adhesive strength for successful chip peeling-off, i.e., $\Gamma_{high} = (G^c_{adh})_{max}/\sigma^c_{chip}$. The strength of the interfacial adhesive should be reduced to get easier chip peeling-off. However, the smaller adhesive strength may lead to serious interruption among neighboring chips. When picking up one chip, the peeling-off of adjacent chips from the substrate should be prevented. Therefore, there exists

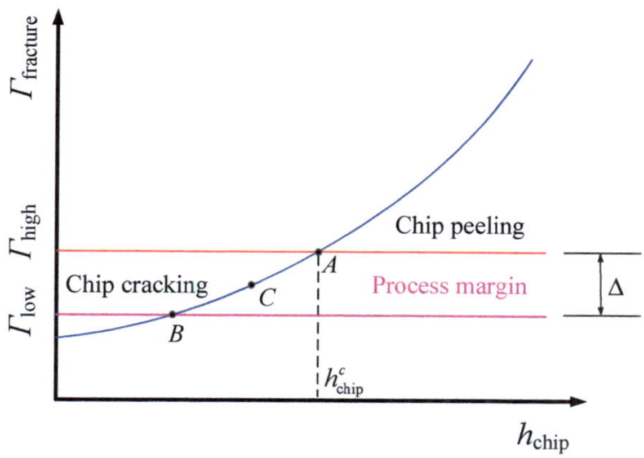

Fig. 5.20 Determination of the process margin. © 2012 IEEE. Reprinted, with permission, from Ref. [8]

a minimum interfacial adhesive strength satisfying $\Gamma_{\text{low}} = (G^c_{\text{adh}})_{\min}/\sigma^c_{\text{chip}}$. For a given silicon chip, the process margin of the chip peeling-off can be defined as: $\Gamma_{\text{process}} = \Gamma_{\text{high}} - \Gamma_{\text{low}}$. An illustration of the determination of the proposed process margin is shown in Fig. 5.20. The points A and B represent the upper limit and lower limit, respectively.

The selection of the adhesive should be on the premise of positive process margin, i.e., $\Gamma_{\text{process}} = \Gamma_{\text{high}} - \Gamma_{\text{low}} > 0$ for a successful chip peeling-off process. To choose suitable adhesive is equivalent to making corresponding competing index located within the process margin, that is, $\Gamma_{\text{low}}\sigma^c_{\text{chip}} \leq G^c_{\text{adh}} \leq \Gamma_{\text{high}}\sigma^c_{\text{chip}}$ e.g., the point C in Fig. 5.20.

(b) The determination of whether the multiple ejector needles should be adopted.

When the chip becomes larger and thinner, the process margin will get narrower under the condition of the single-needle ejecting. Once the process margin turns to be nonpositive, that is, $\Gamma_{\text{process}} = \Gamma_{\text{high}} - \Gamma_{\text{low}} \leq 0$, the single-needle ejecting process reaches the limit, and properly distributed multiple ejector needles should be adopted to guarantee a positive process margin. It should also be noted that by changing the boundary and loading conditions, the competing index and corresponding evaluation method are also applicable to the case of multiple ejector needles process.

(c) The promotion of innovative chip peeling-off process.

The approach proposed here is also beneficial to the promotion of new peeling-off processes for ultra-thin chips. Once $\Gamma_{\text{process}} = \Gamma_{\text{high}} - \Gamma_{\text{low}} \leq 0$, process parameters or configurations of the needles should be adjusted. If it is difficult or costly to obtain a positive process margin, e.g., multiple ejector needles may be unfeasible for tiny and ultra-thin chips, novel peeling-off process such as the laser-assisted peeling technique can be considered [38].

5.5 Impact Effect of Needle Peeling-off Process

5.5.1 *Failed Sample Observation*

Marks of the crack and scratch on the backside of the chip are commonly detected in the microelectronics, which leads to the failure easily in practical service. The failed samples of the radio frequency identification (RFID) chip used in our equipment observed by the Olympus confocal laser scanning microscope, LEXT OLS4000 are depicted in Fig. 5.21. The conical breakage can be found on the backside of the chip at the localized area around the contact point, as shown in Fig. 5.21b, c.

There are two main reasons leading to the local breakage. One is the contact-impact effect by the ejector needle, which gives rise to the local stress concentration. The other is the excessive static contact force caused by improper process parameters such as the excessive displacement of the ejector needle. The effect of the contact impact will be investigated in this section.

5.5.2 *Numerical Model*

The mechanical model of the contact impact is shown in Fig. 5.22. The model involves the chip, adhesive, substrate layers, and an ejector needle tip representing the whole needle module. Since the adhesive layer is very thin, i.e., $h_{\mathrm{adh}} \ll h_{\mathrm{chip}}, h_{\mathrm{adh}} \ll h_{\mathrm{sub}}$, the adhesive layer is thus regarded here as a zero-thickness interface layer with certain physical properties based on the theory of interfacial mechanics [40, 41], and the interfacial toughness is measured by experimental tests [42–44]. The effect of contact impact can be treated as an interaction between the ejector needle tip and the chip-on-substrate structure.

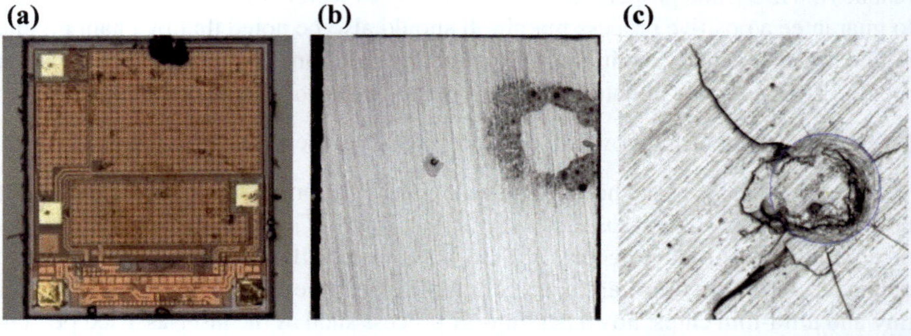

Fig. 5.21 RFID silicon chip with defect: **a** topside, **b** backside with local conical crack, **c** conical crack observed by Olympus LEXT OLS4000. © 2011 IEEE. Reprinted, with permission, from Ref. [39]

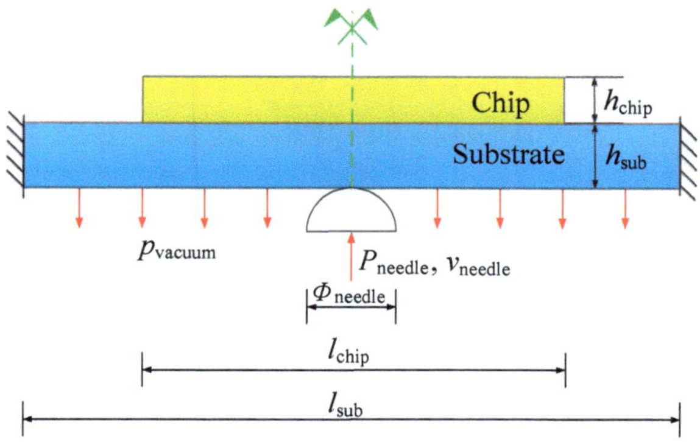

Fig. 5.22 The mechanical model of contact impact based on the chip-on-substrate structure

Table 5.2 Geometric dimensions and loading conditions of the mechanical model

Item	l_{sub}	l_{chip}	h_{chip}	h_{sub}	Φ_{needle}	v_{needle}	p_{vacuum}	m_{needle}
Unit	mm	mm	mm	mm	mm	m/s	kPa	g
Value	3	1	0.1	0.1	0.05	0.01–0.2	−60	10

Table 5.3 Material properties of the mechanical model

Material (−)	Young's modulus (MPa)	Poisson's ratio (−)	Density (kg/m^3)
Chip	1.29×10^5	0.28	2330
Substrate	3.00×10^3	0.38	1370
Needle	2.00×10^5	0.3	7900

The impact analysis is conducted using the explicit dynamic finite element analysis. The chip, substrate, and ejector needle are all treated as linear elastic materials. Because of symmetry, only a quarter of the chip-on-substrate structure in Fig. 5.22 is modeled. The nonlinear geometric deformation of the chip-on-substrate structure is considered here. An explicit three-dimensional solid stress element in Abaqus software, C3D8R, is adopted. A predefined speed along the z-axis is applied to the tip of the ejector needle. The boundary conditions are as follows: fixation at the end of the substrate, symmetry about the y-z plane, and symmetry about the x-z plane. The pressure is defined at the bottom surface of the substrate to represent the vacuum adsorption. A point mass, m_{needle}, is attached to the tip of ejector needle to represent the whole ejector needle. A general contact between the tip of ejector needle and the chip-on-substrate structure is also defined. The geometry dimensions, loading conditions, and material properties of the finite element model are listed in Tables 5.2 and 5.3, which represent a common case of the chip-on-substrate structure for the peeling-off process in electronics assembly.

In consideration of the industrial practice, there may exist two different situations for the analysis of the impact effect by the needle ejecting, that is, without and with penetration of the substrate. The finite element model for these two cases are depicted

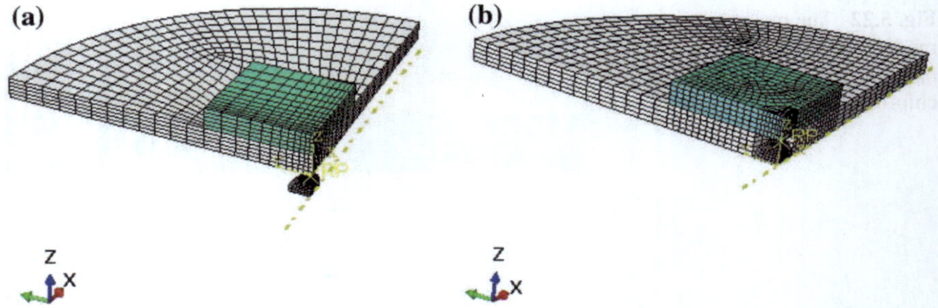

Fig. 5.23 Finite element model for the impact effect analysis of the needle ejecting process: **a** without penetration of the substrate, **b** with through-penetration of the substrate. © 2011 IEEE. Reprinted, with permission, from Ref. [39]

in Fig. 5.23. Local regions surrounding the contact points are refined to ensure the accuracy of numerical results.

5.5.3 Effect of Impact Speed

The dynamic impact force acts directly on the chip in the through-penetration case, resulting in a high instantaneous stress concentration at the local contact area on the backside of the chip. The impact speed is an important factor affecting the stress level. The Von Mises stress contour of the chip subjected to the impact force induced by the ejector needle at a speed of 0.1 m/s under the through-penetration condition is shown in Fig. 5.24. It indicates clearly that the most severe stress occurs at the back center of the chip. However, the high stress is only restricted to a very limited region surrounding the contact center, and most of other regions are nearly in stress-free state. Moreover, the stress due to needle impacting can be 8–10 times larger than static stress, and the stress decreases remarkably as the distance from the contact center enlarges. For the free-penetration case, a direct contact exists between the ejector needle and the substrate, and the impact effect is transmitted through the substrate to the chip.

In both the through-penetration and free-penetration cases, numerical simulations are conducted with the impact speed varying from 0.01 m/s to 0.2 m/s, which cover the most ranges of the impact speed in industrial chip peeling-off process. Since the silicon chip typically possesses a brittle nature, and the contact area generally undergoes a compressive load, the third principle stress is therefore adopted to analyze the chip cracking problem.

The maximum values of the third principle stress versus the impact speed at the bottom center of the chip layer are shown in Fig. 5.25. The line marked by triangles and squares represents the through-penetration and free-penetration conditions, respectively. A nearly linear increase of the maximum third principle stress with

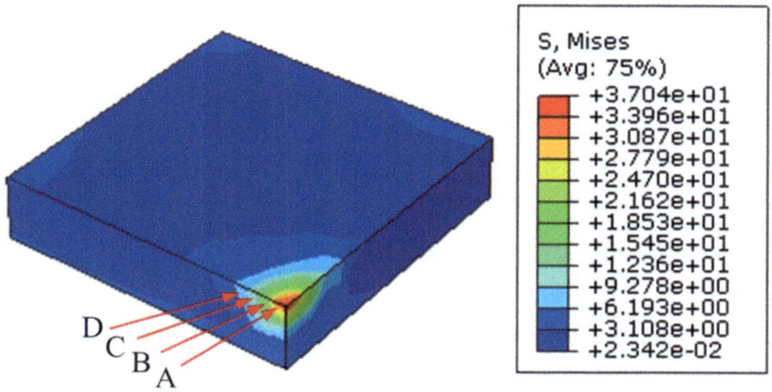

Fig. 5.24 Stress contour of the chip subjected to the impact effect induced by the ejector needle at a speed of 0.1 m/s under the through-penetration condition. © 2011 IEEE. Reprinted, with permission, from Ref. [39]

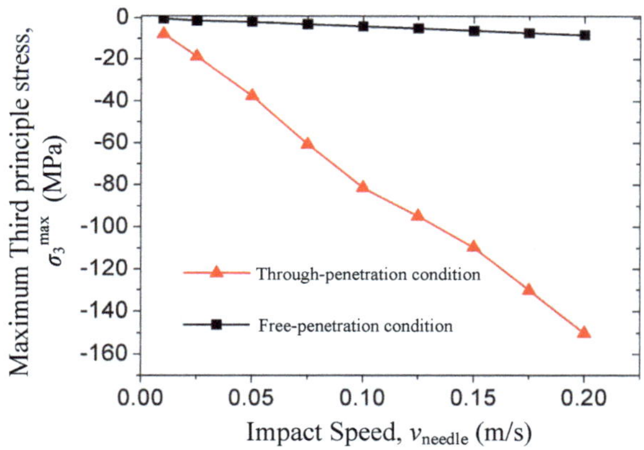

Fig. 5.25 Variation of the maximum third principle stress with the impact speed at the contact area under the through-penetration and free-penetration conditions, separately. © 2011 IEEE. Reprinted, with permission, from Ref. [39]

the increase in impact speed can be found in both cases. Moreover, for the same impact speed, the maximum third principle stress under the through-penetration condition is about 15–20 times larger than that under the free-penetration condition. The maximum third principle stress under the through-penetration condition increases much faster than that under the free-penetration condition as the ejector needle speed increases, which means the through-penetration case is more sensitive to the impact effect. Specially, when the speed of ejector needle reaches about 0.15 m/s, the maximum third principle stress attains the ultimate strength of silicon, 130 MPa [11], which may lead to the cracking at the contact center of the chip.

Therefore, in the free-penetration case, the impact speed is not responsible for the local cracking on the back center of the chip. In contrast, if the substrate is through-penetrated during the peeling-off process, it may lead to excessive local stress. The substrate penetration should be avoided to prevent chip damages. It should be pointed out that the calculation method above mentioned can be used to select suitable impact speed.

5.5.4 *Effect of Substrate Penetration*

The variation of the third principle stresses for four elements around the contact center under the through-penetration condition is depicted in Fig. 5.26. The element A is located at the contact center, the elements B, C, and D are arranged gradually away from the element A, as illustrated in Fig. 5.24. It can be seen that the third principle stress of element A decreases steeply from zero to nearly −38.6 MPa upon the contact, and then the stress level is reduced obviously. Subsequently, a second stress valley is observed at the element A due to the second impact effect of the ejector needle, despite that the stress level is lower than the first valley. In addition, the impact effect on the elements B, C, and D is insignificant, which can be derived from the relatively small variation of the third principle stress. Therefore, the needle impact takes effect only within local contact areas, where the corresponding stresses may be enhanced by ten times, and the impact effect can be neglected as the distance from the contact center enlarges.

The variation of the third principle stresses for elements around the contact center under the free-penetration condition is shown in Fig. 5.27. The positions of elements A′, B′, C′, and D′ are the same as those of elements A, B, C, and D. No high stress exists for all the selected elements during the needle impact process. Additionally, small fluctuation of the third principle stress can be observed among the elements, which indicates an unobvious influence of the ejector needle impact. Therefore, it

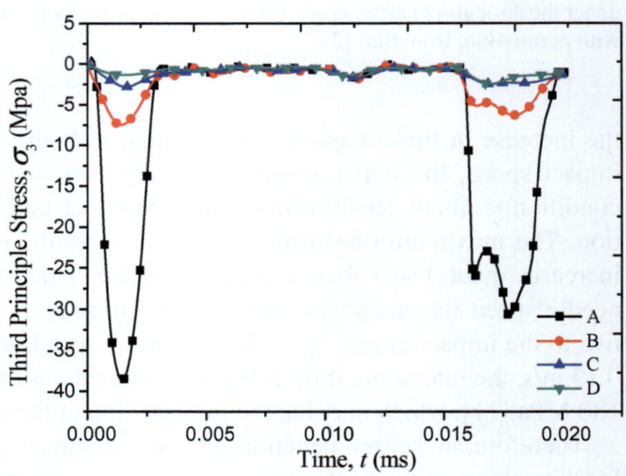

Fig. 5.26 Variation of the third principle stress with time for elements A, B, C, and D under the through-penetration condition with an impact speed of 0.05 m/s. © 2011 IEEE. Reprinted, with permission, from Ref. [39]

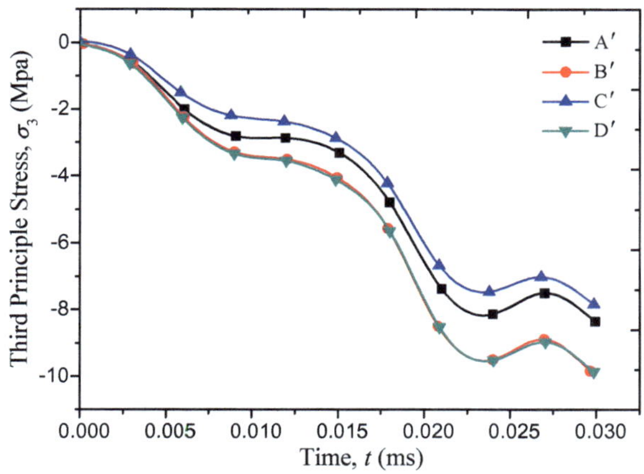

Fig. 5.27 Variation of the third principle stress with time for elements A′, B′, C′, and D′ under the free-penetration condition with an impact speed of 0.05 m/s. © 2011 IEEE. Reprinted, with permission, from Ref. [39]

can be concluded that the impact effect on the local chip cracking introduced by the ejector needle can be neglected under the free-penetration condition.

In conclusion, the chip intactness can be generally maintained in case no penetration of the substrate is detected. In consideration of the chip peeling-off process, the substrate penetration should be avoided to protect the chips, which means that the high-toughness substrate, suitable distribution of the ejector needle, proper needle speed, and reasonable needle displacement are demanded.

5.6 Summary

The single-needle peeling-off process is applied extensively in the flip chip of advanced packaging. Here, theoretical models were established to disclose the interfacial peeling mechanism combining the theory of interfacial fracture mechanics. Advices of choosing process parameters and materials were given for the chip peeling-off process to enhance the success rate especially for thin chips.

(1) The adhesive peel stress, the adhesive shear stress, and the ERR of interfacial peeling during the peeling-off process by a single ejector needle were obtained analytically, and validated by finite element simulations with the VCCT. In particular, the mechanism of the chip peeling-off was disclosed as well as the tunable peeling mode, considering effects of several key factors including geometric dimensions of the chip layer and material properties of the adhesive and substrate layers. Theoretical results showed that the initial peeling at the end of the chip was the toughest, and the crack could propagate by itself after the initiation of the chip peeling until the convergent debonding occurred. The chip peeling-off could be promoted by using a more compliant adhesive tape. The controllable prestress could achieve the tunability of mode II-dominated

peeling, which was of significance to improve the initial peeling for thin and large chips with single ejector needle. A fundamental insight into the mixed-mode peeling behavior was proved useful for the design/optimization of process parameters for electronics assembling.

(2) A competing relationship between the chip peeling and the chip cracking was constructed. A competing index was defined to characterize the competing fracture behavior, and the mechanism of the competing fracture was disclosed fundamentally. Based on the competing index, effects of geometric dimensions of the chip on the competing fracture behavior were investigated. There existed a critical thickness or length separating the chip peeling from the chip cracking. Chips were more difficult to be peeled off for longer ones. The critical chip thickness and critical chip length representing the limit of the peeling-off process were determined by using the critical competing index.

(3) The dynamic impact effect was studied numerically considering key factors of the impact speed, distance from the contact center, and substrate penetration. The impact effect would lead to damage on the backside of the chip during the peeling-off process if the substrate was through-penetrated and the impact speed exceeds a critical value. The contact impact took effect only within limited area around the contact region and could enhance the local stress by 8–10 times, which may result in local damage of chips. Without penetration of the substrate, the impact effect did not contribute to the local chip cracking. The penetration of the substrate penetration should be prevented to achieve an ideal chip peeling-off, which required a tough substrate, suitable ejector needle, proper impact speed, and reasonable ejector needle displacement.

In order to achieve high success rate and prevent damage for thin or large chips, it is suggested that a more compliant and thinner substrate, a weaker adhesive strength, and multiple needles are more preferred. Compared with the more simplified chip-on-substrate system, material properties of the adhesive layer are included in the chip-adhesive-substrate system and corresponding analytical solutions are more reliable, which indicate that the chip-adhesive-substrate system is more applicable than the chip-on-substrate system. In addition, the mechanism of competing fracture disclosed is available for the needle ejecting process with various chip sizes and material properties. The determined critical thickness and length of the chip are useful for optimization of the peeling-off process especially for thin chips. The methodologies proposed here can be used to derive design rules for the peeling-off process of advanced IC packages.

References

1. Rogers JA, Someya T, Huang Y (2010) Materials and mechanics for stretchable electronics. Science 327(5973):1603–1607
2. Feil M, Adler C, Hemmetzberger D, Konig M, Bock K (2004) The challenge of ultra thin chip assembly. In: 54th Electronic Components and Technology Conference (IEEE Cat. No.04CH37546), 1–4 June 2004, pp 35–40
3. Liu Z, Huang Y, Xiao L, Tang P, Yin Z (2015) Nonlinear characteristics in fracture strength test of ultrathin silicon die. Semicond Sci Technol 30(4):045005
4. Chen J, Liu H, Huang Y, Yin Z (2016) High-rate roll-to-roll stack and lamination of multilayer structured membrane electrode assembly. J Manuf Process 23:175–182
5. Huang Y, Liu H, Xu Z, Chen J, Yin Z (2018) Conformal peeling of device-on-substrate system in flexible electronic assembly. IEEE Trans Compon Packag Manuf Technol 8(8):1496–1506
6. Xu Z, Liu Z, Huang Y, Chen J, Liu H, Yin Z (2015) Vacuum-based picking-up of thin chip from adhesive tape. J Adhes Sci Technol 29(13):1315–1329
7. Xu Z, Liu Z, Liu H, Yin Z, Huang Y, Chen J (2015) Analytical evaluation of interfacial crack propagation in vacuum-based picking-up process. IEEE Trans Compon Packag Manuf Technol 5(11):1700–1708
8. Peng B, Huang Y, Yin Z, Xiong Y (2012) Competing fracture modeling of thin chip pick-up process. IEEE Trans Compon Packag Manuf Technol 2(7):1217–1225
9. Liu Z, Huang YA, Liu H, Chen J, Yin Z (2014) Reliable peeling of ultrathin die with multineedle ejector. IEEE Trans Compon Packag Manuf Technol 4(9):1545–1554
10. Liu Z, Wan X, Huang Y, Chen J, Yin Z (2018) Theoretical and experimental studies of competing fracture for flexible chip-adhesive-substrate composite structure. IEEE Trans Compon Packag Manuf Technol 8(1):57–64
11. Cheng T-H, Du C-C, Tseng C-H (2006) Study in IC chip failure during pick-up process by using experimental and finite element methods. J Mater Process Technol 172(3):407–416
12. Liu Z, Huang Y, Chen J, Yin Z (2014) Tunable peeling technique and mechanism of thin chip from compliant adhesive tapes. IEEE Trans Compon Packag Manuf Technol 4(4):560–568
13. da Silva LF, das Neves PJ, Adams R, Spelt J (2009) Analytical models of adhesively bonded joints—Part I: literature survey. Int J Adhes Adhesiv 29(3):319–330
14. Goland M, Reissner E (1944) The stresses in cemented joints. J Appl Mech 11(1):A17–A27
15. Adams R, Peppiatt N (1974) Stress analysis of adhesive-bonded lap joints. J Strain Anal Eng Des 9(3):185–196
16. Delale F, Erdogan F, Aydinoglu M (1981) Stresses in adhesively bonded joints: a closed-form solution. J Compos Mater 15(3):249–271
17. Shahin K, Kember G, Taheri F (2008) An asymptotic solution for evaluation of stresses in balanced and unbalanced adhesively bonded joints. Mech Adv Mater Struct 15(2):88–103
18. Wang J, Zeng S (2008) Thermal stresses analysis in adhesive/solder bonded bimaterial assemblies. J Appl Phys 104(11):113508
19. Bennati S, Colleluori M, Corigliano D, Valvo PS (2009) An enhanced beam-theory model of the asymmetric double cantilever beam (ADCB) test for composite laminates. Compos Sci Technol 69(11):1735–1745
20. Liu Z, Valvo PS, Huang Y, Yin Z (2013) Cohesive failure analysis of an array of IC chips bonded to a stretched substrate. Int J Solids Struct 50(22–23):3528–3538
21. Thouless M (2009) The effects of transverse shear on the delamination of edge-notch flexure and 3-point bend geometries. Compos B Eng 40(4):305–312
22. Peng B, Huang Y, Yin Z, Xiong Y (2011) Analysis of interfacial peeling in IC chip pick-up process. J Appl Phys 110(7):073508
23. Edde F, Verreman Y (1992) On the fracture parameters in a clamped cracked lap shear adhesive joint. Int J Adhes Adhes 12(1):43–48
24. Krenk S (1992) Energy release rate of symmetric adhesive joints. Eng Fract Mech 43(4):549–559

25. He M, Evans A, Hutchinson J (1997) Convergent debonding of films and fibers. Acta Mater 45(8):3481–3489
26. Lin Y-J, Hwang S-J (2005) Static analysis of the die picking process. IEEE Trans Electron Packag Manuf 28(2):142–149
27. Chen W, Nelson C (1979) Thermal stress in bonded joints. IBM J Res Dev 23(2):179–188
28. Saiki N, Inaba K, Kishimoto K, Seno H, Ebe K (2010) Study on peeling behavior in pick-up process of IC chip with adhesive tapes. J Solid Mech Mater Eng 4(7):1051–1060
29. D-h Chen, Nisitani H (1993) Singular stress field near the corner of jointed dissimilar materials. J Appl Mech 60(3):607–613
30. Lu N, Yoon J, Suo Z (2007) Delamination of stiff islands patterned on stretchable substrates. Int J Mater Res 98(8):717–722
31. Chong ACM, Cheung YM (2003) Finite element stress analysis of thin die detachment process. In: Fifth international conference on electronic packaging technology proceedings, 28–30 Oct. 2003, pp 44–51
32. Cheng T-H, Tseng C-H, Hung C-H (2008) Application of a genetic algorithm associated with adhesive joint analysis to the IC chip pick-up process. J Adhes Sci Technol 22(10–11):1057–1072
33. Feil M, Alder C, Klink G, König M, Landesberger C, Scherbaum S, Schwinn G, Spöhrle H (2003) Ultra thin ICs and MEMS elements: techniques for wafer thinning, stress-free separation, assembly and interconnection. Microsyst Technol 9(3):176–182
34. Zimmermann M, Burghartz JN, Appel W, Remmers N, Burwick C, Wurz R, Tobail O, Schubert M, Palfinger G, Werner J (2006) A seamless ultra-thin chip fabrication and assembly process. In: 2006 International electron devices meeting, 11–13 Dec. 2006. IEEE, pp 1–3
35. Burghartz JN, Appel W, Rempp HD, Zimmermann M (2009) A new fabrication and assembly process for ultrathin chips. IEEE Trans Electr Devices 56(2):321–327
36. Richter H, Rempp HD, Hassan M, Harendt C, Wacker N, Zimmermann M, Burghartz JN (2009) Technology and design aspects of ultra-thin silicon chips for bendable electronics. In: 2009 IEEE international conference on IC design and technology, 18–20 May 2009. IEEE, pp 149–154
37. Burghartz JN, Appel W, Harendt C, Rempp H, Richter H, Zimmermann M (2010) Ultra-thin chip technology and applications, a new paradigm in silicon technology. Solid-State Electron 54(9):818–829
38. Marinov V, Swenson O, Miller R, Sarwar F, Atanasov Y, Semler M, Datta S (2012) Laser-enabled advanced packaging of ultrathin bare dice in flexible substrates. IEEE Trans Compon Packag Manuf Technol 2(4):569–577
39. Peng B, Huang Y, Yin Z, Xiong Y (2011) On the analysis of dynamic effect in the die pick-up process. In: 12th international conference on electronic packaging technology and high density packaging, 8–11 Aug. 2011. IEEE, pp 1–4
40. Nied HF (2003) Mechanics of interface with applications in electronic packaging. IEEE Trans Device Mater Reliab 3(4):129–143
41. Yuuki R (1993) Mechanics of interface. Baifukan, Tokyo, Japan
42. Cheung YM, Chong ACM, Huang B (2006) Determination of the interfacial fracture toughness of laminated silicon die on adhesive dicing tape from stud pull measurement. In: International conference on electronic materials and packaging, 11–14 Dec. 2006, pp 1–10
43. Suo Z, Hutchinson JW (1989) Sandwich test specimens for measuring interface crack toughness. Mater Sci Eng A 107:135–143
44. Volinsky A, Moody N, Gerberich W (2002) Interfacial toughness measurements for thin films on substrates. Acta Mater 50(3):441–466

Chapter 6
Multi-needle Peeling

6.1 Introduction

Nowadays, the thickness of commercial silicon chip has been reduced to thinner than 50 μm in the laboratory [1, 2]. Based on the production experience, the thinner and larger chip is more prone to failure during the chip peeling/transferring [3, 4]. The single-needle peeling-off process may reach its limit for a chip thinner than 50 μm due to the vulnerability and flexibility [5–7]. In general, the chip peeling along the adhesive layer should be promoted and the chip cracking across the chip layer should be prevented for a successful chip peeling-off. To lift the chip more effectively, the adhesive tape is recommended to be ejected by more needles [5, 8]. A practical peeling-off process of a chip from the adhesive tape through four-needle ejecting is shown in Fig. 6.1, from which it is clearly observed that the initial peeling starts from the adhesive ends and propagates along the adhesive layer. The needle configuration can be conveniently arranged to meet the technological requirement, as shown in Fig. 6.1a. The chip cracking can be prevented in this way. It is thus technically desired to understand fundamentally the mechanism of the multi-needle peeling-off process and design the process window for an effective chip peeling-off [9–12]. A mechanical model is also essential to predict whether the ultra-thin chip can be peeled off nondestructively through multi-needle ejecting. Besides, investigations on the comparison between the single-needle and the multi-needle peeling-off processes are also very meaningful.

This chapter is outlined as follows. An analytical model was first introduced to address the multi-needle peeling-off process. A competing fracture model was then established to investigate the limitation of the needle peeling-off process and improve the peelability of the ultra-thin chip. The chip peeling-off behavior was analyzed combining the bending normal stress of the chip layer, the ERR of interfacial peeling, the chip peeling index, and the chip cracking index. An experimental platform is finally designed for the needle peeling-off process.

Y. Huang et al., *Modeling and Application of Flexible Electronics Packaging*,
https://doi.org/10.1007/978-981-13-3627-0_6

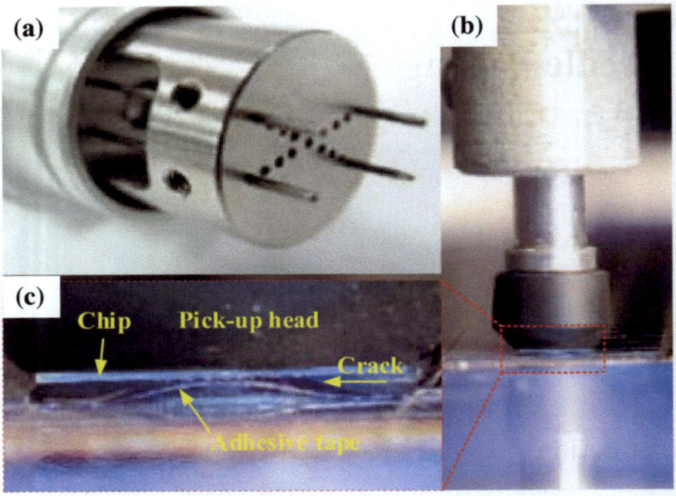

Fig. 6.1 Typical peeling-off process of a chip from the adhesive tape through four-needle ejecting: **a** typical four-needle configuration, **b** pick-up head, **c** peeling-off propagation along the adhesive layer. © 2014 IEEE. Reprinted, with permission, from Ref. [13]

6.2 Mechanical Model of Multi-needle Peeling-off

6.2.1 Model Description

A two-dimensional mechanical model under plane strain condition is constructed considering both the characteristic of practical peeling process and simplification of the analytical model. A schematic description of the mechanical model for the double-needle peeling-off process is shown in Fig. 6.2, in which the ejector needle force is denoted by P_{needle}, and the needle distance is represented by d_{needle}. Both of the substrate ends are clamped. All layers are assumed to be linear elastic and isotropic. For more specific meanings of the symbols in Fig. 6.2, Appendix A can be referred.

It is assumed that the crack will initiate from the free end of the adhesive layer and propagate along the mid-plane, where the initial crack length and virtual propagation length are denoted by a and Δa, respectively. A 1/2 model is constructed due to the structural symmetry. Two stages described in Fig. 6.2b, c, namely stages I and II, are discussed here. At stage I, namely $0 < a < (l_{\text{chip}} - d_{\text{needle}})/2$, the 1/2 model is divided into three regions with lengths $l_{\text{reg-1}} = d_{\text{needle}}/2$, $l_{\text{reg-2}} = (l_{\text{chip}} - d_{\text{needle}})/2 - a$, and $l_{\text{reg-3}} = (l_{\text{sub}} - l_{\text{chip}})/2 + a$, separately. At stage II, namely $(l_{\text{chip}} - d_{\text{needle}})/2 < a < l_{\text{chip}}/2$, the lengths are separately $l_{\text{reg-1}} = l_{\text{chip}}/2 - a$, $l_{\text{reg-2}} = a - (l_{\text{chip}} - d_{\text{needle}})/2$, and $l_{\text{reg-3}} = (l_{\text{sub}} - d_{\text{needle}})/2$. For segment S_i ($i = 1, 2, \ldots, 5$) in Region j ($j = 1, 2, 3$), the local coordinate system (x_j, y_i, z_i) is adopted with the origin at mid-point of the

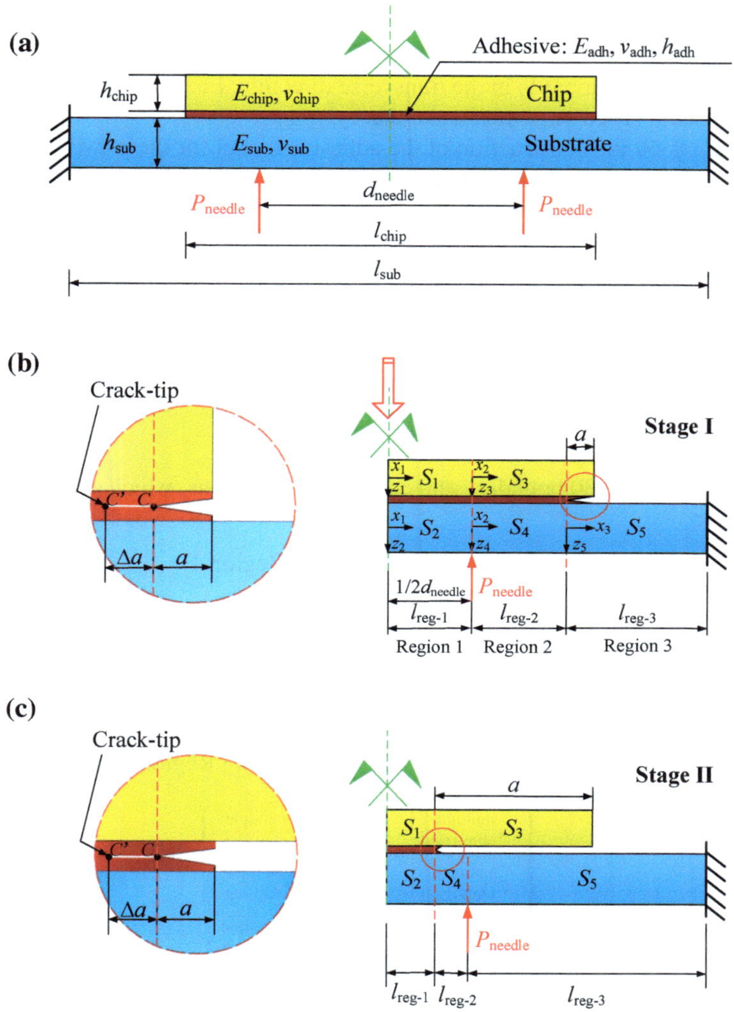

Fig. 6.2 Schematic illustration of the mechanical model for double-needle peeling-off process: **a** overall model, in which the substrate is symmetrically ejected by P_{needle} and the needle distance is denoted by d_{needle}, **b** 1/2 model at stage I, **b** 1/2 model at stage II. A magnified view of the crack-tip is also illustrated in **b** and **c**, where a and Δa denote the adhesive crack length and virtual propagation length, respectively. © 2014 IEEE. Reprinted, with permission, from Ref. [13]

left-side end of the segment S_i. Local value of x_j denotes the distance from left-side end of Region j in the axial direction. Local value of z_i denotes the distance from the mid-plane of the segment S_i in the transverse direction. Each segment is modeled as an elastic beam based on Timoshenko theory.

For details on the construction of the adhesive model for these two stages Chap. 2 can be referred. Schematic diagrams in Fig. 6.2b, c show that the divided regions can be classified as two kinds of mechanical structures, one of which is the chip-adhesive-substrate structure including Regions 1 and 2 at stage I, and Region 1 at stage II, and the other is a common beam including the segment S_5 at stage I, and S_3, S_4 and S_5 at stage II.

6.2.2 Boundary and Continuity Conditions

At stage I, there are 30 boundary and continuity conditions, which can be described as follows:

(a) Symmetric conditions at the left-side end of Region 1:

$$u_{\text{chip-reg-1}}^{\text{multiple}}\Big|_{1x_1=0} = 0,\ \phi_{\text{chip-reg-1}}^{\text{multiple}}\Big|_{x_1=0} = 0,\ Q_{\text{chip-reg-1}}^{\text{multiple}}\Big|_{x_1=0} = 0 \tag{6.1}$$

$$u_{\text{sub-reg-1}}^{\text{multiple}}\Big|_{x_1=0} = 0,\ \phi_{\text{sub-reg-1}}^{\text{multiple}}\Big|_{x_1=0} = 0,\ Q_{\text{sub-reg-1}}^{\text{multiple}}\Big|_{x_1=0} = 0 \tag{6.2}$$

(b) Continuity conditions at cross-sections connecting Regions 1 and 2:

$$\begin{cases} u_{\text{chip-reg-1}}^{\text{multiple}}\Big|_{x_1=l_{\text{reg-1}}} = u_{\text{chip-reg-2}}^{\text{multiple}}\Big|_{x_2=0} \\ \phi_{\text{chip-reg-1}}^{\text{multiple}}\Big|_{x_1=l_{\text{reg-1}}} = \phi_{\text{chip-reg-2}}^{\text{multiple}}\Big|_{x_2=0} \\ w_{\text{chip-reg-1}}^{\text{multiple}}\Big|_{x_1=l_{\text{reg-1}}} = w_{\text{chip-reg-2}}^{\text{multiple}}\Big|_{x_2=0} \end{cases} \tag{6.3}$$

$$\begin{cases} N_{\text{chip-reg-1}}^{\text{multiple}}\Big|_{x_1=l_{\text{reg-1}}} = N_{\text{chip-reg-2}}^{\text{multiple}}\Big|_{x_2=0} \\ M_{\text{chip-reg-1}}^{\text{multiple}}\Big|_{x_1=l_{\text{reg-1}}} = M_{\text{chip-reg-2}}^{\text{multiple}}\Big|_{x_2=0} \\ Q_{\text{chip-reg-1}}^{\text{multiple}}\Big|_{x_1=l_{\text{reg-1}}} = Q_{\text{chip-reg-2}}^{\text{multiple}}\Big|_{x_2=0} = 0 \end{cases} \tag{6.4}$$

$$\begin{cases} u_{\text{sub-reg-1}}^{\text{multiple}}\Big|_{x_1=l_{\text{reg-1}}} = u_{\text{sub-reg-2}}^{\text{multiple}}\Big|_{x_2=0} \\ \phi_{\text{sub-reg-1}}^{\text{multiple}}\Big|_{x_1=l_{\text{reg-1}}} = \phi_{\text{sub-reg-2}}^{\text{multiple}}\Big|_{x_2=0} \\ w_{\text{sub-reg-1}}^{\text{multiple}}\Big|_{x_1=l_{\text{reg-1}}} = w_{\text{sub-reg-2}}^{\text{multiple}}\Big|_{x_2=0} \end{cases} \tag{6.5}$$

$$\begin{cases} N_{\text{sub-reg-1}}^{\text{multiple}}\Big|_{x_1=l_{\text{reg-1}}} = N_{\text{sub-reg-2}}^{\text{multiple}}\Big|_{x_2=0} \\ M_{\text{sub-reg-1}}^{\text{multiple}}\Big|_{x_1=l_{\text{reg-1}}} = M_{\text{sub-reg-2}}^{\text{multiple}}\Big|_{x_2=0} \\ Q_{\text{sub-reg-1}}^{\text{multiple}}\Big|_{x_1=l_{\text{reg-1}}} = 0 \\ Q_{\text{sub-reg-2}}^{\text{multiple}}\Big|_{x_2=0} = P_{\text{needle}} \end{cases} \tag{6.6}$$

(c) Free end conditions at the right-side end of the segment S_3:

$$N_{\text{chip-reg-2}}^{\text{multiple}}\Big|_{x_2=l_{\text{reg-2}}} = 0,\ Q_{\text{chip-reg-2}}^{\text{multiple}}\Big|_{x_2=l_{\text{reg-2}}} = 0,\ M_{\text{chip-reg-2}}^{\text{multiple}}\Big|_{x_2=l_{\text{reg-2}}} = 0 \tag{6.7}$$

(d) Continuity conditions at cross-sections connecting segments S_4 and S_5:

$$\begin{cases} u_{\text{sub-reg-2}}^{\text{multiple}}\Big|_{x_2=l_{\text{reg-2}}} = u_{\text{sub-reg-3}}^{\text{multiple}}\Big|_{x_3=0} \\ \phi_{\text{sub-reg-2}}^{\text{multiple}}\Big|_{x_2=l_{\text{reg-2}}} = \phi_{\text{sub-reg-3}}^{\text{multiple}}\Big|_{x_3=0} \\ w_{\text{sub-reg-2}}^{\text{multiple}}\Big|_{x_2=l_{\text{reg-2}}} = w_{\text{sub-reg-3}}^{\text{multiple}}\Big|_{x_3=0} \end{cases} \tag{6.8}$$

$$\begin{cases} N_{\text{sub-reg-2}}^{\text{multiple}}\Big|_{x_2=l_{\text{reg-2}}} = N_{\text{sub-reg-3}}^{\text{multiple}}\Big|_{x_3=0} \\ M_{\text{sub-reg-2}}^{\text{multiple}}\Big|_{x_2=l_{\text{reg-2}}} = M_{\text{sub-reg-3}}^{\text{multiple}}\Big|_{x_3=0} \\ Q_{\text{sub-reg-2}}^{\text{multiple}}\Big|_{x_2=l_{\text{reg-2}}} = Q_{\text{sub-reg-3}}^{\text{multiple}}\Big|_{x_3=0} \end{cases} \tag{6.9}$$

(e) Displacement conditions at the right-side end of the segment S_5:

$$u_{\text{sub-reg-3}}^{\text{multiple}}\Big|_{x_3=l_{\text{reg-3}}} = 0,\ \phi_{\text{sub-reg-3}}^{\text{multiple}}\Big|_{x_3=l_{\text{reg-3}}} = 0,\ w_{\text{sub-reg-3}}^{\text{multiple}}\Big|_{x_3=l_{\text{reg-3}}} = 0 \tag{6.10}$$

At stage II, there are 24 boundary and continuity conditions, which can be described as follows:

(a) Symmetric conditions at the left-side end of Region 1:

$$u_{\text{chip-reg-1}}^{\text{multiple}}\Big|_{x_1=0} = 0,\ \phi_{\text{chip-reg-1}}^{\text{multiple}}\Big|_{x_1=0} = 0,\ Q_{\text{chip-reg-1}}^{\text{multiple}}\Big|_{x_1=0} = 0 \tag{6.11}$$

$$u_{\text{sub-reg-1}}^{\text{multiple}}\Big|_{x_1=0} = 0,\ \phi_{\text{sub-reg-1}}^{\text{multiple}}\Big|_{x_1=0} = 0,\ Q_{\text{sub- reg-1}}^{\text{multiple}}\Big|_{x_1=0} = 0 \tag{6.12}$$

(b) Loading conditions at the right-side end of the segment S_1:

$$N_{\text{chip-reg-1}}^{\text{multiple}}\Big|_{x_1=l_{\text{reg-1}}} = 0,\ Q_{\text{chip-reg-1}}^{\text{multiple}}\Big|_{x_1=l_{\text{reg-1}}} = 0,\ M_{\text{chip-reg-1}}^{\text{multiple}}\Big|_{x_1=l_{\text{reg-1}}} = 0 \tag{6.13}$$

(c) Continuity conditions at cross-sections connecting segments S_2 and S_4:

$$\begin{cases} u^{\text{multiple}}_{\text{sub-reg-1}}\Big|_{x_1=l_{\text{reg-1}}} = u^{\text{multiple}}_{\text{sub-reg-2}}\Big|_{x_2=0} \\ \phi^{\text{multiple}}_{\text{sub-reg-1}}\Big|_{x_1=l_{\text{reg-1}}} = \phi^{\text{multiple}}_{\text{sub-reg-2}}\Big|_{x_2=0} \\ w^{\text{multiple}}_{\text{sub-reg-1}}\Big|_{x_1=l_{\text{reg-1}}} = w^{\text{multiple}}_{\text{sub-reg-2}}\Big|_{x_2=0} \end{cases} \tag{6.14}$$

$$\begin{cases} N^{\text{multiple}}_{\text{sub-reg-1}}\Big|_{x_1=l_{\text{reg-1}}} = N^{\text{multiple}}_{\text{sub-reg-2}}\Big|_{x_2=0} \\ M^{\text{multiple}}_{\text{sub-reg-1}}\Big|_{x_1=l_{\text{reg-1}}} = M^{\text{multiple}}_{\text{sub-reg-2}}\Big|_{x_2=0} \\ Q^{\text{multiple}}_{\text{sub-reg-1}}\Big|_{x_2=0} = Q^{\text{multiple}}_{\text{sub-reg-2}}\Big|_{x_2=0} = 0 \end{cases} \tag{6.15}$$

(d) Continuity conditions at cross-sections connecting segments S_4 and S_5:

$$\begin{cases} u^{\text{multiple}}_{\text{sub-reg-2}}\Big|_{x_2=l_{\text{reg-2}}} = u^{\text{multiple}}_{\text{sub-reg-3}}\Big|_{x_3=0} \\ \phi^{\text{multiple}}_{\text{sub-reg-2}}\Big|_{x_2=l_{\text{reg-2}}} = \phi^{\text{multiple}}_{\text{sub-reg-3}}\Big|_{x_3=0} \\ w^{\text{multiple}}_{\text{sub-reg-2}}\Big|_{x_2=l_{\text{reg-2}}} = w^{\text{multiple}}_{\text{sub-reg-3}}\Big|_{x_3=0} \\ N^{\text{multiple}}_{\text{sub-reg-2}}\Big|_{x_2=l_{\text{reg-2}}} = N^{\text{multiple}}_{\text{sub-reg-3}}\Big|_{x_3=0} \\ M^{\text{multiple}}_{\text{sub-reg-2}}\Big|_{x_2=l_{\text{reg-2}}} = M^{\text{multiple}}_{\text{sub-reg-3}}\Big|_{x_3=0} \end{cases} \tag{6.16}$$

(e) Zero shear force at the right-side end of the segment S_4, and shear force at the left-side end of the segment S_5:

$$Q^{\text{multiple}}_{\text{sub-reg-2}}\Big|_{x_2=l_{\text{reg-2}}} = 0, \quad Q^{\text{multiple}}_{\text{sub-reg-3}}\Big|_{x_3=0} = P_{\text{needle}} \tag{6.17}$$

(f) Displacement conditions at the right-side end of the segment S_5:

$$u^{\text{multiple}}_{\text{sub-reg-3}}\Big|_{x_3=l_{\text{reg-3}}} = 0,\ \phi^{\text{multiple}}_{\text{sub-reg-3}}\Big|_{x_3=l_{\text{reg-3}}} = 0,\ w^{\text{multiple}}_{\text{sub-reg-3}}\Big|_{x_3=l_{\text{reg-3}}} = 0 \tag{6.18}$$

6.2.3 Determination of Integration Constants

At stage I, there are a total of 45 integration constants, which can be determined by using the 15 relationships among them in combination with the 30 boundary and continuity conditions. However, not all of them are concerned here, except 14 integration constants namely $F_{1,1}$, $F_{1,2}$, $F_{1,3}$, $F_{1,4}$, $F_{1,5}$, $F_{1,6}$, $F_{1,7}$, $F_{2,1}$, $F_{2,2}$, $F_{2,3}$, $F_{2,4}$, $F_{2,5}$, $F_{2,6}$, and $F_{2,7}$. These constants are not all independent of each other. $F_{1,7} = 0$ can be obtained, and the remaining integration constants can be obtained through following linear equation sets:

$$\sum_{n=1}^{6} f_{1,n}\lambda_n = 0 \tag{6.19}$$

$$\sum_{n=1}^{6} f_{1,n}\frac{1}{\lambda_n} = 0 \tag{6.20}$$

$$\sum_{n=1}^{6} f_{1,n}\frac{1}{\lambda_n^3} = 0 \tag{6.21}$$

$$\sum_{n=1}^{6} f_{2,n}\frac{1}{\lambda_n} = -f_{2,9} \tag{6.22}$$

$$\sum_{n=1}^{6} f_{2,n}\frac{1}{\lambda_n}\exp(\lambda_n l_{\text{reg-2}}) = -f_{2,9} \tag{6.23}$$

$$\sum_{n=1}^{6}\left[f_{1,n}\exp(\lambda_n l_{\text{reg-1}}) - f_{2,n}\right] = 0 \tag{6.24}$$

$$\sum_{n=1}^{6}\left[f_{1,n}\exp(\lambda_n l_{\text{reg-1}}) - f_{2,n}\right]\lambda_n^2 = 0 \tag{6.25}$$

$$\sum_{n=1}^{6}\left[f_{1,n}\exp(\lambda_n l_{\text{reg-1}}) - f_{2,n}\right]\frac{1}{\lambda_n^2} = 0 \tag{6.26}$$

$$\sum_{n=1}^{6}\left[f_{1,n}\exp(\lambda_n l_{\text{reg-1}}) - f_{2,n}\right]\frac{1}{\lambda_n^3} = \alpha_1 \tag{6.27}$$

$$\sum_{n=1}^{6}\left[f_{1,n}\exp(\lambda_n l_{\text{reg-1}}) - f_{2,n}\right]\left(\frac{\lambda_n}{k_\sigma} - \frac{c_{\text{chip}} + c_{\text{sub}}}{\lambda_n}\right) = -\left(c_{\text{chip}} f_{2,9} + c_{\text{sub}} f_{2,12}\right) \tag{6.28}$$

$$\sum_{n=1}^{6} f_{2,n}\left[\beta_1\left(\frac{\lambda_n^2}{k_\sigma} - c_{\text{chip}} - c_{\text{sub}}\right) - \left(\frac{1}{\lambda_n^3} - \frac{\beta_2}{\lambda_n^2} + \frac{\beta_3}{\lambda_n}\right)\right]\exp(\lambda_n l_{\text{reg-2}}) = \beta_4 + \alpha_1 \tag{6.29}$$

$$\sum_{n=1}^{6} f_{2,n}\left[\gamma_1\left(\frac{\lambda_n^2}{k_\sigma} - c_{\text{chip}} - c_{\text{sub}}\right) + \frac{\lambda_n}{k_\sigma} + \frac{\gamma_2}{\lambda_n^2}\right]\exp(\lambda_n l_{\text{reg-2}}) = \gamma_3 \tag{6.30}$$

where $f_{j,\text{n}} = F_{j,\text{n}}/P_{\text{needle}}$. The constant coefficients of $f_{2,7}$, $f_{2,9}$, and $f_{2,12}$ can be determined by Eqs. (6.31)–(6.33).

$$f_{2,7} = -\frac{d_{\text{chip}} d_{\text{sub}}\left(h_{\text{chip-half}} + h_{\text{sub-half}}\right)\left(d_{\text{chip}} h_{\text{chip-half}} - d_{\text{sub}} h_{\text{sub-half}}\right)}{d_{\text{chip}} d_{\text{sub}}\left(h_{\text{chip-half}} + h_{\text{sub-half}}\right)^2 + \left(a_{\text{chip}} + a_{\text{sub}}\right)\left(d_{\text{chip}} + d_{\text{sub}}\right)} \tag{6.31}$$

$$f_{2,9} = -\frac{d_{\text{sub}}\left[d_{\text{chip}}h_{\text{chip-half}}\left(h_{\text{chip-half}} + h_{\text{sub-half}}\right) + \left(a_{\text{chip}} + a_{\text{sub}}\right)\right]}{d_{\text{chip}}d_{\text{sub}}\left(h_{\text{chip-half}} + h_{\text{sub-half}}\right)^2 + \left(a_{\text{chip}} + a_{\text{sub}}\right)\left(d_{\text{chip}} + d_{\text{sub}}\right)} \tag{6.32}$$

$$f_{2,12} = \frac{d_{\text{chip}}\left[d_{\text{sub}}h_{\text{sub-half}}\left(h_{\text{chip-half}} + h_{\text{sub-half}}\right) + \left(a_{\text{chip}} + a_{\text{sub}}\right)\right]}{d_{\text{chip}}d_{\text{sub}}\left(h_{\text{chip-half}} + h_{\text{sub-half}}\right)^2 + \left(a_{\text{chip}} + a_{\text{sub}}\right)\left(d_{\text{chip}} + d_{\text{sub}}\right)} \tag{6.33}$$

and $\alpha_1 = [f_{2,7}/k_\tau + (a_{\text{chip}} + a_{\text{sub}} + d_{\text{chip}}h^2_{\text{chip-half}} + d_{\text{sub}}h^2_{\text{sub-half}})(c_{\text{chip}}f_{2,9} + c_{\text{sub}}f_{2,12})]/[(a_{\text{chip}} + a_{\text{sub}})(d_{\text{chip}} + d_{\text{sub}}) + d_{\text{chip}}d_{\text{sub}}(h_{\text{chip-half}} + h_{\text{sub-half}})^2]$, $\beta_1 = (l_{\text{reg-1}} + l_{\text{reg-2}} + l_{\text{reg-3}})[a_{\text{chip}}d_{\text{sub}}h_{\text{sub-half}} + a_{\text{sub}}d_{\text{chip}}h_{\text{chip-half}} + d_{\text{chip}}d_{\text{sub}}h_{\text{chip-half}}h_{\text{sub-half}}(h_{\text{chip-half}} + h_{\text{sub-half}})]/[a_{\text{sub}}d_{\text{sub}}(d_{\text{chip}}h_{\text{chip-half}} - d_{\text{sub}}h_{\text{sub-half}})]$, $\beta_2 = l_{\text{reg-1}} + l_{\text{reg-2}} + h_{\text{sub-half}}(l_{\text{reg-1}} + l_{\text{reg-2}} + l_{\text{reg-3}})[(a_{\text{chip}} + a_{\text{sub}})(d_{\text{chip}} + d_{\text{sub}}) + d_{\text{chip}}d_{\text{sub}}(h_{\text{chip-half}} + h_{\text{sub-half}})^2]/a_{\text{sub}}(d_{\text{chip}}h_{\text{chip-half}} - d_{\text{sub}}h_{\text{sub-half}})$, $\beta_3 = l_{\text{reg-1}}l_{\text{reg-2}} + 0.5l^2_{\text{reg-2}}$, $\beta_4 = -l_{\text{reg-1}}l_{\text{reg-2}} - 0.5l^2_{\text{reg-2}} + 0.5l^2_{\text{reg-3}}$, $\gamma_1 = l_{\text{reg-3}} + d_{\text{chip}}(l_{\text{reg-1}} + l_{\text{reg-2}} + l_{\text{reg-3}})[a_{\text{chip}}d_{\text{sub}}(h_{\text{chip-half}} + h_{\text{sub-half}}) + a_{\text{sub}}h_{\text{chip-half}}(d_{\text{chip}} + d_{\text{sub}}) + d_{\text{chip}}d_{\text{sub}}h_{\text{chip-half}}(h_{\text{chip-half}} + h_{\text{sub-half}})^2]/a_{\text{sub}}d_{\text{sub}}(d_{\text{chip}}h_{\text{chip-half}} - d_{\text{sub}}h_{\text{sub-half}})$, $\gamma_2 = d_{\text{chip}}(h_{\text{chip-half}} + h_{\text{sub-half}})(l_{\text{reg-1}} + l_{\text{reg-2}} + l_{\text{reg-3}})[(a_{\text{chip}} + a_{\text{sub}})(d_{\text{chip}} + d_{\text{sub}}) + d_{\text{chip}}d_{\text{sub}}(h_{\text{chip-half}} + h_{\text{sub-half}})^2]/a_{\text{sub}}(d_{\text{chip}}h_{\text{chip-half}} - d_{\text{sub}}h_{\text{sub-half}})$, $\gamma_3 = c_{\text{sub}} - d_{\text{chip}}l_{\text{reg-1}}l_{\text{reg-2}} - 0.5d_{\text{chip}}l^2_{\text{reg-2}} + 0.5(d_{\text{chip}} + d_{\text{sub}})l^2_{\text{reg-3}}$.

At stage II, there are a total of 33 integration constants, which can be determined by combining the 9 internal relationships among them with the 24 boundary and continuity conditions in Eqs. (6.11)–(6.18). Attentions are only paid to the seven integration constants namely $F_{1,1}$, $F_{1,2}$, $F_{1,3}$, $F_{1,4}$, $F_{1,5}$, $F_{1,6}$, and $F_{1,7}$. $F_{1,7} = 0$ can be obtained, and the remaining six integration constants can be derived by following linear equation sets:

$$\sum_{n=1}^{6} f_{1,n}\lambda_n = 0 \tag{6.34}$$

$$\sum_{n=1}^{6} f_{1,n}\frac{1}{\lambda_n} = 0 \tag{6.35}$$

$$\sum_{n=1}^{6} f_{1,n}\frac{1}{\lambda_n^3} = 0 \tag{6.36}$$

$$\sum_{n=1}^{6} f_{1,n}\frac{1}{\lambda_n}\exp\left(\lambda_n l_{\text{reg-1}}\right) = 0 \tag{6.37}$$

$$\sum_{n=1}^{6} f_{1,n}\left[\beta_1\left(\frac{\lambda_n^2}{k_\sigma} - c_{\text{chip}} - c_{\text{sub}}\right) - \frac{1}{\lambda_n^3} + \frac{\beta_2}{\lambda_n^2}\right]\exp\left(\lambda_n l_{\text{reg-1}}\right) = \beta_3 \tag{6.38}$$

$$\sum_{n=1}^{6} f_{1,n}\left[\gamma_1\left(\frac{\lambda_n^2}{k_\sigma} - c_{\text{chip}} - c_{\text{sub}}\right) + \frac{\lambda_n}{k_\sigma} + \frac{\gamma_2}{\lambda_n^2}\right]\exp\left(\lambda_n l_{\text{reg-1}}\right) = \gamma_3 \tag{6.39}$$

where $f_{1,n} = F_{1,n}/P_{\text{needle}}$, $\beta_1 = (l_{\text{reg-1}} + l_{\text{reg-2}} + l_{\text{reg-3}})[d_{\text{sub}}h_{\text{sub-half}}(a_{\text{chip}} + d_{\text{chip}}h^2_{\text{chip-half}}) + d_{\text{chip}}h_{\text{chip-half}}(a_{\text{sub}} + d_{\text{sub}}h^2_{\text{sub-half}})]/[a_{\text{sub}}d_{\text{sub}}(d_{\text{chip}}h_{\text{chip-half}} - d_{\text{sub}}h_{\text{sub-half}})]$, $\beta_2 = l_{\text{reg-1}} + [\eta_3 h_{\text{sub-half}}(l_{\text{reg-1}} + l_{\text{reg-2}} + l_{\text{reg-3}})]/[k_\tau k_\sigma a_{\text{sub}}(d_{\text{chip}}h_{\text{chip-half}} - d_{\text{sub}}h_{\text{sub-half}})]$, $\beta_3 = 0.5l^2_{\text{reg-3}}$, $\gamma_1 = l_{\text{reg-1}} + l_{\text{reg-2}} - d_{\text{chip}}(l_{\text{reg-1}} + l_{\text{reg-2}} + l_{\text{reg-3}})[\eta_3 h_{\text{chip-half}} + k_\tau k_\sigma a_{\text{chip}}(d_{\text{chip}}h_{\text{chip-half}} - d_{\text{sub}}h_{\text{sub-half}})]/[k_\tau k_\sigma a_{\text{sub}}d_{\text{sub}}(d_{\text{chip}}h_{\text{chip-half}} - d_{\text{sub}}h_{\text{sub-half}})]$, $\gamma_2 = [\eta_3 d_{\text{chip}}(h_{\text{chip-half}} + h_{\text{sub-half}})(l_{\text{reg-1}} + l_{\text{reg-2}} + l_{\text{reg-3}})]/[k_\tau k_\sigma a_{\text{sub}}(d_{\text{chip}}h_{\text{chip-half}} - d_{\text{sub}}h_{\text{sub-half}})]$, and $\gamma_3 = 0.5(d_{\text{chip}} + d_{\text{sub}})l^2_{\text{reg-3}}$.

MATLAB software is used here to solve the linear equation sets. Analytical expressions of the adhesive stresses are obtained by substituting the integration constants into the general expressions in Chap. 2.

6.3 Competing Fracture Model of Chip-Adhesive-Substrate Structure

6.3.1 Modeling of Competing Fracture Behavior

A schematic description of the competing fracture behavior considering the chip-adhesive-substrate structure is shown in Fig. 6.3. The structure is fixed at both ends and ejected by single or multiple needles at the bottom surface of the substrate, as illustrated in Fig. 6.3a. A local magnification of the chip peeling and the chip cracking is shown in Fig. 6.3b. The chip peeling is intended to propagate along the adhesive layer. The chip cracking generally initiates at the top surface center of the chip and propagates across the chip layer. Generally, the mismatch of the bending deformation between the chip and substrate layers results in final initiation of the chip peeling or chip cracking. Complete peel-off of the chip from the substrate is generally expected. However, one of the main drawbacks of the needle peeling-off process is the susceptibility to the chip cracking, especially for ultra-thin chips.

In order to analyze the competing fracture behavior, four basic mechanical parameters are considered here, i.e., $G^{\text{crack-tip}}_{\text{adh}}$, G^c_{adh}, $\sigma^{\max}_{\text{chip}}$, and σ^c_{chip}. To be specific, $G^{\text{crack-tip}}_{\text{adh}}$ is the ERR of interfacial peeling at crack tip of the adhesive layer. G^c_{adh} is the adhesive fracture energy of the adhesive layer. $\sigma^{\max}_{\text{chip}}$ refers to the maximum tensile stress of the chip layer. σ^c_{chip} represents the ultimate tensile strength of the chip layer.

According to Griffith fracture criterion, the chip peeling requires that

$$G^{\text{crack-tip}}_{\text{adh}} > G^c_{\text{adh}}, \tag{6.40}$$

and can be rewritten as

$$G^{\text{crack-tip}}_{\text{adh}}/G^c_{\text{adh}} > 1 \tag{6.41}$$

According to the classical strength theory, the chip uncracking requires that

$$\sigma_{\text{chip}}^{\max} < \sigma_{\text{chip}}^{c}, \tag{6.42}$$

and can be rewritten as

$$\sigma_{\text{chip}}^{\max} / \sigma_{\text{chip}}^{c} < 1 \tag{6.43}$$

The criterion of the chip peelability, namely necessary and sufficient conditions for nondestructive chip peeling-off, can thus be represented by

$$\frac{\sigma_{\text{chip}}^{\max}}{\sigma_{\text{chip}}^{c}} < 1 < \frac{G_{\text{adh}}^{\text{crack-tip}}}{G_{\text{adh}}^{c}} \tag{6.44}$$

Two indexes are defined here to describe the competing fracture behavior. The chip peeling index is termed as $\Gamma_{\text{peel}} = G_{\text{adh}}^{\text{crack-tip}} / G_{\text{adh}}^{c}$, which is used to illustrate the extent of the chip peeling. The chip peeling occurs when the chip peeling index exceeds the switch level $\Gamma_{\text{switch}} = 1$. Similarly, the chip cracking index is defined as $\Gamma_{\text{crack}} = \sigma_{\text{chip}}^{\max} / \sigma_{\text{chip}}^{c}$. Different from the chip peeling index, the chip cracking index should always be lower than the switch level. Efforts should thus be focused on reducing the chip cracking index while maintaining a high level of the chip peeling index.

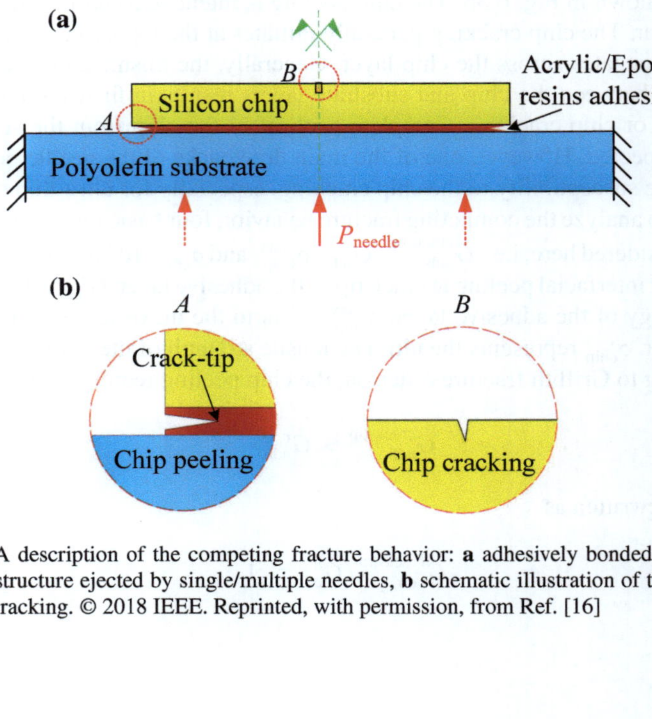

Fig. 6.3 A description of the competing fracture behavior: **a** adhesively bonded chip-adhesive-substrate structure ejected by single/multiple needles, **b** schematic illustration of the chip peeling and chip cracking. © 2018 IEEE. Reprinted, with permission, from Ref. [16]

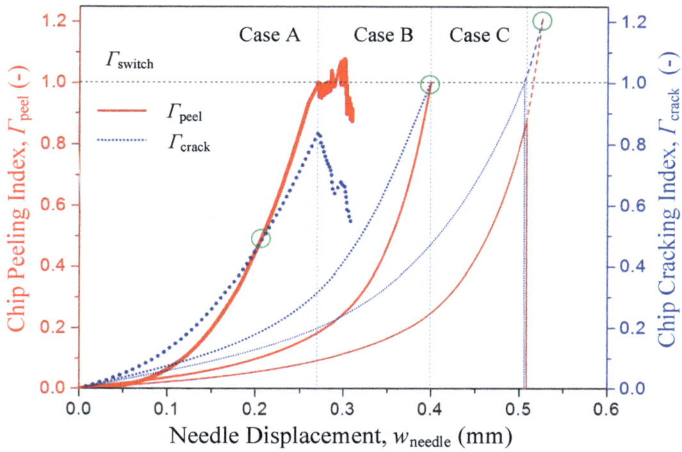

Fig. 6.4 A typical schematic of the competing fracture behavior in the needle peeling-off process, where Case A is a practical simulation of successful chip peeling-off, and Cases B and C refer to the transition state and the chip cracking, respectively.

The criterion of the chip peelability can then be rewritten as

$$\Gamma_{crack} < \Gamma_{switch} < \Gamma_{peel} \tag{6.45}$$

Based on preliminary researches [6, 13–15], the competing relationship can be schematically illustrated in Fig. 6.4. The horizontal dash-line represents the switch level. The horizontal coordinate represents a certain parameter like the needle displacement affecting the chip peeling and chip cracking indexes. In practice, there may exist three typical curve combinations after analyzing the variation of the chip peeling and chip cracking indexes, as shown in Fig. 6.4. The conditions with intersection points of the chip peeling and chip cracking indexes below, on, and over the switch level are classified as Cases A, B, and C, respectively. Case A results from a group of practical simulations by Abaqus software, in which the chip peeling index firstly reaches the switch level. In Case B, the chip peeling and chip cracking indexes simultaneously reach the switch level. In Case C, the chip cracking index first reaches the switch level.

Case A indicates the occurrence of the chip peeling. With the increase in the needle displacement, the crack-tip moves forward along the adhesive layer. Meanwhile, the maximum chip peeling index jitters around the switch level, and the chip cracking index decreases in the subsequent stage. Case C indicates that the chip cracking occurs first. The chip peeling index decreases steeply once the chip cracking occurs. The needle peeling-off process is failed then. Case B denotes that the chip peeling or the chip cracking may occur finally, depending on practical disturbance. Therefore, Case B can be treated as a transition state between Case A and Case C.

Table 6.1 Geometric dimensions, material properties and boundary conditions of the chip-adhesive-substrate structure

Parameter	Unit	Chip	Adhesive	Substrate
Material	–	Silicon	Propylene/epoxy resin	Polyolefin/polyester film
Young's modulus	MPa	1.69×10^5	20	110
Poisson's ratio	–	0.28	0.40	0.45
Fracture strength	MPa	52–415	–	–
Adhesive fracture energy	N/mm	–	0.008–0.035	–
Thickness	mm	0.01–0.15	0.005	0.07
Length	mm	5–10	5–10	7.5–15
Displacement boundary	–	Free end	Free end	Fixed end
Load boundary	–	Free end	Free end	Needle force

The chip peeling and chip cracking indexes are generally affected by geometric dimensions and material properties of the chip-adhesive-substrate system. In order to determine suitable ranges of process parameters under different conditions of needle peeling-off process, effects of key factors including the chip thickness, chip length, and needle force on the index values should be analyzed. The process window can thus be established to determine which fracture mode will occur finally.

According to Eq. (6.45), the four basic fracture parameters namely $G_{\text{adh}}^{\text{crack-tip}}$, G_{adh}^{c}, $\sigma_{\text{chip}}^{\max}$, and σ_{chip}^{c} should be determined. $G_{\text{adh}}^{\text{crack-tip}}$ and $\sigma_{\text{chip}}^{\max}$ can both be obtained through theoretical analysis and finite element approach, details can be found in Chaps. 4 and 5 [13, 15]. G_{adh}^{c} and σ_{chip}^{c} can be obtained through experiments, details have already been given in Chap. 3. Estimation on G_{adh}^{c} is conducted here using a homemade experimental peeling platform. The power of the ultraviolet (UV) radiation in the peeling test is selected between 60 W and 120 W, and the radiation time is varied from 20 s to 50 s. G_{adh}^{c} can be finally determined as 0.008–0.035 N/mm. σ_{chip}^{c} is determined by the three-point bending test combining the nonlinear theory. It can be deduced that the minimum σ_{chip}^{c} of the chip with thickness of 0.01–0.15 mm is roughly at 52–415 MPa, and a thinner chip corresponds to a larger σ_{chip}^{c}. The geometric dimensions, material properties, and boundary conditions of all layers involved in the analytical model, numerical model, and experimental verification are summarized in Table 6.1.

6.4 Analysis of Bending Normal Stress and ERR of Interfacial Peeling

6.4.1 Analysis of Bending Normal Stress

Based on Timoshenko theory on linear elastic beam, the bending normal stress at the top surface of the chip layer in the double-needle peeling-off process can be described by the following piecewise function [13]:

$$\sigma_{\text{chip}}^{\text{top-surf}} = \begin{cases} \dfrac{3M_{\text{chip-reg-1}}^{\text{multiple}}(x_1)}{2h_{\text{chip-half}}^2} - \dfrac{N_{\text{chip-reg-1}}^{\text{multiple}}(x_1)}{2h_{\text{chip-half}}}, & 0 \le x_1 \le l_{\text{reg-1}} \\ \dfrac{3M_{\text{chip-reg-2}}^{\text{multiple}}(x_2)}{2h_{\text{chip-half}}^2} - \dfrac{N_{\text{chip-reg-2}}^{\text{multiple}}(x_2)}{2h_{\text{chip-half}}}, & 0 \le x_2 \le l_{\text{reg-2}} \end{cases} \tag{6.46}$$

Both terms of Eq. (6.46) are applicable to the stage I in Fig. 6.2. Only the first term of Eq. (6.46) is applicable to stage II in Fig. 6.2. The bending normal stress can be found only related to the bending moment and axial force in the chip layer, and there exist

$$N_{\text{chip-reg-}j}^{\text{multiple}}(x_j)/P_{\text{needle}} = \frac{\sum_{n=1}^{6} f_{j,n}\left[\begin{array}{l}\dfrac{\lambda_n^2}{k_\sigma} - (c_{\text{chip}} + c_{\text{sub}}) \\ + \dfrac{d_{\text{chip}} + d_{\text{sub}}}{\lambda_n^2}\end{array}\right]\exp(\lambda_n x_j) + f_{j,7}x_j + f_{j,8}}{d_{\text{chip}}h_{\text{chip-half}} - d_{\text{sub}}h_{\text{sub-half}}} \tag{6.47}$$

$$\begin{aligned} M_{\text{chip-reg - }j}^{\text{multiple}}(x_j)/P_{\text{needle}} = {} & \frac{1}{d_{\text{chip}}h_{\text{chip-half}} - d_{\text{sub}}h_{\text{sub-half}}} \\ & \times \left\{\begin{array}{l} \sum\limits_{n=1}^{6} f_{j,n}\left[\begin{array}{l}\dfrac{h_{\text{chip-half}}\lambda_n^2}{k_\sigma} - h_{\text{chip-half}}(c_{\text{chip}} + c_{\text{sub}}) \\ + \dfrac{d_{\text{sub}}(h_{\text{chip-half}} + h_{\text{sub-half}})}{\lambda_n^2}\end{array}\right]\exp(\lambda_n x_j) \\ + \left[h_{\text{chip-half}}f_{j,7} - (d_{\text{chip}}h_{\text{chip-half}} - d_{\text{sub}}h_{\text{sub-half}})f_{j,9}\right]x_j \\ \qquad + h_{\text{chip-half}}f_{j,10} \end{array}\right\} \end{aligned} \tag{6.48}$$

where $j = 1$ and 2.

At stage I, the integration constants can be determined by Eqs. (6.49)–(6.54):

$$f_{1,7} = 0 \tag{6.49}$$

$$f_{1,9} = 0 \tag{6.50}$$

$$f_{1,8} = -\sum_{n=1}^{6} f_{2,n}\left[\frac{\lambda_n^2}{k_\sigma} - (c_{\text{chip}} + c_{\text{sub}}) + \frac{d_{\text{chip}} + d_{\text{sub}}}{\lambda_n^2}\right]\exp(\lambda_n l_{\text{reg-2}}) - l_{\text{reg-2}} f_{2,7} \tag{6.51}$$

$$f_{2,8} = -\sum_{n=1}^{6} f_{2,n}\left[\frac{\lambda_n^2}{k_\sigma} - (c_{\text{chip}} + c_{\text{sub}}) + \frac{d_{\text{chip}} + d_{\text{sub}}}{\lambda_n^2}\right]\exp(\lambda_n l_{\text{reg-2}}) - l_{\text{reg-2}} f_{2,7} \tag{6.52}$$

$$\begin{aligned} f_{1,10} = & -\sum_{n=1}^{6} f_{2,n}\left[\frac{\lambda_n^2}{k_\sigma} - (c_{\text{chip}} + c_{\text{sub}}) + \frac{d_{\text{sub}}(h_{\text{chip-half}} + h_{\text{sub-half}})}{h_{\text{chip-half}}\lambda_n^2}\right]\exp(\lambda_n l_{\text{reg-2}}) \\ & + l_{\text{reg-2}} f_{2,7} - \frac{l_{\text{reg-2}} f_{2,9}(d_{\text{chip}} h_{\text{chip-half}} - d_{\text{sub}} h_{\text{sub-half}})}{h_{\text{chip-half}}} \end{aligned} \tag{6.53}$$

$$\begin{aligned} f_{2,10} = & -\sum_{n=1}^{6} f_{2,n}\left[\frac{\lambda_n^2}{k_\sigma} - (c_{\text{chip}} + c_{\text{sub}}) + \frac{d_{\text{sub}}(h_{\text{chip-half}} + h_{\text{sub-half}})}{h_{\text{chip-half}}\lambda_n^2}\right]\exp(\lambda_n l_{\text{reg-2}}) \\ & + l_{\text{reg-2}} f_{2,7} - \frac{l_{\text{reg-2}} f_{2,9}(d_{\text{chip}} h_{\text{chip-half}} - d_{\text{sub}} h_{\text{sub-half}})}{h_{\text{chip-half}}} \end{aligned} \tag{6.54}$$

At stage II, the integration constants can be determined by Eqs. (6.55)–(6.58):

$$f_{1,7} = 0 \tag{6.55}$$

$$f_{1,9} = 0 \tag{6.56}$$

$$f_{1,8} = -\sum_{n=1}^{6} f_{1,n}\left[\frac{\lambda_n^2}{k_\sigma} - (c_{\text{chip}} + c_{\text{sub}}) + \frac{d_{\text{chip}} + d_{\text{sub}}}{\lambda_n^2}\right]\exp(\lambda_n l_{\text{reg-1}}) \tag{6.57}$$

$$f_{1,10} = -\sum_{n=1}^{6} f_{1,n}\left[\frac{\lambda_n^2}{k_\sigma} - (c_{\text{chip}} + c_{\text{sub}}) + \frac{d_{\text{sub}}(h_{\text{chip-half}} + h_{\text{sub-half}})}{h_{\text{chip-half}}\lambda_n^2}\right]\exp(\lambda_n l_{\text{reg-1}}) \tag{6.58}$$

Distribution of the bending normal stress at the top surface of the chip layer away from the center is shown in Fig. 6.5, where $l_{\text{chip}} = 8$ mm and $l_{\text{sub}}/l_{\text{chip}} = 1.5$. Effects of the crack length and needle distance are included in Fig. 6.5a, b, respectively. Good agreement can be found between analytical predictions and finite element results. The bending normal stress vanishes gradually toward the free end, which means that the maximum tensile stress always locates near the center of the chip layer.

It is assigned that $d_{\text{needle}}/l_{\text{chip}} = 0.4$ in Fig. 6.5a. The bending normal stress between needles can be found significantly larger than that close to the free end, which may explain why in practice the local breakage on the chip backside always occurs near the central region. As the chip peeling propagates, a rapid decrease in the bending normal stress can be observed. It can be seen that the maximum tensile stress reduces about 50% from the initial peeling to the process end. The chip cracking is more likely

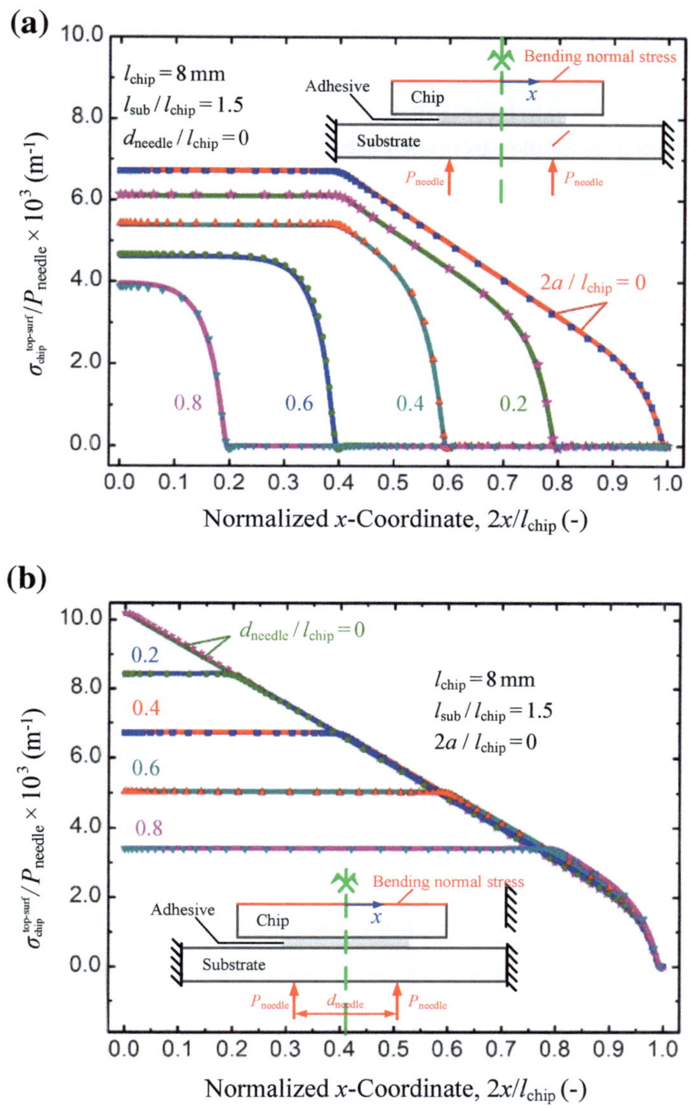

Fig. 6.5 Distribution of the bending normal stress at the top surface of the chip layer: **a** the effect of the crack length, **b** the effect of the needle distance. Here, analytical predictions and finite element results are represented by solid lines and scatter plots, respectively. © 2014 IEEE. Reprinted, with permission, from Ref. [13]

to occur at the initial stage. Once the chip peeling initiates, the occurrence of the chip cracking will be suspended.

The effect of the needle distance on the distribution of the bending normal stress at the top surface of the chip layer is shown in Fig. 6.5b. It is obvious that the maximum value of the bending normal stress decreases with the increase of $d_{\text{needle}}/l_{\text{chip}}$. In particular, for $d_{\text{needle}}/l_{\text{chip}} = 0$, which is equivalent to the normal single-needle peeling-off process, the maximum value of the bending normal stress under the same process condition reaches the ultimate tensile strength of the silicon chip. It quantitatively makes clear that reasonable increase in the needle distance has a contribution to the reduction in the bending normal stress, thereby improving the success rate of the chip peeling-off process.

6.4.2 Analysis of ERR of Interfacial Peeling

At stage I, the normalized crack-tip is supposed at $x_2 = l_{\text{reg-2}}$, where the ERR of interfacial peeling can be calculated as

$$\frac{G_{\text{I}}}{P_{\text{needle}}^2} = \frac{\left[\sum_{n=1}^{6} f_{2,n} \exp\left(\lambda_n l_{\text{reg-2}}\right)\right]^2}{2k_\sigma} \tag{6.59}$$

$$\frac{G_{\text{II}}}{P_{\text{needle}}^2} = \frac{\left[\sum_{n=1}^{6} f_{2,n}\lambda_n\left(c_{\text{chip}} + c_{\text{sub}} - \lambda_n^2/k_\sigma\right)\exp\left(\lambda_n l_{\text{reg-2}}\right) - d_{\text{sub}}\right]^2}{2k_\tau\left(d_{\text{chip}}h_{\text{chip-half}} - d_{\text{sub}}h_{\text{sub-half}}\right)^2} \tag{6.60}$$

At stage II, there exist

$$\frac{G_{\text{I}}}{P_{\text{needle}}^2} = \frac{\left[\sum_{n=1}^{6} f_{1,n} \exp\left(\lambda_n l_{\text{reg-1}}\right)\right]^2}{2k_\sigma} \tag{6.61}$$

$$\frac{G_{\text{II}}}{P_{\text{needle}}^2} = \frac{\left[\sum_{n=1}^{6} f_{1,n}\lambda_n\left(c_{\text{chip}} + c_{\text{sub}} - \lambda_n^2/k_\sigma\right)\exp\left(\lambda_n l_{\text{reg-1}}\right)\right]^2}{2k_\tau\left(d_{\text{chip}}h_{\text{chip-half}} - d_{\text{sub}}h_{\text{sub-half}}\right)^2} \tag{6.62}$$

The variation of the ERR of interfacial peeling and the mode-mixity angle with the crack length and the needle distance is shown in Fig. 6.6. It can be seen that predictions of the analytical model show good agreement with finite element results. The position of the ejector needle plays an important role in the crack propagation, as illustrated in Fig. 6.6a. A clear watershed can be observed near the position of the ejector needle, i.e., $2a/l_{\text{chip}} = 0.6$. Obviously, the ERR of interfacial peeling increases with crack propagation from the initial peeling to the end of the stage I. The ERR of interfacial peeling decreases gradually at stage II until a convergent debonding occurs after $2a/l_{\text{chip}} = 0.9$. That is, once the chip peeling initiates, it can propagate

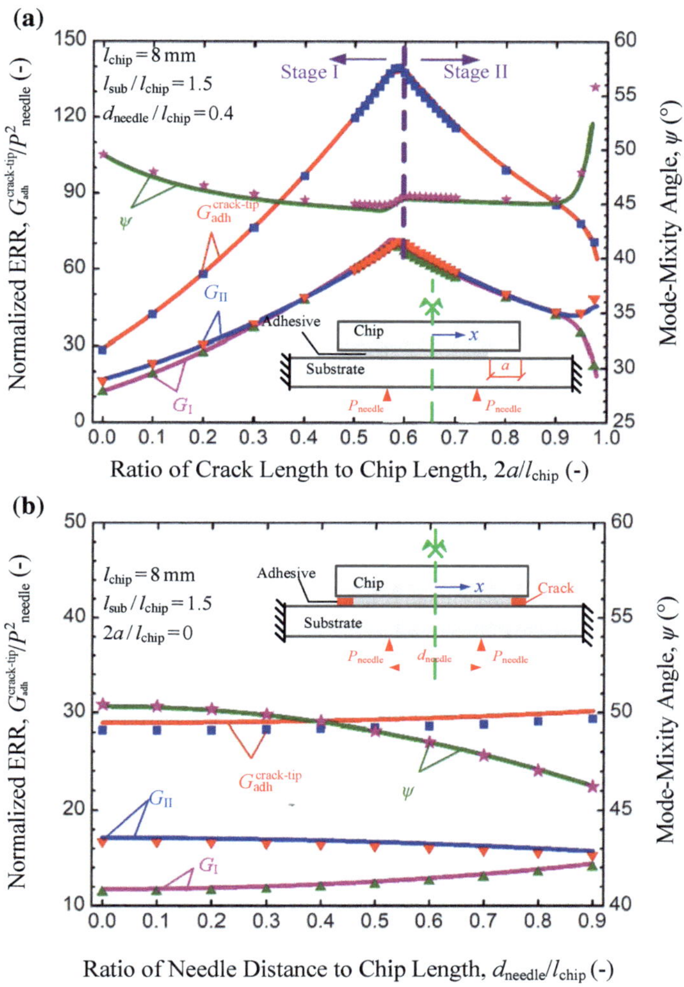

Fig. 6.6 The variation of the ERR of interfacial peeling and the mode-mixity angle: **a** the effect of the crack length, **b** the effect of the needle distance. Analytical predictions and finite element results are represented by solid lines and scatter plots, respectively. © 2014 IEEE. Reprinted, with permission, from Ref. [13]

by itself at stage I under the same process condition. When moving into stage II, the chip peeling stops in case the total ERR of interfacial peeling is smaller than the adhesive fracture energy. Further scrutiny from Fig. 6.6b reveals that the total ERR of interfacial peeling increases slightly as the needle distance increases, which indicates that the improvement on the initial ERR of interfacial peeling is inapparent just by changing the needle distance.

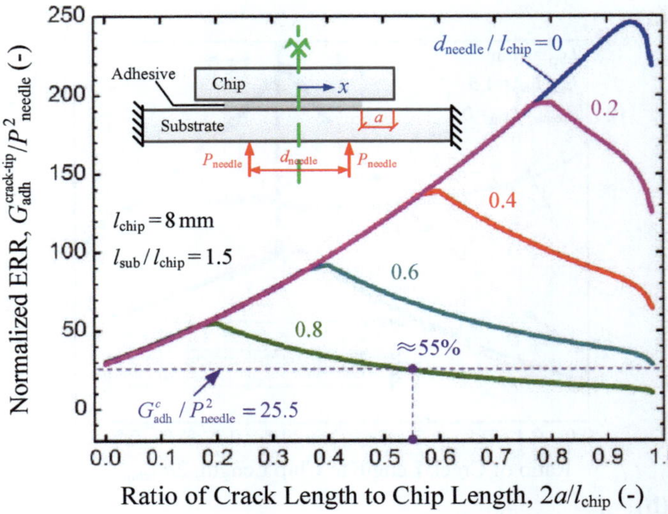

Fig. 6.7 The total ERR of interfacial peeling versus the crack length under different conditions of ejector needle distribution, including $d_{needle}/l_{chip} = 0, 0.2, 0.4, 0.6,$ and 0.8. In particular, $d_{needle}/l_{chip} = 0$ represents the single-needle peeling-off process. Here, all results are obtained from analytical predictions. © 2014 IEEE. Reprinted, with permission, from Ref. [13]

The total ERR of interfacial peeling versus the crack length under different conditions of the ejector needle distribution is shown in Fig. 6.7. It can be obviously found that the larger needle distance leads to the lower overall level of the total ERR of interfacial peeling (namely, it is harder to maintain the crack propagation, especially at stage II). Here, the normalized adhesive fracture energy is assumed to be $G_{adh}^c/P_{needle}^2 = 25.5$. The total length that is peeled off from the adhesive tape may reach only about 55% under the condition of $d_{needle}/l_{chip} = 0.8$, which means the failure of the peeling-off process.

6.5 Effect of Chip Geometry on Competing Fracture Behavior

6.5.1 Effect of Chip Thickness

The effect of the chip thickness on the chip peeling and chip cracking indexes in the single-needle peeling-off process is shown in Fig. 6.8. The horizontal line represents the switch level. The ejector needle force is set as 0.015 N/mm. The chip length is maintained at 10 mm. The adhesive fracture energy is calculated as 0.025 N/mm. Unless specified otherwise, all process parameters are kept constant except the chip

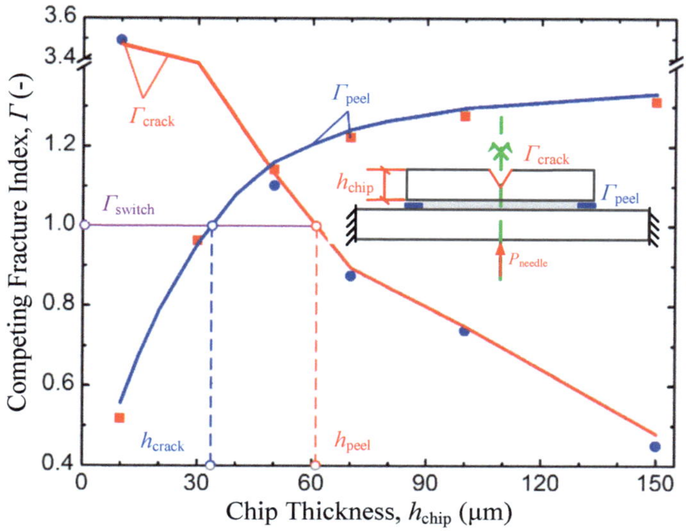

Fig. 6.8 The effect of the chip thickness on the chip peeling and chip cracking indexes under the condition of the single-needle peeling-off process. The solid curve is obtained through theoretical analysis. The discrete points are obtained by using the finite element simulation. © 2018 IEEE. Reprinted, with permission, from Ref. [16]

thickness. As can be seen, a larger chip peeling index can be found for a thicker chip, which indicates an improvement on the chip peelability. A rapid increase in the chip peeling index can be observed when the chip thickness is smaller than 60 μm. The chip peeling index converges with a further increase in the chip thickness. The peeling-off behavior of the ultra-thin chip can be found sensitive to the chip thickness. Meanwhile, the chip cracking index decreases as the chip thickness increases, which means that the thicker chip is less susceptible to the chip cracking.

As depicted in Fig. 6.8, curves of the chip peeling and chip cracking indexes are intersected with the switch level at different chip thicknesses, which are defined as the critical chip peeling thickness (h_{peel}) and critical chip cracking thickness (h_{crack}), separately. The chip cracking should not occur when $h_{peel} > h_{crack} \approx 61$ μm. As indicated previously, a nondestructive chip peeling-off requires that Eq. (6.45) must be satisfied. Therefore, the chip thicker than 61 μm can achieve the purpose of nondestructive chip peeling-off by the single-needle ejecting.

Generally, the chip peeling index can be improved by increasing the ejector needle force. However, the chip cracking index can be found increased meanwhile. As indicated by Liu et al. [13], the chip peelability can be improved through multi-needle adoption. The maximum tensile stress of the chip layer can be reduced obviously in the multi-needle peeling-off process. The ERR of interfacial peeling increases meanwhile since the multi-needle adoption is equivalent to the reduction in the chip length.

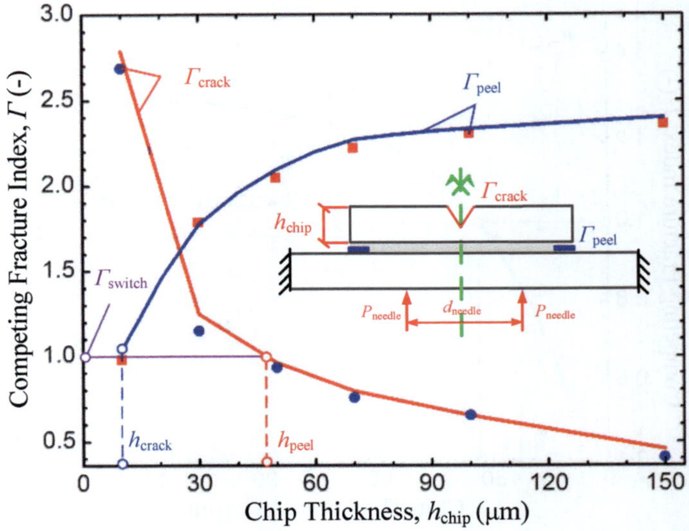

Fig. 6.9 The effect of the chip thickness on the chip peeling and chip cracking indexes under the condition of the double-needle peeling-off process. The solid curve is obtained through theoretical analysis. The discrete points are obtained by using the finite element simulation. © 2018 IEEE. Reprinted, with permission, from Ref. [16]

The effect of the chip thickness on the chip peeling and chip cracking indexes under the condition of the double-needle ejecting process is depicted in Fig. 6.9. The ejector needle force is selected as 0.02 N/mm. The needle distance is set as 4 mm. The horizontal line $\Gamma = \Gamma_{switch}$ represents the critical level. Unless specified otherwise, all process parameters are maintained constant except the chip thickness. As can be seen, although the ejector needle force is roughly 130% of that in the single-needle peeling-off process, the corresponding chip cracking index is only about 70% of that in the single-needle peeling-off process. $\Gamma_{peel} > \Gamma_{switch}$ is already satisfied when the chip thickness is only about 10 μm. The critical chip peeling thickness in the double-needle ejecting process is only about 48 μm. The multi-needle peeling-off process is proved beneficial for the ultra-thin chip peeling-off.

6.5.2 *Effect of Chip Length*

The effect of the chip length on the chip peeling and chip cracking indexes under the condition of the single-needle peeling-off process is given in Fig. 6.10. The chip thickness is maintained at 0.03 mm. All process parameters are kept constant except the chip length. Based on the conclusions obtained by Liu et al. [13], the energy required for a shorter chip to be peeled off nondestructively is less than that for a

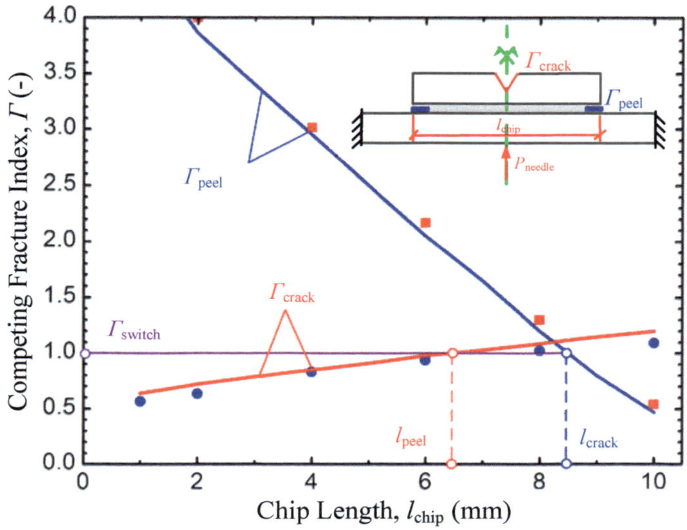

Fig. 6.10 The effect of the chip length on the chip peeling and chip cracking indexes under the condition of the single-needle peeling-off process. The solid curve is obtained through theoretical analysis. The discrete points are obtained by using the finite element simulation. © 2018 IEEE. Reprinted, with permission, from Ref. [16]

longer one. As depicted in Fig. 6.8, the switch level can already be attained when the ejector needle force is 0.015 N/mm for the chip with dimensions of 10 mm × 10 mm × 0.03 mm. Therefore, the chip peeling index for a smaller chip should be higher in this case. The ejector needle force is thus selected here as 0.01 N/mm to reflect the effect of the chip length more reasonably.

As can be seen from Fig. 6.10, the success rate of the chip peeling-off is higher for a shorter chip. For a chip with dimensions of 1 mm × 1 mm × 0.03 mm, the chip peeling index is much higher than the switch level, and the chip cracking index is lower than the switch level. The successful chip peeling-off is easy to be achieved then. A rapid decrease in the chip peeling index and a slow increase in the chip cracking index can be observed as the chip length increases. The chip cracking is found easier to occur for a longer chip. The successful chip peeling-off can be achieved when $l_{chip} \approx 6.5$ mm, considering that Eq. (6.45) is satisfied. Moreover, the chip cracking index can be found close to the switch level when the chip length is between 4.0 and 6.5 mm. In practice, the chip cracking may still occur even when the chip length is around 4.0 mm. Therefore, the theoretically established process window should be adjusted to improve the system stability. Additionally, an improvement on the success rate of the chip peeling-off through the multi-needle peeling-off process is similar to the conclusion obtained from Sect. 6.5.1.

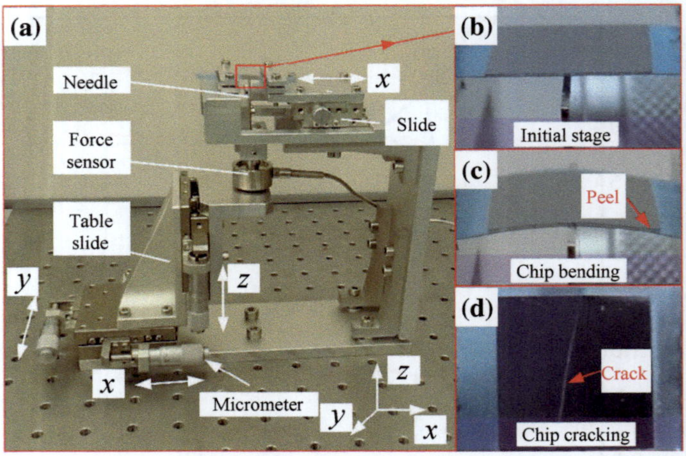

Fig. 6.11 Experimental system designed for the needle peeling-off process: **a** overall setup, **b** initial state, **c** chip peeling, **d** chip cracking. © 2018 IEEE. Reprinted, with permission, from Ref. [16]

6.6 Experimental Validation

6.6.1 Experimental System

An experimental system designed for the needle peeling-off process is shown in Fig. 6.11, which can be used to conduct the single-needle and the multi-needle peeling-off processes. The system is mainly composed of the rectangular ejector needle, force sensor, and table slide. Platform movement along x, y, and z axes is driven by micrometers. Platform movement along z axis is used as the needle displacement. The ejector needle force can be extracted from the force sensor. In addition, position adjustment of the chip-adhesive-substrate structure along x axis is controlled by a table slide with a single degree of freedom.

The chip dimension in Fig. 6.11 is selected as 10 mm × 10 mm × 0.03 mm. Structural bending of the chip-adhesive-substrate system as a whole can be observed at the initial stage, as seen in Fig. 6.11b. Bending stiffness of traditional thick chip is much larger than that of the substrate. With the increase in the ejector needle force, an initial chip peeling can be triggered easily due to the deformation mismatch between chip and substrate. However, flexibility of the ultra-thin chip is more significant than that of a traditional thick one. Bending deformation of the ultra-thin chip remains consistent with the substrate even when the ejector needle force is set as 0.02 N/mm, as depicted in Fig. 6.11c. The chip cracking even occurs when the ejector needle force further increases to about 0.04 N/mm, as shown in Fig. 6.11d. Experimental results of the single-needle peeling-off process are basically in accordance with theoretical predictions in previous sections. Moreover, a successful chip peeling-off can already

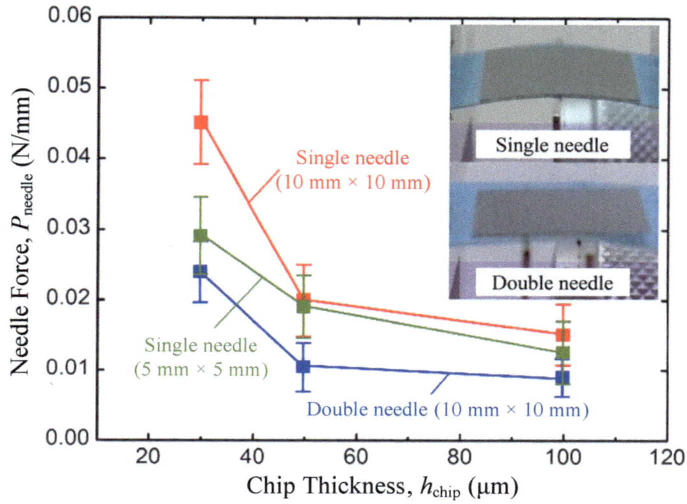

Fig. 6.12 Threshold ejector needle force needed for the chip peeling or the chip cracking under different process conditions. © 2018 IEEE Reprinted, with permission, from Ref. [16]

be attained experimentally when the ejector needle force is only about 0.011 N/mm under the condition of the double-needle ejecting process.

The threshold value of the ejector needle force demanded for the chip peeling or the chip cracking under different process conditions is given in Fig. 6.12. Parameters of the chip length, chip thickness, and ejector needle number are included. A total of 26, 34, and 36 samples are used for chips with dimensions of 10 mm × 10 mm × 0.03 mm, 10 mm × 10 mm × 0.05 mm, and 10 mm × 10 mm × 0.1 mm, respectively. Extra 11, 14, and 20 samples with dimensions of 5 mm × 5 mm × 0.03 mm, 5 mm × 5 mm × 0.05 mm, 5 mm × 5 mm × 0.1 mm are also adopted as a supplement. Experimental results show that a larger ejector needle force is generally needed for a thinner chip to attain the chip peeling or the chip cracking. The multi-needle peeling-off process can simultaneously increase the chip peeling index and decrease the chip cracking index. The threshold value of the ejector needle force for a 10 mm × 10 mm chip under the condition of the multi-needle peeling-off process is much smaller than that in the single-needle peeling-off process. The chip with dimensions of 10 mm × 10 mm × 0.03 mm can hardly be peeled off in the single-needle peeling-off process, while the success rate of the chip peeling-off can reach 76% and 94% for samples with thicknesses of 0.05 mm and 0.1 mm, respectively. In comparison, the success rate of the chip peeling-off in the double-needle ejecting process can increase up to 77%, 100%, and 100% for samples with thicknesses of 0.03 mm, 0.05 mm, and 0.1 mm, respectively. The rate of the chip cracking is roughly at 18% for samples with dimensions of 5 mm × 5 mm × 0.03 mm under the condition of the single-needle peeling-off process, while a nearly 100% success rate of the chip peeling-off can be achieved for samples with a thickness of 0.05 and 0.1 mm.

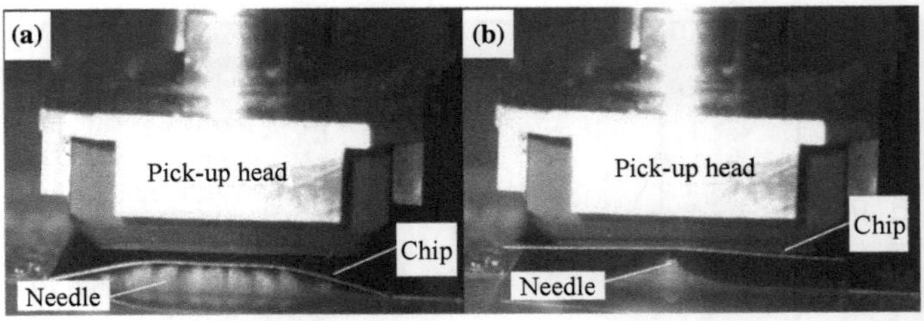

Fig. 6.13 Experimental attempt on the ultra-thin and large chip peeling-off through a modified multi-needle peeling-off process: **a** initial chip peeling ejected by multiple needles, **b** final chip peeling-off ejected by a single needle. © 2018 IEEE. Reprinted, with permission, from Ref. [16]

As mentioned in literature [5, 8], the single-needle peeling-off process has already reached its limit for chips thinner than 50 μm, under which condition the maximum tensile stress of the chip layer for a chip larger than 5 mm × 5 mm should be very high. The multi-needle peeling-off process can then be used to prevent excessive bending of the chip-adhesive-substrate structure. However, although the application of the multi-needle peeling-off process is indeed helpful to promote an initial chip peeling, the crack propagation along the adhesive layer is gradually restricted due to the distributed ejector needles [13].

An experimental attempt on the ultra-thin and large chip (20 mm × 20 mm × 0.02 mm) peeling-off process through a modified multi-needle peeling-off process is given in Fig. 6.13. Multiple ejector needles are employed at the initial stage, as shown in Fig. 6.13a. Afterward, the needles are removed sequentially during the subsequent stage as the crack propagates along the adhesive layer. Finally, the chip peeling-off is driven by a single ejector needle, as depicted in Fig. 6.13b. Then the chip can be picked up after a nearly complete chip peeling-off. The modified approach is supposed theoretically feasible, although a very high precision of automatic control is required for the experimental equipment.

6.7 Summary

This chapter first introduced a mechanical model to address the reliable peeling of an ultra-thin chip from the adhesive tape through the multi-needle peeling-off process. The competing fracture behavior in the single/multiple needle peeling-off process based on the chip-adhesive-substrate structure was then investigated. The criterion for the nondestructive chip peeling-off was established.

A practical case of the ultra-thin chip peeling-off through the multi-needle peeling-off process was addressed. The chip cracking was found more likely to occur at the initial stage and the maximum tensile stress of the chip layer was always located near the chip central region. The bending normal stress at the top surface of the chip layer could be reduced obviously through multi-needle ejecting, while an improvement on the ERR of interfacial peeling was not obvious just by changing the needle distribution. The multi-needle peeling-off process could be used to prevent the chip from cracking, but the promotion on initial chip peeling was not obvious. The chip peeling may not have occurred for an ultra-thin and large chip unless increasing the multi-needle force.

As to the investigation on the competing fracture behavior between the chip peeling and the chip cracking, effects of chip geometries and single/multiple needles were considered. Based on the analysis of the chip peeling and chip cracking indexes, the nondestructive chip peeling-off was found difficult to be achieved for the chip thinner than 50 μm and larger than 5 mm under the condition of the single-needle peeling-off process. An experimental platform was also designed for the single/multiple needle peeling-off process. A potential solution strategy was finally proposed by applying the modified multi-needle peeling-off process with independently controlled needles.

References

1. Jeon EB, Park JD, Song JH, Lee HJ, Kim HS (2012) Bi-axial fracture strength characteristic of an ultra-thin flash memory chip. J Micromech Microeng 22(10):105014
2. Endler S, Rempp H, Harendt C, Burghartz JN (2012) Compensation of externally applied mechanical stress by stacking of ultrathin chips. Solid-State Electron 74:102–107
3. Cheng T-H, Du C-C, Tseng C-H (2006) Study in IC chip failure during pick-up process by using experimental and finite element methods. J Mater Process Technol 172(3):407–416
4. Liu Z, Huang Y, Xiao L, Tang P, Yin Z (2015) Nonlinear characteristics in fracture strength test of ultrathin silicon die. Semicond Sci Technol 30(4):045005
5. Feil M, Adler C, Hemmetzberger D, Konig M, Bock K (2004) The challenge of ultra thin chip assembly. In: 54th electronic components and technology conference (IEEE Cat. No.04CH37546), 1–4 June 2004, pp 35–40
6. Peng B, Huang Y, Yin Z, Xiong Y (2012) Competing fracture modeling of thin chip pick-up process. IEEE Trans Compon Packag Manuf Technol 2(7):1217–1225
7. Peng B, Huang Y, Yin Z, Xiong Y (2011) On the analysis of dynamic effect in the die pick-up process. In: 12th international conference on electronic packaging technology and high density packaging, 8–11 Aug 2011. IEEE, pp 1–4
8. Medding J, Stalder R, Niederhauser M, Stoessel P (2004) Thin die bonding techniques. In: IEEE/CPMT/SEMI 29th international electronics manufacturing technology symposium (IEEE Cat. No.04CH37585), 14–16 July 2004, pp 68–73
9. Malachowski K, Gonzales D, Miller RA, Schnegg F, Arzberger A, Hajdarevic Z, LaManna A (2011) An investigation into damage-free thin die pick and place for 3D stacking. In: IEEE 13th electronics packaging technology conference, 7–9 Dec 2011, pp 819–823
10. Schnegg F, Kostner H, Bock G, Engensteiner S (2011) Handling of thin dies with emphasis on chip-to-wafer bonding. In: Burghartz J (ed) Ultra-thin chip technology and applications. Springer New York, New York, NY, pp 167–183
11. Yong L, Irving S, Desbiens D, Timwah L, Qiuxiao Q (2006) Simulation and analysis for typical package assembly manufacture. In: EuroSime 2006—7th international conference on thermal, mechanical and multiphysics simulation and experiments in micro-electronics and micro-systems, 24–26 Apr 2006, pp 1–10
12. Saiki N, Inaba K, Kishimoto K, Seno H, Ebe K (2010) Study on peeling behavior in pick-up process of IC chip with adhesive tapes. J Solid Mech Mater Eng 4(7):1051–1060
13. Liu Z, Huang YA, Liu H, Chen J, Yin Z (2014) Reliable peeling of ultrathin die with multineedle ejector. IEEE Trans Compon Packag Manuf Technol 4(9):1545–1554

14. Peng B, Huang Y, Yin Z, Xiong Y (2011) Analysis of interfacial peeling in IC chip pick-up process. J Appl Phys 110(7):073508
15. Liu Z, Valvo PS, Huang Y, Yin Z (2013) Cohesive failure analysis of an array of IC chips bonded to a stretched substrate. Int J Solids Struct 50(22–23):3528–3538
16. Liu Z, Wan X, Huang Y, Chen J, Yin Z (2018) Theoretical and experimental studies of competing fracture for flexible chip-adhesive-substrate composite structure. IEEE Trans Compon Packag Manuf Technol 8(1):57–64

Chapter 7
Conformal Peeling

7.1 Introduction

Flexible electronic devices are generally constituted by multiple thin films with various material properties and geometric dimensions, and the fabrication of which are increasingly integrated with the R2R process [1–4] considering the advantages of high efficiency and low cost. The conformal peeling of flexible devices from the donor substrate through the R2R system is crucially important in the flexible electronics assembly. For instance, in the packaging of the radio-frequency identification (RFID) tag, the silicon-based integrated circuit (IC) chip is first bonded onto the antenna to form an Inlay. The antenna integrated with the chip is adhered to the cover layer and the substrate layer. Finally, the RFID is transferred from the donor substrate to the acceptor substrate for specific application purpose through the conformal peeling achieved by the R2R process.

During the conformal peeling, the left substrate moves at a certain speed under the tension condition. The left substrate will then be forced to bend when passing the peel blade. If the adhesion strength is not enough to make the device layer conformally shaped with the left substrate, the device layer will be peeled from the left substrate. With the moving/sliding of the left substrate on the blade, the device layer will be peeled continuously, then placed onto the right substrate. The process parameters including the substrate tension, roller press, substrate speed, and bending stress, can tune the conformal peeling performance. Many researches have been carried out on the interface fracture through theoretical and experimental methods considering the peel force [5], peel angle [6, 7], peel speed [8], adhesive layer thickness [9], etc.

The layout of this chapter is as follows: A device–adhesive–substrate model was first established to address the analytical solutions of the conformal peeling process. The conformal peeling mechanism was revealed considering the effects of material properties, geometric dimensions, and process parameters. Optimal design of the device thickness, substrate thickness, peel blade, and substrate tension was conducted combining the competing fracture behavior between the device peeling and device

Y. Huang et al., *Modeling and Application of Flexible Electronics Packaging*,
https://doi.org/10.1007/978-981-13-3627-0_7

cracking. The measurement of the adhesive fracture energy was achieved under different conditions of peel speed. Experimental validation on the optimal design was finally made by using the multilayer R2R manufacturing system.

7.2 Modeling of Conformal Peeling for Multilayer Structure

A schematic illustration of the mechanical model for the conformal peeling process is shown in Fig. 7.1, in which the peel radius, conformal angle, substrate speed, and substrate tension are represented by r_{blade}, $\theta_{\text{conformal}}$, v_{sub}, and T_{sub}, respectively. The following assumptions are adopted in the mechanical model: the top and the bottom layers are treated as plates that bend according to Kirchhoff–Love theory; the plane strain hypothesis is adopted; the half thicknesses of the device, adhesive, and substrate layers are, respectively, represented by $h_{\text{dev-half}}$, $h_{\text{adh-half}}$, $h_{\text{sub-half}}$, and much smaller than the corresponding lengths and widths; the material properties of each layer are isotropic, homogeneous, and linearly elastic; and the interfacial peeling always propagates along the adhesive layer.

The theoretical model is established through solving the equilibrium, kinematic, and constitutive equations, together with the boundary conditions. The horizontal, vertical, and moment equilibrium of the system are formulated by using a free-body element, as shown in Fig. 7.2. The positive directions are indicated by the arrows. For each layer, there exist the following equations:

For the adhesive layer:

$$\frac{M_{\text{adh}}}{\mathrm{d}x} = h_{\text{adh-half}}\tau_{\text{adh}} \tag{7.1}$$

where M_{adh} and τ_{adh} are the bending moment and shear stress of the adhesive layer, respectively.

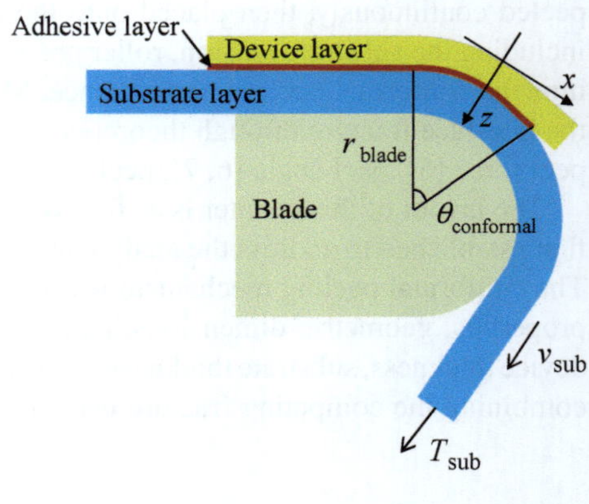

Fig. 7.1 Schematic illustration of the conformal peeling process

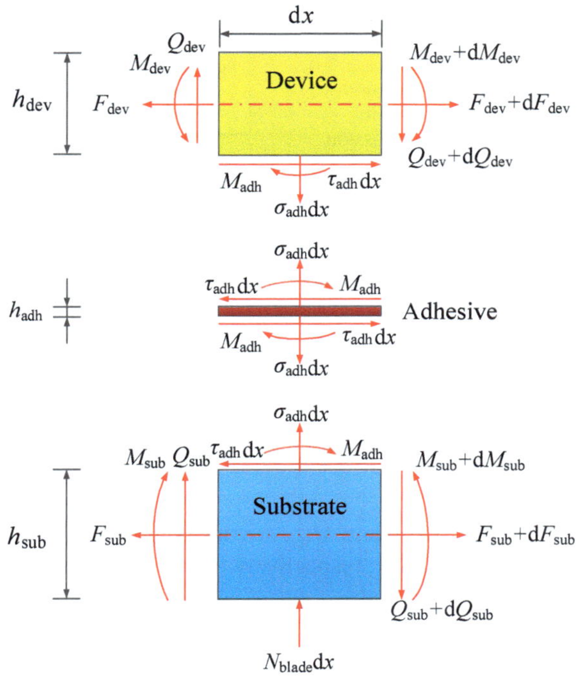

Fig. 7.2 Mechanical model of the bending-induced internal forces and moments acting on a free-body element with unit width at the overlapped region of the system. © 2018 IEEE. Reprinted, with permission, from Ref. [10]

For the device layer:

$$\frac{dF_{dev}}{dx} + \tau_{adh} = 0 \tag{7.2}$$

$$\frac{dQ_{dev}}{dx} + \sigma_{adh} = 0 \tag{7.3}$$

$$\frac{dM_{dev}}{dx} + Q_{dev} + (h_{adh\text{-}half} - h_{dev\text{-}half})\tau_{adh} = 0 \tag{7.4}$$

where F_{dev}, Q_{dev}, and M_{dev} are the axial force, shear force, and bending moment of the device layer, respectively. σ_{adh} is the normal (or peel) stress of the adhesive layer.

For the substrate layer:

$$\frac{dF_{sub}}{dx} - \tau_{adh} = 0 \tag{7.5}$$

$$\frac{dQ_{sub}}{dx} - \sigma_{adh} - N_{blade} = 0 \tag{7.6}$$

$$\frac{dM_{sub}}{dx} + Q_{sub} + (h_{adh\text{-}half} - h_{sub\text{-}half})\tau_{adh} = 0 \tag{7.7}$$

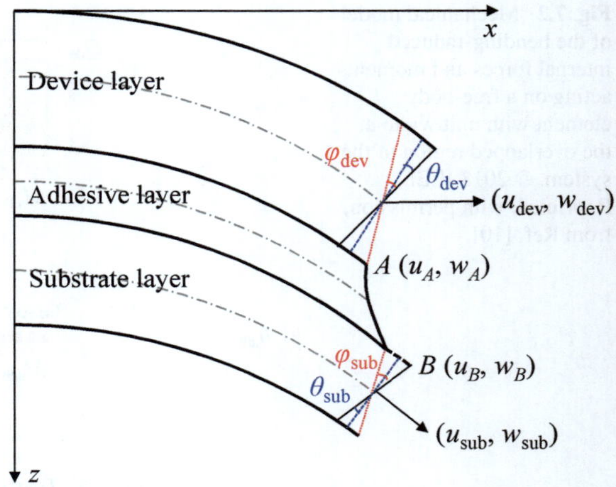

Fig. 7.3 Schematic illustration of the deformation of the device–adhesive–substrate structure. © 2018 IEEE. Reprinted, with permission, from Ref. [10]

where F_{sub}, Q_{sub}, and M_{sub} are the axial force, shear force, and bending moment of the substrate layer, respectively. N_{blade} is the supporting force of the peel blade on the substrate.

Combined with constitutive equations:

$$F_i = \frac{1}{c_i}\frac{\mathrm{d}u_i}{\mathrm{d}x} \tag{7.8}$$

$$M_i = \frac{1}{d_i}\frac{\mathrm{d}^2 w_i}{\mathrm{d}x^2} \tag{7.9}$$

Here, u_i and w_i ($i =$ dev, adh, sub) represent the displacement in x and z directions, respectively. Under the plane strain condition, c_i and d_i can be represented by $(1 - \nu_i^2)/2E_i h_{i\text{-half}}$ and $3(1 - \nu_i^2)/2E_i h_{i\text{-half}}^2$, respectively.

During the conformal peeling, the transverse displacements compatibility is considered here. A schematic illustration of the deformation of the device–adhesive—substrate structure is shown in Fig. 7.3, where θ_i and φ_i denote the rotation angle and the shear angle, respectively. The displacement of the top surface of the adhesive layer keeps consistent with that of the bottom surface of the device layer. Similarly, the displacement of the bottom surface of the adhesive layer is the same as that of the top surface of the substrate layer.

There exists

$$\frac{2h_{\text{adh-half}}\left(1 - \nu_{\text{adh}}^2\right)}{E_{\text{adh}}}\sigma_{\text{adh}} = w_B - w_A \tag{7.10}$$

where $w_A = w_{dev} + [h_{\text{dev-half}}(1 - \nu_{dev}^2)/E_{dev}]\sigma_{adh}$, $w_B = w_{sub} - [h_{\text{sub-half}}(1 - \nu_{sub}^2)/E_{sub}]\sigma_{adh}$. The normal stress of the adhesive layer can then be obtained as

$$\sigma_{adh} = f_\sigma(w_{dev} - w_{sub}) \tag{7.11}$$

where $f_\sigma = E_{dev}E_{adh}E_{sub}/[E_{adh}E_{sub}h_{dev\text{-}half}(1-\nu_{dev}^2)+E_{dev}E_{adh}h_{sub\text{-}half}(1-\nu_{sub}^2)+2E_{dev}E_{sub}h_{adh\text{-}half}(1-\nu_{adh}^2)]$.

By differentiating Eq. (7.11) with respect to x for two times, one can get

$$\frac{d^2\sigma_{adh}}{dx^2} = f_\sigma\left(\frac{d^2w_{sub}}{dx^2} - \frac{d^2w_{dev}}{dx^2}\right) \tag{7.12}$$

During the whole procedure of the conformal peeling, the radius of the peel blade is constant, which means $d^2w_{sub}/dx^2 = 1/(r_{blade}+h_{sub\text{-}half})$. By differentiating Eq. (7.11) with respect to x for four times, then

$$\frac{d^4\sigma_{adh}}{dx^4} = -f_\sigma\frac{d^4w_{dev}}{dx^4} \tag{7.13}$$

By differentiating Eq. (7.4) with respect to x for four times, and by combining Eq. (7.13), one can get

$$\frac{d^4\sigma_{adh}}{dx^4} + f_\sigma d_{dev}\sigma_{adh} - f_\sigma d_{dev}(h_{adh\text{-}half} - h_{dev\text{-}half})\frac{d\tau_{adh}}{dx} = 0 \tag{7.14}$$

The transverse displacements compatibility requires

$$\frac{2h_{adh\text{-}half}}{\overline{G}_{adh}}\tau_{adh} = u_B - u_A \tag{7.15}$$

where $\overline{G}_{adh} = E_{adh}/2(1+\nu_{adh})$ is the shear modulus of the adhesive layer under the plane strain condition. Moreover, it can be seen from Fig. 7.3 that $u_B = u_{sub} - h_{sub\text{-}half}(\varphi_{sub} - \theta_{sub})$, $u_A = u_{dev} + h_{dev\text{-}half}(\varphi_{dev} - \theta_{dev})$, $\varphi_{sub} = \tau_{adh}/4\overline{G}_{sub} + Q_{sub}/2\overline{G}_{sub}h_{sub\text{-}half}$, $\varphi_{dev} = \tau_{adh}/4\overline{G}_{dev} + Q_{dev}/2\overline{G}_{dev}h_{dev\text{-}half}$, $\theta_{sub} = dw_{sub}/dx$, and $\theta_{dev} = dw_{dev}/dx$ [11]. Then Eq. (7.15) becomes

$$\tau_{adh} = \frac{u_{sub} - u_{dev} + h_{dev\text{-}half}\dfrac{dw_{dev}}{dx} + h_{sub\text{-}half}\dfrac{dw_{sub}}{dx} + \dfrac{1}{2\overline{G}_{dev}}\dfrac{dM_{dev}}{dx} + \dfrac{1}{2\overline{G}_{sub}}\dfrac{dM_{sub}}{dx}}{\dfrac{2h_{adh\text{-}half}}{\overline{G}_{adh}} + \dfrac{3h_{dev\text{-}half} - 2h_{adh\text{-}half}}{4\overline{G}_{dev}} + \dfrac{3h_{sub\text{-}half} - 2h_{adh\text{-}half}}{4\overline{G}_{sub}}} \tag{7.16}$$

where $\overline{G}_{dev} = E_{dev}/2(1+\nu_{dev})$ and $\overline{G}_{sub} = E_{sub}/2(1+\nu_{sub})$ are the shear modulus of the device layer and substrate layer under the plane strain condition. If $f_\tau = 4\overline{G}_{dev}\overline{G}_{adh}G_{sub}/[8\overline{G}_{dev}\overline{G}_{sub}h_{adh\text{-}half} + \overline{G}_{adh}\overline{G}_{sub}(3h_{dev\text{-}half} - 2h_{adh\text{-}half}) + \overline{G}_{dev}\overline{G}_{adh}(3h_{sub\text{-}half} - 2h_{adh\text{-}half})]$, then Eq. (7.16) can be simplified as

$$\tau_{\text{adh}} = f_\tau \left(\begin{array}{l} u_{\text{sub}} - u_{\text{dev}} + h_{\text{dev-half}} \dfrac{\text{d}w_{\text{dev}}}{\text{d}x} + h_{\text{sub-half}} \dfrac{\text{d}w_{\text{sub}}}{\text{d}x} \\ -\dfrac{1}{2\overline{G}_{\text{dev}}} \dfrac{\text{d}M_{\text{dev}}}{\text{d}x} + \dfrac{1}{2\overline{G}_{\text{sub}}} \dfrac{\text{d}M_{\text{sub}}}{\text{d}x} \end{array} \right) \tag{7.17}$$

By differentiating Eq. (7.17) with respect to x, one can obtain

$$\frac{\text{d}\tau_{\text{adh}}}{\text{d}x} = f_\tau \left(\begin{array}{l} \dfrac{\text{d}u_{\text{sub}}}{\text{d}s} - \dfrac{\text{d}u_{\text{dev}}}{\text{d}x} + h_{\text{dev-half}} \dfrac{\text{d}^2 w_{\text{dev}}}{\text{d}x^2} \\ + h_{\text{sub-half}} \dfrac{\text{d}^2 w_{\text{sub}}}{\text{d}x^2} + \dfrac{1}{2\overline{G}_{\text{dev}} d_{\text{dev}}} \dfrac{\text{d}^4 w_{\text{dev}}}{\text{d}x^4} \end{array} \right) \tag{7.18}$$

By differentiating Eq. (7.17) with respect to x for three times, then

$$\frac{\text{d}^3\tau_{\text{adh}}}{\text{d}x^3} - f_\tau(c_{\text{dev}} + c_{\text{sub}}) \frac{\text{d}\tau_{\text{adh}}}{\text{d}x} + \frac{f_\tau h_{\text{dev}}}{f_\sigma} \frac{\text{d}^4\sigma_{\text{adh}}}{\text{d}x^4} + \frac{f_\tau}{2\overline{G}_{\text{dev}} d_{\text{dev}} f_\sigma} \frac{\text{d}^6\sigma_{\text{adh}}}{\text{d}x^6} = 0 \tag{7.19}$$

Combining Eqs. (7.14) and (7.19), one can get

$$\frac{\text{d}^6\sigma_{\text{adh}}}{\text{d}x^6} + P_1 \frac{\text{d}^4\sigma_{\text{adh}}}{\text{d}x^4} + P_2 \frac{\text{d}^2\sigma_{\text{adh}}}{\text{d}x^2} + P_3\sigma_{\text{adh}} = 0 \tag{7.20}$$

where $P_1 = -2\overline{G}_{\text{dev}} f_\tau [c_{\text{dev}} + c_{\text{sub}} - d_{\text{dev}} h_{\text{dev-half}} (h_{\text{adh-half}} - h_{\text{dev-half}})]/[2\overline{G}_{\text{dev}} + f_\tau (h_{\text{adh-half}} - h_{\text{dev-half}})]$, $P_2 = 2\overline{G}_{\text{dev}} f_\sigma d_{\text{dev}}/[2\overline{G}_{\text{dev}} + f_\tau (h_{\text{adh-half}} - h_{\text{dev-half}})]$, $P_3 = -2\overline{G}_{\text{dev}} f_\sigma f_\tau d_{\text{dev}} (c_{\text{dev}} + c_{\text{sub}})/[2\overline{G}_{\text{dev}} + f_\tau (h_{\text{adh-half}} - h_{\text{dev-half}})]$.

The solution for Eq. (7.20) can be determined as

$$\sigma_{\text{adh}}(x) = \sum_{n=1}^{6} C_n \exp(\lambda_n x) \tag{7.21}$$

The shear stress can be obtained as

$$\tau_{\text{adh}}(x) = \frac{1}{f_\sigma d_{\text{dev}} (h_{\text{adh-half}} - h_{\text{dev-half}})} \left\{ \sum_{n=1}^{6} C_n \left[(\lambda_n)^3 + \frac{f_\sigma d_{\text{dev}}}{\lambda_n} \right] \exp(\lambda_n x) + C_7 \right\} \tag{7.22}$$

The characteristic equation of Eq. (7.20) can be written as

$$\lambda^6 + P_1\lambda^4 + P_2\lambda^2 + P_3 = 0 \tag{7.23}$$

By regarding λ^2 as a whole, one can convert Eq. (7.23) into a cubic equation about λ^2, the roots of which can be determined easily by $\left[(P_1^2 - 3P_2)/9\right]^3 \left[(2P_1^3 - 9P_1P_2 + 27P_3)\right]/54]^2$. Considering that the adhesive layer is very thin, the characteristic roots of Eq. (7.23) can be obtained as

$$\begin{cases} \lambda_1 = \sqrt{\frac{\sqrt{3}}{3}\sqrt{|P_1| - 2\sqrt{P_1^2 - 3P_2}\cos\left(\frac{\theta}{3}\right)}} \\ \lambda_2 = -\sqrt{\frac{\sqrt{3}}{3}\sqrt{|P_1| - 2\sqrt{P_1^2 - 3P_2}\cos\left(\frac{\theta}{3}\right)}} \\ \lambda_3 = \sqrt{\frac{\sqrt{3}}{3}\sqrt{|P_1| + \sqrt{P_1^2 - 3P_2}\left[\cos\left(\frac{\theta}{3}\right) + \sqrt{3}\sin\left(\frac{\theta}{3}\right)\right]}} \\ \lambda_4 = -\sqrt{\frac{\sqrt{3}}{3}\sqrt{|P_1| + \sqrt{P_1^2 - 3P_2}\left[\cos\left(\frac{\theta}{3}\right) + \sqrt{3}\sin\left(\frac{\theta}{3}\right)\right]}} \\ \lambda_5 = \sqrt{\frac{\sqrt{3}}{3}\sqrt{|P_1| + \sqrt{P_1^2 - 3P_2}\left[\cos\left(\frac{\theta}{3}\right) - \sqrt{3}\sin\left(\frac{\theta}{3}\right)\right]}} \\ \lambda_6 = -\sqrt{\frac{\sqrt{3}}{3}\sqrt{|P_1| + \sqrt{P_1^2 - 3P_2}\left[\cos\left(\frac{\theta}{3}\right) - \sqrt{3}\sin\left(\frac{\theta}{3}\right)\right]}} \end{cases} \tag{7.24}$$

where $\theta = \arccos\left\{\left[(2P_1^3 - 9P_1P_2 + 27P_3)/54\right]/\left[(P_1^2 - 3P_2)/9\right]^{1.5}\right\}$.

The multilayer structure is symmetric about z axis. A substrate tension T_{sub} and a moment M_{sub} are applied at the right-side end of the substrate layer. The boundary conditions in the middle portion of the three layers can be summarized as follows:

$$u_{\text{dev}}|_{x=0} = 0, \quad \theta_{\text{dev}}|_{x=0} = 0, \quad Q_{\text{dev}}|_{x=0} = 0 \tag{7.25}$$

$$u_{\text{sub}}|_{x=0} = 0, \quad \theta_{\text{sub}}|_{x=0} = 0, \quad Q_{\text{sub}}|_{x=0} = 0 \tag{7.26}$$

The boundary conditions at the right side end of the three layers are as follows:

$$M_{\text{dev}}|_{x=l_{\text{sub-half}}} = 0, \quad F_{\text{dev}}|_{x=l_{\text{sub-half}}} = 0, \quad Q_{\text{dev}}|_{x=l_{\text{sub-half}}} = 0 \tag{7.27}$$

$$M_{\text{sub}}|_{x=l_{\text{sub-half}}} = M_{\text{sub}}, \quad F_{\text{sub}}|_{x=l_{\text{sub-half}}} = T_{\text{sub}}, \quad Q_{\text{sub}}|_{x=l_{\text{sub-half}}} = 0 \tag{7.28}$$

where $l_{\text{sub-half}}$ is the half length of the substrate layer.

The relationship among the constants C_1, C_2, C_3, C_4, C_5, C_6, and C_7 in Eq. (7.22) is

$$\begin{cases} \sum_{i=1}^{6} \frac{C_i}{\lambda_i} e^{l_{\text{dev-half}}\lambda_i} = 0 \\ \sum_{i=1}^{6} C_i\lambda_i = 0 \\ \sum_{n=1}^{6} C_n\lambda_n e^{l_{\text{sub-half}}\lambda_n} = \frac{f_\sigma}{r_{\text{blade}} + h_{\text{sub-half}}} \\ \sum_{n=1}^{6} C_n\lambda_n^3 = 0 \\ \sum_{n=1}^{6} C_n\lambda_n^5 = 0 \\ \sum_{n=1}^{6} \left[C_n\lambda_n^4 + \frac{2\overline{G}_{\text{dev}} f_\sigma d_{\text{dev}}}{2\overline{G}_{\text{dev}} + f_\tau(h_{\text{adh-half}} - h_{\text{dev-half}})(r_{\text{blade}} + h_{\text{sub}})} C_n\lambda_n \right] e^{\lambda_n l_{\text{sub-half}}} \\ = \frac{2 f_\tau \overline{G}_{\text{dev}} h_{\text{sub-half}} f_\sigma d_{\text{dev}} (h_{\text{adh-half}} - h_{\text{dev-half}})(r_{\text{blade}} T_{\text{sub}} + h_{\text{sub}} T_{\text{sub}} + 1)}{2\overline{G}_{\text{dev}} + f_\tau(h_{\text{adh-half}} - h_{\text{dev-half}})(r_{\text{blade}} + h_{\text{sub-half}})} \\ \frac{1}{f_\sigma d_{\text{dev}}(h_{\text{adh-half}} - h_{\text{dev-half}})} \left\{ \sum_{n=1}^{6} C_n \left[(\lambda_n)^3 + \frac{f_\sigma d_{\text{dev}}}{\lambda_n} \right] + C_7 \right\} = 0 \end{cases} \tag{7.29}$$

After solving Eq. (7.29), one can get

$$\begin{cases} C_1 = C_2 = \frac{f_\sigma - f_\tau h_{\text{sub-half}}(h_{\text{adh-half}} - h_{\text{dev-half}})(r_{\text{blade}} T_{\text{sub}} + h_{\text{sub-half}} T_{\text{sub}} + 1)(r_{\text{blade}} + h_{\text{sub-half}})}{\lambda_1 \left[\exp(\lambda_1 l_{\text{sub-half}}) - \exp(-\lambda_1 l_{\text{sub-half}}) \right] (r_{\text{blade}} + h_{\text{sub-half}})} \\ C_3 = C_4 = \frac{f_\tau h_{\text{sub-half}}(h_{\text{adh-half}} - h_{\text{dev-half}}) \begin{pmatrix} r_{\text{balde}} T_{\text{sub}} \\ + h_{\text{sub}} T_{\text{sub}} + 1 \end{pmatrix} \begin{bmatrix} (r_{\text{blade}} + h_{\text{sub-half}})\lambda_1^2\lambda_3^2 \\ + \begin{pmatrix} 1 - r_{\text{blade}} \\ - h_{\text{sub-half}} \end{pmatrix} \lambda_1^2\lambda_5^2 \\ - \lambda_3^2\lambda_5^2 \end{bmatrix} + f_\sigma\lambda_3\lambda_5^2}{\lambda_1^2\lambda_3 \left[\exp(\lambda_3 l_{\text{sub-half}}) - \exp(-\lambda_3 l_{\text{sub-half}}) \right] (\lambda_3^2 - \lambda_5^2)(r_{\text{blade}} + h_{\text{sub-half}})} \\ C_5 = C_6 = \frac{f_\tau h_{\text{sub-half}}(h_{\text{adh-half}} - h_{\text{dev-half}})(r_{\text{blade}} T_{\text{sub}} + h_{\text{sub-half}} T_{\text{sub}} + 1)(\lambda_3^2 - \lambda_1^2)\lambda_5 - f_\sigma\lambda_3\lambda_5}{\lambda_1^2 \left[\exp(\lambda_5 l_{\text{sub-half}}) - \exp(-\lambda_5 l_{\text{sub-half}}) \right] (\lambda_3^2 - \lambda_5^2)(r_{\text{blade}} + h_{\text{sub-half}})} \\ C_7 = 0 \end{cases} \tag{7.30}$$

The maximum tensile stress of the device layer can be determined as

$$\sigma_{\text{dev}}^{\max} = \frac{F_{\text{dev}}}{2h_{\text{dev-half}}} + \frac{3M_{\text{dev}}}{2h_{\text{dev-half}}^2} \tag{7.31}$$

To ensure that the device layer can be peeled-off safely during the conformal peeling process, the maximum tensile stress of the device layer must be smaller than the corresponding fracture strength, i.e., $\sigma_{\text{dev}}^{\max} < \sigma_{\text{dev}}^{c}$.

Combining Eqs. (7.9) and (7.12), the bending moment of the device layer under the conformal condition can be written as

$$M_{\text{dev}}(x) = \frac{1}{f_\sigma d_{\text{dev}}} \left(\frac{f_\sigma}{r_{\text{blade}} + h_{\text{adh-half}}} - \frac{\mathrm{d}^2\sigma_{\text{adh}}}{\mathrm{d}x^2} \right), \tag{7.32}$$

and the axial force of the device layer is

$$F_{\mathrm{dev}}(x) = \int_0^x [-\tau_{\mathrm{adh}}(x)\mathrm{d}x] \tag{7.33}$$

Then one can get

$$N_{\mathrm{blade}}(x) = \frac{\mathrm{d}Q_{\mathrm{sub}}}{\mathrm{d}x} - \sigma_{\mathrm{adh}} \tag{7.34}$$

Under the condition of the conformal peeling, the substrate is in close contact with the surface of the peel blade, namely $\mathrm{d}M_{\mathrm{sub}}/\mathrm{d}x = 0$, then $Q_{\mathrm{sub}} = (h_{\text{sub-half}} - h_{\text{adh-half}})\tau_{\mathrm{adh}}$. One can get

$$N_{\mathrm{blade}}(x) = (h_{\text{sub-half}} - h_{\text{adh-half}})\frac{\mathrm{d}\tau_{\mathrm{adh}}}{\mathrm{d}x} - \sigma_{\mathrm{adh}} \tag{7.35}$$

The conformal peeling exists under the condition of $N_{\mathrm{blade}}(x) > 0$. The status of $N_{\mathrm{blade}}(x) < 0$ denotes that there exists tension between the blade surface and the substrate. However, there is no adhesive strength between these two surfaces, which means that $N_{\mathrm{blade}}(x) < 0$ indicates the non-conformal state.

As for the adhesive layer, the ERR of interfacial peeling can be obtained by using the linear elastic fracture mechanics theory as follows:

$$G_{\mathrm{adh}}^{\text{crack-tip}} = G_{\mathrm{I}} + G_{\mathrm{II}} \tag{7.36}$$

Under the plain strain condition, the above Eq. (7.36) can be rewritten as

$$G_{\mathrm{adh}}^{\text{crack-tip}} = \frac{h_{\text{adh-half}}\left(1 - \nu_{\mathrm{adh}}^2\right)}{E_{\mathrm{adh}}}\left(\sigma_{\mathrm{adh}}^{\max}\right)^2 + \frac{h_{\text{adh-half}}}{\overline{G}_{\mathrm{adh}}}\left(\tau_{\mathrm{adh}}^{\max}\right)^2 \tag{7.37}$$

The successful peeling-off of the device layer from the substrate layer requires that

$$G_{\mathrm{adh}}^{\text{crack−tip}} > G_{\mathrm{adh}}^{c} \tag{7.38}$$

where G_{adh}^{c} is the adhesive fracture energy of the adhesive layer.

Here, the mode-mixity angle of interfacial peeling can be calculated as $\psi_{\mathrm{adh}}^{\text{crack-tip}} = \arctan(\sqrt{G_{\mathrm{II}}/G_{\mathrm{I}}}) = \sqrt{2}\tau_{\mathrm{adh}}^{\max}/(\sqrt{1+\nu_{\mathrm{adh}}}\sigma_{\mathrm{adh}}^{\max})$. $\psi_{\mathrm{adh}}^{\text{crack-tip}} = 0$ means that it is the normal stress of the adhesive layer separating the device layer from the substrate layer, while $\psi_{\mathrm{adh}}^{\text{crack-tip}} = 90°$ denotes that it is the shear stress of the adhesive layer which separates the device layer from the substrate layer. Obviously, $\psi_{\mathrm{adh}}^{\text{crack-tip}} = 90°$ should be avoided during the conformal peeling process considering that the device layer will be separated from the substrate layer in an uncontrolled manner.

7.3 Effect of Material Properties and Geometric Size

The competition between the maximum tensile stress of the device layer and the ERR of interfacial peeling will affect the peeling process. The positive peel force results from the strain mismatch between the device layer and the substrate layer increases the ERR of interfacial peeling. When the peel force is large enough, the device layer will be separated from the substrate layer. The maximum strain, the normal stress, and the shear stress will also increase with the increase of the Young's modulus of the device layer. Generally, the interfacial peeling occurs at the free end of the layered structure when Eq. (7.38) is satisfied. Meanwhile, the device layer should be kept intact, which requires that $\sigma_{\text{dev}}^{\max} < \sigma_{\text{dev}}^{c}$. Based on Eqs. (7.31)–(7.33), the maximum tensile stress of the device layer can be written as

$$\sigma_{\text{dev}}^{\max} = \frac{1}{2h_{\text{dev-half}}} \int_0^{l_{\text{dev-half}}} [-\tau_{\text{adh}}(x)\mathrm{d}x] + \frac{3}{2f_\sigma d_{\text{dev}} h_{\text{dev-half}}^2} \left(\frac{f_\sigma}{r_{\text{blade}} + h_{\text{sub-half}}} - \frac{\mathrm{d}^2\sigma_{\text{adh}}}{\mathrm{d}x^2} \right) \tag{7.39}$$

By using Eqs. (7.21) and (7.22), the maximum tensile stress of the device layer can be obtained as

$$\begin{aligned} \sigma_{\text{dev}}^{\max} = & \frac{3}{2d_{\text{dev}} h_{\text{dev-half}}^2 (r_{\text{blade}} + h_{\text{sub-half}})} \\ & + \frac{3f_\sigma d_{\text{dev}}(h_{\text{adh-half}} - h_{\text{dev-half}}) - h_{\text{dev-half}}}{2h_{\text{dev-half}}^2 f_\sigma d_{\text{dev}}(h_{\text{dev-half}} - h_{\text{adh-half}})} \sum_{n=1}^{6} C_n \lambda_n^2 \exp(\lambda_n l_{\text{dev-half}}) \\ & - \frac{3}{2h_{\text{dev-half}}(h_{\text{adh-half}} - h_{\text{dev-half}})} \sum_{n=1}^{6} \frac{C_n}{\lambda_n^2} \exp(\lambda_n l_{\text{dev-half}}) \end{aligned} \tag{7.40}$$

It can be found from Eq. (7.40) that there are several factors affecting the maximum tensile stress of the device layer, including the thickness of the device, adhesive, and substrate layers, the radius of the peel blade, and even the device length.

The substrate tension can be written as

$$\begin{aligned} T_{\text{sub}} &= \int_0^{l_{\text{sub-half}}} \tau_{\text{adh}}(x)\mathrm{d}x \\ &= \frac{1}{f_\sigma d_{\text{dev}}(h_{\text{adh-half}} - h_{\text{dev-half}})} \sum_{n=1}^{6} C_n \left(\lambda_n^2 + \frac{f_\sigma d_{\text{dev}}}{\lambda_n^2} \right) \exp(\lambda_n l_{\text{sub-half}}) \end{aligned} \tag{7.41}$$

7.3.1 *Effect of Young's Modulus of Device Layer*

The effect of the Young's modulus of the device layer and $(E_{\text{dev}})^{0.5}$ on the maximum tensile stress of the device layer is shown in Fig. 7.4. Here, the material parameters

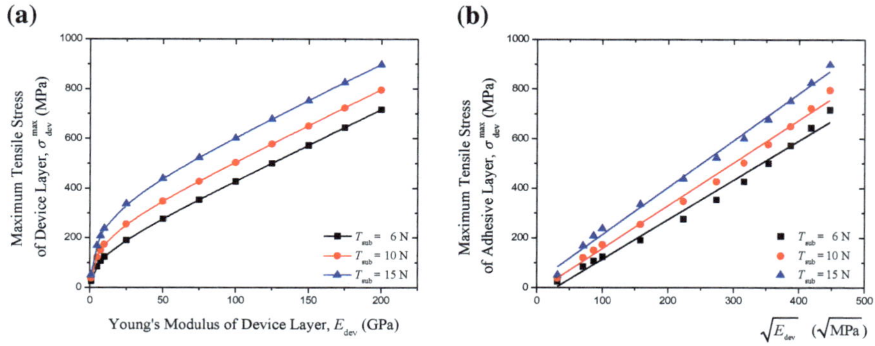

Fig. 7.4 Results of the maximum tensile stress of the device layer: **a** the effect of the Young's modulus of the device layer, **b** the effect of $(E_{dev})^{0.5}$. © 2018 IEEE. Reprinted, with permission, from Ref. [10]

Table 7.1 Material properties of different layers

Layer (–)	Thickness (mm)	Young's modulus (MPa)	Poisson's ratio (–)	Material (–)
Device	0.05	129,000	0.28	Silicon
Adhesive	0.01	3600	0.39	Epotek301-2
Substrate	0.05	5370	0.32	Kapton

of different layers are listed in Table 7.1. The radius of the peel blade is set as 10 mm, and the angle of the conformal peeling is adopted as 0.1 rad. As can be seen from Fig. 7.4a, the maximum tensile stress of the device layer increases monotonously with the increase of the Young's modulus of the device layer despite different substrate tension conditions. It can be observed from Fig. 7.4b that the maximum tensile stress of the device layer changes linearly with $(E_{dev})^{0.5}$ under the conditions of the substrate tension of 6 N, 10 N, and 15 N.

The effect of the Young's modulus of the device layer on the maximum normal stress and the maximum shear stress of the adhesive layer is shown in Fig. 7.5. It can be seen that the maximum normal stress and the maximum shear stress increase sharply and then converge at $E_{dev} \approx 20$ GPa. Moreover, the ERR of interfacial peeling is influenced by both the substrate tension and the Young's modulus of the device layer. The effect of the Young's modulus of the device layer on the ERR of interfacial peeling is shown in Fig. 7.6. As can be found, the ERR of interfacial peeling increases obviously with the substrate tension increasing. Similarly, the ERR of interfacial peeling increases with the Young's modulus of the device layer, and converges when $E_{dev} \approx 50$ GPa. The results denote that the substrate tension can be used to tune the peelability.

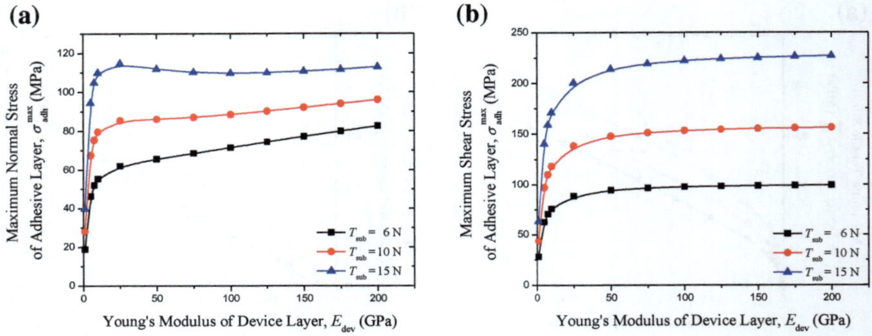

Fig. 7.5 The effect of the Young's modulus of the device layer on the maximum normal stress and the maximum shear stress of the adhesive layer. © 2018 IEEE. Reprinted, with permission, from Ref. [10]

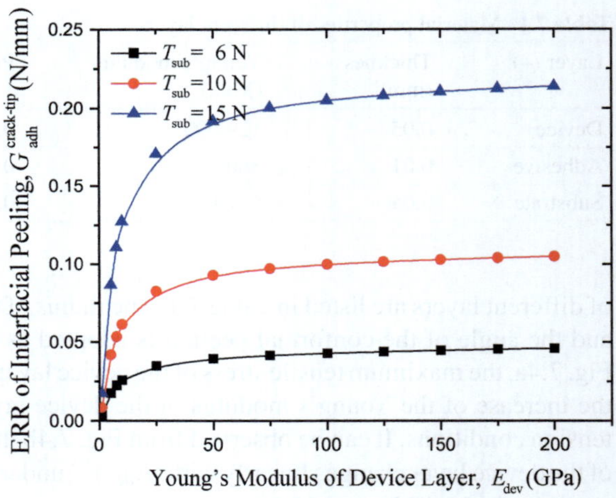

Fig. 7.6 The effect of the Young's modulus of the device layer on the ERR of interfacial peeling. © 2018 IEEE. Reprinted, with permission, from Ref. [10]

7.3.2 *Effect of Young's Modulus of Adhesive Layer*

Although theoretical analysis shows that the change of the normal stress and the shear stress of the adhesive layer can be usually neglected due to the small thickness of the adhesive layer, a significant effect of the Young's modulus of the adhesive layer on the stress distribution can still be found. Through the finite element simulation, similar behavior appears that the adhesive layer shows an obvious effect on the stress distribution and the ERR of interfacial peeling. The effect of the Young's modulus of the adhesive layer on the maximum tensile stress of the device layer is depicted in Fig. 7.7. As can be seen, the maximum tensile stress of the device layer increases with the Young's modulus of the adhesive layer when $E_{adh} < 2$ GPa, and converges when

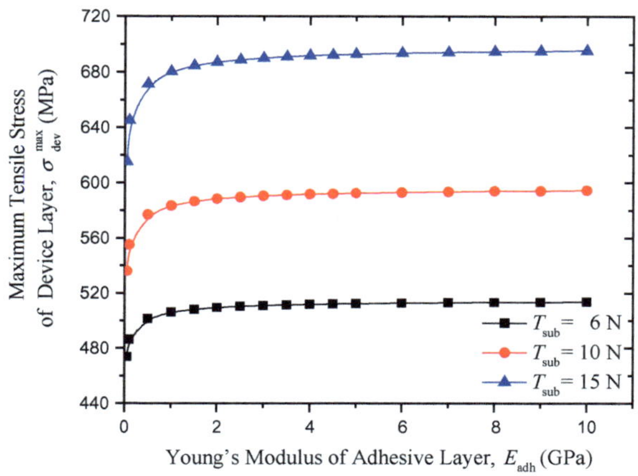

Fig. 7.7 The effect of the Young's modulus of the adhesive layer on the maximum tensile stress of the device layer

$E_{adh} > 2$ GPa. Moreover, the maximum tensile stress of the device layer increases obviously with the increase of the substrate tension.

The effect of the Young's modulus of the adhesive layer on the maximum normal stress and the maximum shear stress of the adhesive layer is shown in Fig. 7.8. Both stresses increase with the increase of the Young's modulus of the adhesive layer and the substrate tension. According to Eq. (7.37), the ERR of interfacial peeling is closely related to the Young's modulus of the adhesive layer. The effect of the Young's modulus of the adhesive layer on the ERR of interfacial peeling is shown in Fig. 7.9. It can be seen that the ERR of interfacial peeling decreases obviously with the increase of the Young's modulus of the adhesive layer. It can also be observed from Figs. 7.8 and 7.9 that the substrate tension promotes the improvement of the maximum normal stress of the adhesive layer, the maximum shear stress of the adhesive layer, and the ERR of interfacial peeling.

According to the definition of the mode-mixity angle, one can get

$$\psi_{\text{adh}}^{\text{crack-tip}} = \frac{1}{f_\sigma d_{\text{dev}}(h_{\text{adh-half}} - h_{\text{dev-half}})}\sqrt{\frac{2}{1+\nu_{\text{adh}}}}\frac{\sum_{n=1}^{6} C_n\left(\lambda_n^3 + \frac{f_\sigma d_{\text{dev}}}{\lambda_n}\right)\exp(\lambda_n)}{\sum_{n=1}^{6} C_n \exp(\lambda_n)} \tag{7.42}$$

The change of the mode-mixity angle with the Young's modulus of the adhesive layer is shown in Fig. 7.10. The mode-mixity angle initially increases and then decreases with the increase of the Young's modulus of the adhesive layer. Additionally, the mode-mixity angle decreases with the substrate tension increasing. Namely, the mode II peeling plays a dominate role under the condition of small substrate tension.

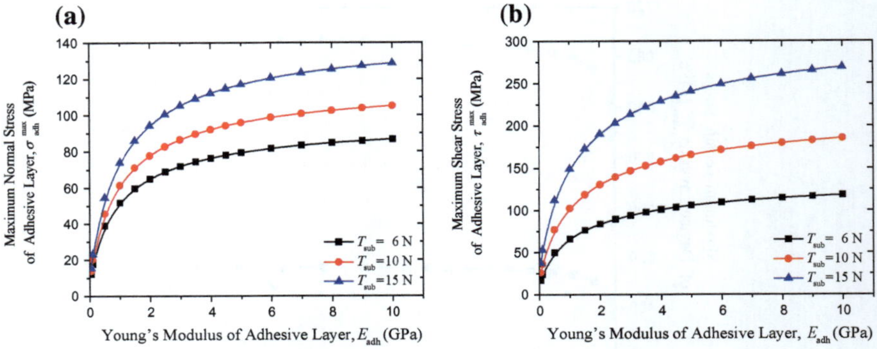

Fig. 7.8 The effect of the Young's modulus of the adhesive layer on: **a** the maximum normal stress of the adhesive layer, **b** the maximum shear stress of the adhesive layer

Fig. 7.9 The effect of the Young's modulus of the adhesive layer on the ERR of interfacial peeling

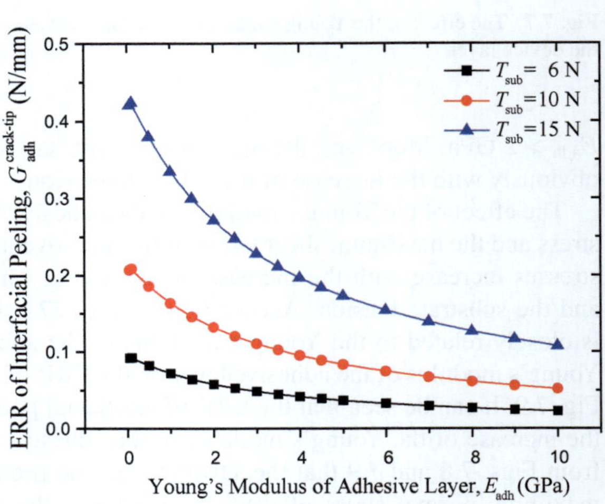

The effect of the Young's modulus of the substrate layer on the maximum tensile stress of the device layer is shown in Fig. 7.11. As can be seen, the maximum tensile stress of the device layer increases with the Young's modulus of the substrate layer and the substrate tension. The variation trend of the ERR of interfacial peeling and the corresponding mode-mixity angle with the Young's modulus of the device layer is shown in Fig. 7.12. The monotonicity of the ERR of interfacial peeling is determined by both the substrate tension and the Young's modulus of the device layer. When the substrate tension is relatively small, e.g., $T_{\mathrm{sub}} = 6$ N, the ERR of interfacial peeling increases with the increase of the Young's modulus of the device layer. It is interesting to see that the ERR of interfacial peeling first decreases and then increases under the condition of $T_{\mathrm{sub}} = 10$ N. When the substrate tension increases further to 15 N, the ERR of interfacial peeling decreases monotonously, and converges at $E_{\mathrm{dev}} \approx 50$ GPa.

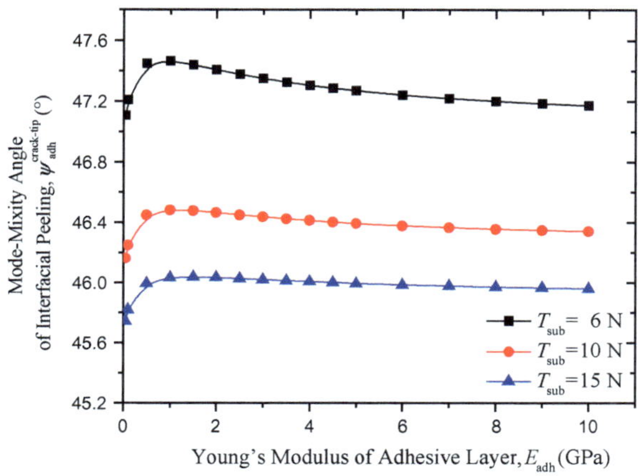

Fig. 7.10 The effect of the Young's modulus of the adhesive layer on the mode-mixity angle of interfacial peeling

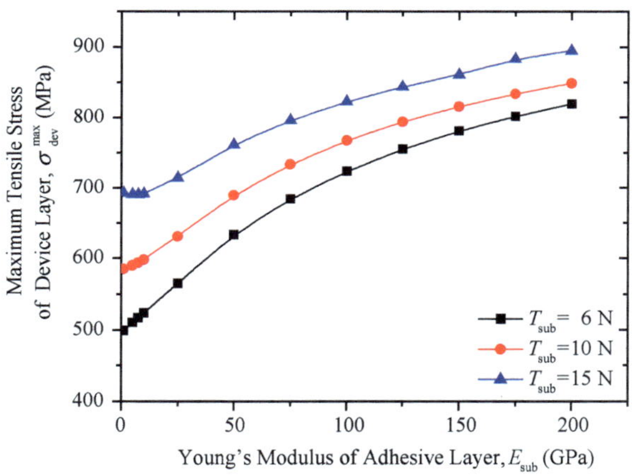

Fig. 7.11 The effect of the Young's modulus of the substrate layer on the maximum tensile stress of the device layer. © 2018 IEEE. Reprinted, with permission, from Ref. [10]

However, the mode-mixity angle shows an identical increasing trend under different substrate tension conditions, and converges at $E_{dev} \approx 50$ GPa.

7.3.3 *Effect of Device Thickness*

RFID tags with the thickness of 0.3 mm were adopted as the device layer in the experiment. The effect of the adhesive thickness is neglected here, since it is much

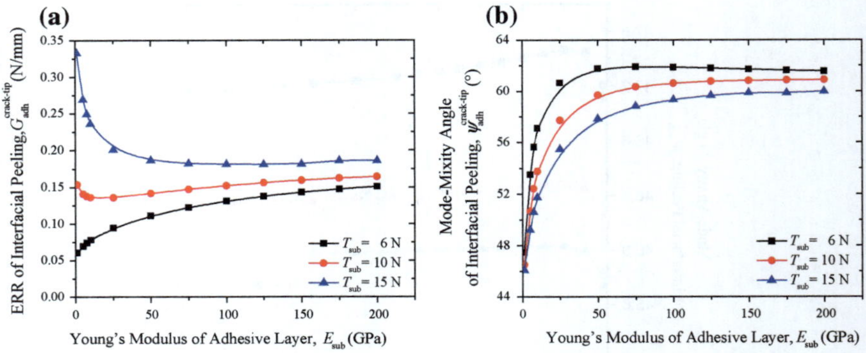

Fig. 7.12 The effect of the Young's modulus of the substrate layer on the ERR and the mode-mixity angle of interfacial peeling. © 2018 IEEE. Reprinted, with permission, from Ref. [10]

smaller than that of the substrate layer and the device layer. Under the bending radius of $r_{\text{blade}} + 2h_{\text{sub-half}} + h_{\text{adh-half}}$, one can rewrite Eq. (7.31) as

$$\sigma_{\text{dev}}^{\max} = \frac{1}{2h_{\text{dev-half}} f_\sigma d_{\text{dev}}(h_{\text{dev-half}} - h_{\text{adh-half}})} \sum_{n=1}^{6} C_n \left(\lambda_n^2 + \frac{f_\sigma d_{\text{dev}}}{\lambda_n^2}\right) \exp(\lambda_n l_{\text{dev-half}}) + \frac{E_{\text{dev}} h_{\text{dev-half}}}{r_{\text{blade}} + 2h_{\text{sub-half}} + h_{\text{adh-half}}} \tag{7.43}$$

It can be seen from Eq. (7.43) that the maximum tensile stress of the device layer is directly related to the substrate thickness and the device thickness, the effect of which on the maximum tensile stress of the device layer is then investigated separately.

The maximum tensile stress of the device layer versus the device thickness under the condition of the peel radius of 10 mm and the conformal angle of 0.1 rad is shown in Fig. 7.13. An obvious increasing trend of the maximum tensile stress of the device layer with the increase of the device thickness can be found. The substrate tension has a relatively large influence on the maximum tensile stress of the device layer when the device thickness is small. The above results agree with Ref. [12]. However, the substrate tension plays a negligible role in the maximum tensile stress of the device layer when the device thickness is relatively large.

The effect of the device thickness on the maximum normal and shear stresses of the adhesive layer is shown in Fig. 7.14. Both stresses increase with the device thickness increasing. Therefore, the force resulting from the stresses of the adhesive layer that makes the device layer bend into the same radius increases with the device thickness. The effect of the device thickness on the ERR of interfacial peeling and the corresponding mode-mixity angle is shown in Fig. 7.15. Obviously, the ERR of interfacial peeling increases with the device thickness increasing, as shown in Fig. 7.15a. The relationship between the mode-mixity angle and the device thickness

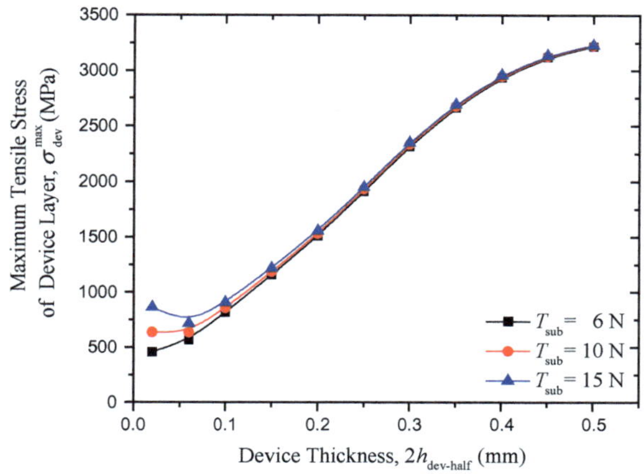

Fig. 7.13 The effect of the device thickness on the maximum tensile stress of the device layer

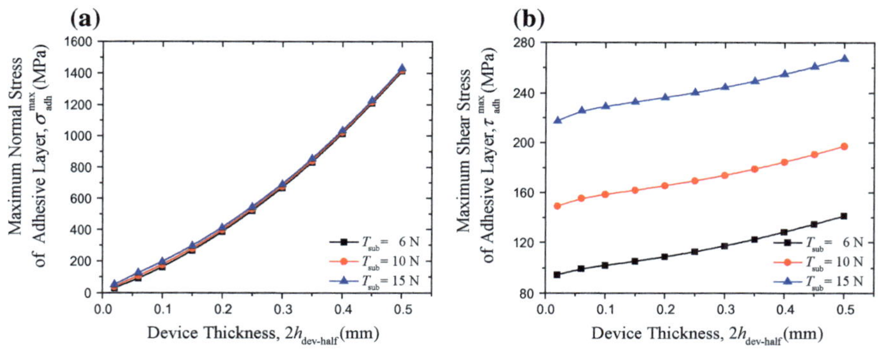

Fig. 7.14 The effect of the device thickness on: **a** the maximum normal stress of the adhesive layer, **b** the maximum shear stress of the adhesive layer

is shown in Fig. 7.15b. The mode-mixity angle gets larger with the increase of the device thickness, but get smaller with the increase of the substrate tension.

7.3.4 Effect of Substrate Thickness

The effect of the substrate thickness on the maximum tensile stress of the device layer under various substrate tension conditions is shown in Fig. 7.16. As can be seen, the maximum tensile stress increases with the increase of the substrate thickness and the substrate tension. The effect of the substrate thickness on the maximum normal stress and the maximum shear stress of the adhesive layer is depicted in Fig. 7.17. It can be found that the maximum normal stress of the adhesive layer decreases with the

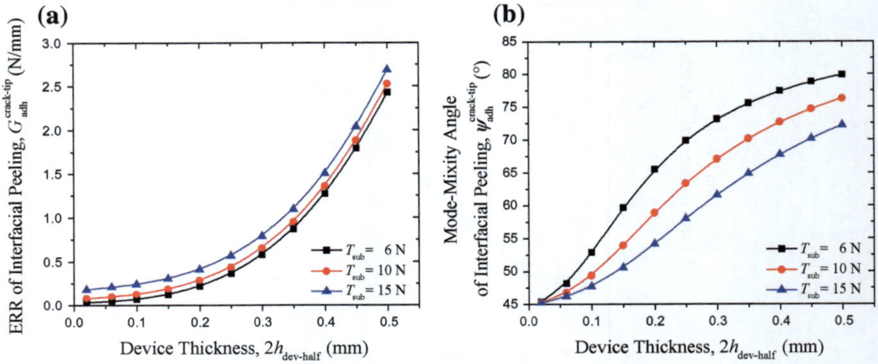

Fig. 7.15 The effect of the device thickness on: **a** the ERR of interfacial peeling, **b** the mode-mixity angle of interfacial peeling

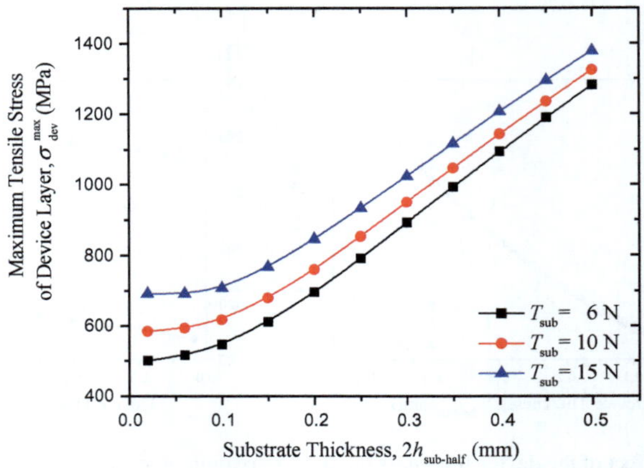

Fig. 7.16 The effect of the substrate thickness on the maximum tensile stress of the device layer

substrate thickness increasing, and converges when the substrate thickness is larger than 0.2 mm. The maximum shear stress of the adhesive layer also decreases with the increase of the substrate thickness. There exists a minimum value of the maximum shear stress when the substrate thickness is about 0.15 mm.

The effect of the substrate thickness on the ERR of interfacial peeling and the corresponding mode-mixity angle is shown in Fig. 7.18. It can be seen that the ERR of interfacial peeling decreases with the increase of the substrate thickness. While the corresponding mode-mixity angle first increases and then decreases with the increase of the substrate thickness.

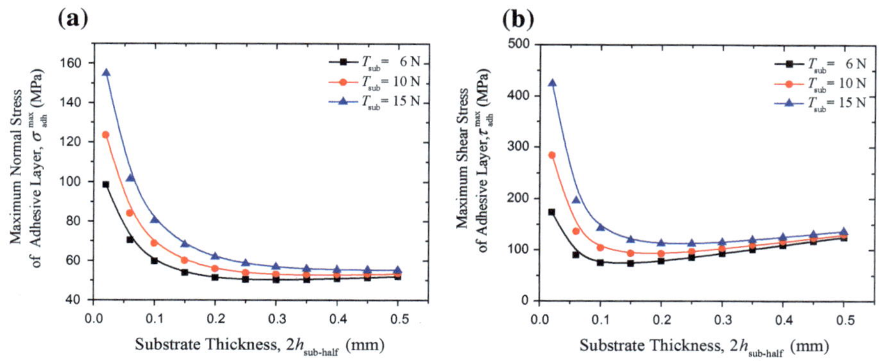

Fig. 7.17 The effect of the substrate thickness on: **a** the maximum normal stress of the adhesive layer, **b** the maximum shear stress of the adhesive layer

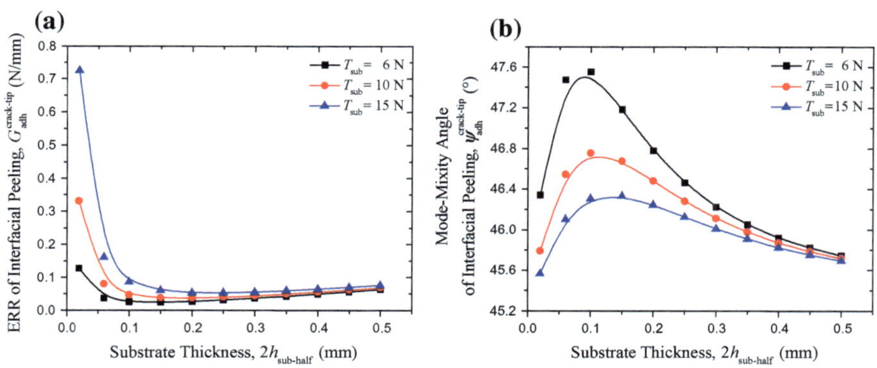

Fig. 7.18 The effect of the substrate thickness on: **a** the ERR of interfacial peeling, **b** the mode-mixity angle of interfacial peeling

7.3.5 Selection of Device Thickness

The following requirements should be satisfied in the conformal peeling process: (1) The device layer should be peeled-off from the substrate safely and reliably considering that there exists a competing relationship between the device peeling and the device cracking, which is determined by the ERR of interfacial peeling and the maximum tensile stress of the device layer. (2) The device layer should be transferred from the left substrate to the right substrate safely and reliably. During the peeling process, three kinds of failure modes may be detected, i.e., the unexpected unpeeling mode, the unexpected device broken mode, and the expected successful peeling-off mode. When the ERR of interfacial peeling is larger than the corresponding critical value, the device peeling can be observed, i.e., the crack propagation along the adhesive layer can be achieved. When the maximum tensile stress of the device

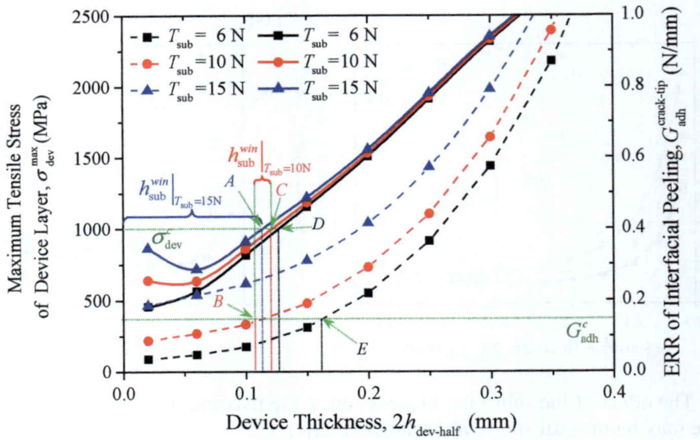

Fig. 7.19 Selection of the device thickness for the successful device peeling-off from the substrate. © 2018 IEEE. Reprinted, with permission, from Ref. [10]

layer is smaller than the fracture strength of the device layer, the device layer will be maintained intact during the peeling process. Only when the above two conditions are simultaneously satisfied, the device layer can be successfully peeled-off from the substrate.

When the material of the device layer and the peel blade are given, the effect of the device thickness on the maximum tensile stress of the device layer and the ERR of interfacial peeling is shown in Fig. 7.19. The solid and dash lines represent the maximum tensile stress of the device layer and the ERR of interfacial peeling, respectively. As mentioned above, to ensure that the device layer can be peeled from the substrate successfully, $\sigma_{\text{dev}}^{\text{max}} < \sigma_{\text{dev}}^{c}$ and $G_{\text{adh}}^{\text{crack-tip}} > G_{\text{adh}}^{c}$ should be simultaneously satisfied. It is supposed here that $\sigma_{\text{dev}}^{c} = 1000\,\text{MPa}$ and $G_{\text{adh}}^{c} = 0.15\,\text{N/mm}$. As can be seen, the substrate tension shows significant effect on the peelability. In case of $T_{\text{sub}} = 1.5\,\text{N}$, all the ERR of interfacial peeling is larger than the adhesive fracture energy, and the device will crack after reaching Point A. It denotes that the device can be peeled-off as long as $h_{\text{sub}} < h_A$, and the window of the substrate thickness is $h_{\text{sub}}^{win}|_{T_{\text{sub}}=10\,\text{N}\in(0,h_A)}$. When the substrate tension decreases to 10 N, the ERR of interfacial peeling is larger than the adhesive fracture energy under the condition of $h_{\text{sub}} > h_B$, and the device will crack at Point C. The results show that the window of peelability reduces to $h_{\text{sub}}^{win}|_{T_{\text{sub}}=10\,\text{N}\in(h_B,\text{h}_C)}$. When the substrate tension further decreases to 6 N, the device cannot be peeled from the substrate since the device breaks before the interfacial peeling occurs because of $h_D < h_E$. Therefore, the maximum tensile stress of the device layer and the ERR of interfacial peeling can be tuned by changing the device thickness, the device peelability can also be tuned by the substrate tension.

7.3.6 Selection of Substrate Thickness

The variation trend of the maximum tensile stress of the device layer and the ERR of interfacial peeling with the substrate thickness is shown in Fig. 7.20. It is also supposed that $\sigma_{\text{dev}}^{c} = 1000$ MPa and $\sigma_{\text{adh}}^{c} = 0.15\,\text{N/mm}$. The solid line represents the maximum tensile stress of the device layer. The dash line denotes the ERR of interfacial peeling. The window of the substrate thickness can be obtained in a similar way as Fig. 7.19. When the substrate tension is 6 N, to ensure the intactness of the peeled device layer, the substrate thickness should be smaller than h_A. Meanwhile, to ensure the interfacial peeling occurs along the adhesive layer, the ERR of interfacial peeling should be larger than the adhesive fracture energy, which requires that the substrate thickness should be smaller than h_D. The window of the substrate thickness can thus be determined as $h_{\text{sub}}^{win}|_{T_{\text{sub}}=6\text{N}} \in (0, h_D)$. When the substrate tension is 10 N, the intactness of the peeled device layer requires that the substrate thickness should be smaller than h_B while the device peeling from the substrate requires that the substrate thickness should be smaller than h_E. One can obtain $h_{\text{sub}}^{win}|_{T_{\text{sub}}=10\text{N}} \in (0, h_E)$ since $h_E < h_B$ under the condition of $T_{\text{sub}} = 10\,\text{N}$. With the increase of the substrate tension, the available substrate thickness gets larger. Similarly, in order to peel the device from the substrate successfully under the condition of $T_{\text{sub}} = 15\,\text{N}$, the substrate thickness should be smaller than both h_C and h_F. Therefore, the window of the substrate thickness can be determined as $h_{\text{sub}}^{win}|_{T_{\text{sub}}=15\text{N}} \in (0, h_F)$ considering that $h_F < h_C$ in the case of $T_{\text{sub}} = 15\,\text{N}$. Additionally, it can be derived through $h_{\text{sub}}^{win}|_{T_{\text{sub}}=6\text{N}} < h_{\text{sub}}^{win}|_{T_{\text{sub}}=10\text{N}} < h_{\text{sub}}^{win}|_{T_{\text{sub}}=15\text{N}}$ that the substrate tension is helpful to improve the window of peelability.

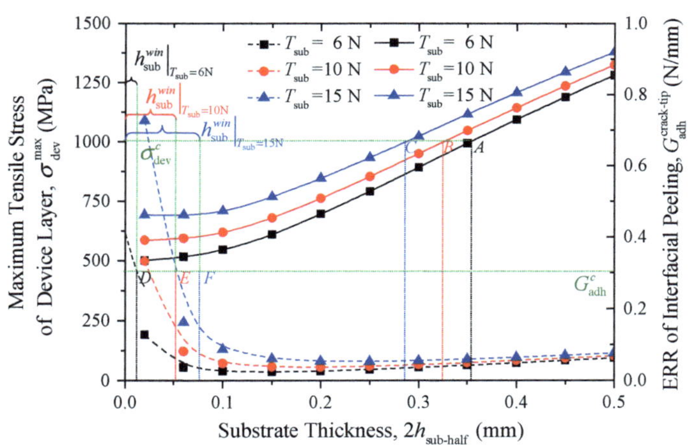

Fig. 7.20 Selection of the substrate thickness for the successful device peeling-off from the substrate. © 2018 IEEE. Reprinted, with permission, from Ref. [10]

7.4 Effects of Process Parameters

7.4.1 Effect of Peel Radius

As for the multilayer structure, only when the strain mismatch among the layers reaches to a certain extent, the interfacial peeling along the adhesive layer can be achieved. Based on Eq. (7.40), the maximum tensile stress of the device layer is not only influenced by the material properties of the multilayer structure, but also by the peel radius. Using the parameters listed in Table 7.1, the effect of the peel radius on the maximum tensile stress of the device layer is shown in Fig. 7.21. As can be seen, the maximum tensile stress of the device layer decreases with the increase of the peel radius. To ensure the intactness of the peeled device, the peel radius should be larger than the critical value. In the case of $\sigma_{\mathrm{dev}}^{c} = 1000$ MPa, the critical peel radius under conditions of $T_{\mathrm{sub}} = 6$ N, 10 N, and 15 N is termed here as $r_{\mathrm{blade}}^{T_{\mathrm{sub}}=6\,\mathrm{N}}|\sigma_{\mathrm{dev}}^{c}$, $r_{\mathrm{blade}}^{T_{\mathrm{sub}}=10\,\mathrm{N}}|\sigma_{\mathrm{dev}}^{c}$, and $r_{\mathrm{blade}}^{T_{\mathrm{sub}}=15\,\mathrm{N}}|\sigma_{\mathrm{dev}}^{c}$, respectively. It can be observed from Fig. 7.21 that $r_{\mathrm{blade}}^{T_{\mathrm{sub}}=6\,\mathrm{N}}|\sigma_{\mathrm{dev}}^{c}, r_{\mathrm{blade}}^{T_{\mathrm{sub}}=15\,\mathrm{N}}|\sigma_{\mathrm{dev}}^{c}$, which means that a larger peel radius is required for a successful device peeling under the condition of higher substrate tension.

7.4.2 Effect of Conformal Angle

Many researches have been conducted to study the effect of the peel angle during peeling multilayer structures [6–8, 13–15]. However, the conformal angle is different from the peel angle. The peel angle is the angel between the device layer and the substrate layer, while the peeling process never starts when the conformal angle is

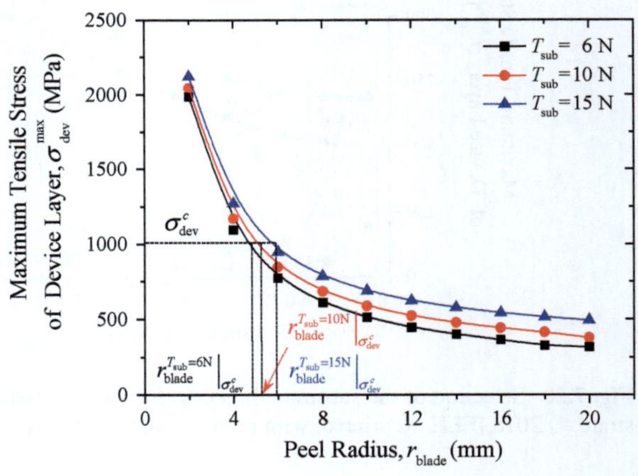

Fig. 7.21 The effect of the peel radius on the maximum tensile stress of the device layer. © 2018 IEEE. Reprinted, with permission, from Ref. [10]

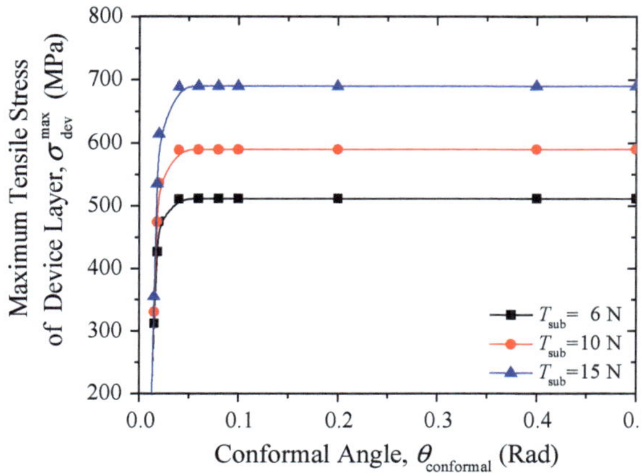

Fig. 7.22 The effect of the conformal angle on the maximum tensile stress of the device layer. © 2018 IEEE. Reprinted, with permission, from Ref. [10]

smaller than the critical angle. In the R2R peeling process, the conformal angle can be tuned by the layout of rollers.

The effect of the conformal angle on the conformal peeling process is different from that of the peel angle. By combining Eq. (7.40) and using the parameters in Table 7.1, the effect of the conformal angle on the maximum tensile stress of the device layer is shown in Fig. 7.22. When the conformal angle is smaller than 0.1 rad, the maximum tensile stress of the device layer increases with the conformal angle increasing. When the conformal angle is larger than 0.1 rad, the maximum tensile stress of the device layer converges. The results agree well with the literature [16]. As a consequence, the effect of the conformal angle on the maximum tensile stress of the device layer is very small when it is larger than the critical value such as 0.1 rad.

7.4.3 Effect of Substrate Tension

The substrate tension has two important roles in the R2R manufacturing process: keeping the R2R running stably and promoting the device peeling from the substrate. The feasible range of the substrate tension is the intersection of both roles. Many researches have been carried out on the effect of the peel force on the peeling process [5, 17–19]. In the R2R transferring process, there is strict requirement for the substrate tension to ensure the intactness of the flexible film. The effect of the substrate tension on the performance of electronic products has already been studied [20–22].

It is supposed here that there are no other rollers between the discharging roller and the receiving roller. The discharging roller sends the substrate with speed $\nu_{\mathrm{sub}}^{\mathrm{left}}$, while the receiving roller withdraws the substrate with speed $\nu_{\mathrm{sub}}^{\mathrm{left}}$. When $\nu_{\mathrm{sub}}^{\mathrm{right}} < \nu_{\mathrm{sub}}^{\mathrm{left}}$, the substrate will be relaxed, and the R2R manufacturing system fails to work. While

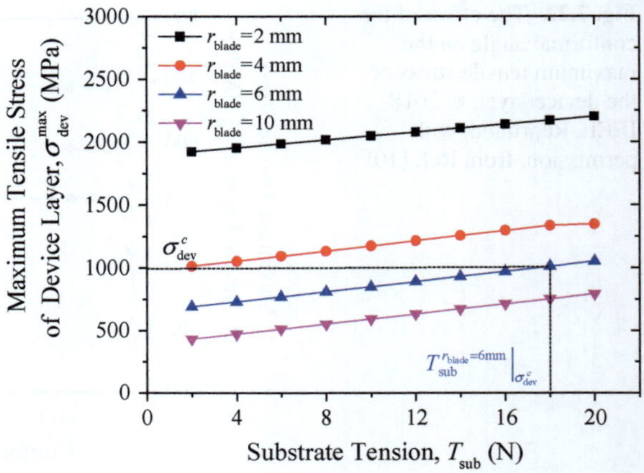

Fig. 7.23 The effect of the substrate tension on the maximum tensile stress of the device layer.

$v_{sub}^{right} > v_{sub}^{left}$, the substrate is stretched and the substrate tension appears. According to Hooke's law, the substrate tension can be written as

$$T_{sub} = \frac{E_{sub}A_{sub}}{L_{sub}} \int_0^t \left(v_{sub}^{right} - v_{sub}^{left}\right) dt \tag{7.44}$$

where E_{sub} is the Young's modulus of the substrate, A_{sub} is the section area, L_{sub} is the distance between the two rollers, t is the time taken for the substrate transferring from the discharging roller to the receiving roller. It can be observed from Eq. (7.44) that the substrate tension is determined by the difference of the speed between the discharging roller and the receiving roller.

Using the parameters listed in Table 7.1, the variation trend of the maximum tensile stress of the device layer with the substrate tension under the condition of the conformal angle of 0.1 rad is shown in Fig. 7.23. It can be seen that the maximum tensile stress of the device layer is linearly proportional to the substrate tension. The maximum tensile stress of the device layer increases with the reduction of the peel radius under the same substrate tension condition. When $\sigma_{dev}^{c} = 1000$ MPa, the device layer will crack under the condition of the peel radius of 2 mm or 4 mm, which means that the device layer cannot be peeled from the substrate layer intactly despite the extent of the substrate tension. While in the case of the peel radius of 6 mm, the substrate tension should be smaller than $T_{sub}^{r_{blade}=6mm}|\sigma_{dev}^{c}$, otherwise the device layer will be broken.

The effect of the substrate tension on the ERR of interfacial peeling and the corresponding mode-mixity angle under different conditions of the peel radius is shown in Fig. 7.24. As can be seen from Fig. 7.24a, the ERR of interfacial peeling increases with the increase of the substrate tension. One can get the following results from Figs. 7.23 and 7.24 under the condition of the adhesive fracture energy $G_{adh}^{c} = 0.1$ N/mm and the fracture strength of the device layer $\sigma_{dev}^{c} = 1000$ MPa. When the peel radius is 2 mm, the device layer is first wrecked, and cannot be peeled from the

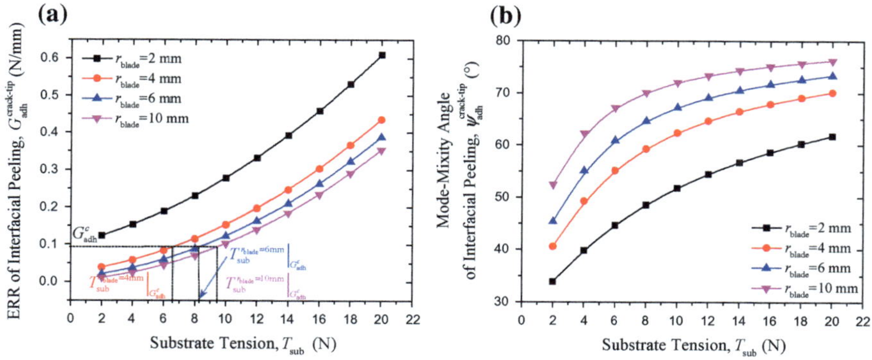

Fig. 7.24 The effect of the substrate tension on: **a** the ERR of interfacial peeling, **b** the mode-mixity angle of interfacial peeling under different conditions of peel radius. © 2018 IEEE. Reprinted, with permission, from Ref. [10]

substrate layer. When the peel radius is 4 mm, the device layer is first wrecked similar to that of $r_{\text{blade}} = 2\,\text{mm}$, and the device layer cannot be peeled until the substrate tension is larger than $T_{\text{sub}}^{r_{\text{blade}}=4\text{mm}}\big|G_{\text{adh}}^{c}$. When the peel radius is 6 mm, the device layer can be peeled when the substrate tension $T_{\text{sub}}^{r_{\text{blade}}=6\text{mm}}\big|\sigma_{\text{dev}}^{c} < T_{\text{sub}} < T_{\text{sub}}^{r_{\text{blade}}=6\text{mm}}\big|G_{\text{adh}}^{c}$. When the peel radius is 10 mm, the proper substrate tension for a successful device peeling is $T_{\text{sub}} > T_{\text{sub}}^{r_{\text{blade}}=10\text{mm}}\big|\sigma_{\text{adh}}^{c}$. The above results indicate that a bigger peel radius leads to a larger scope of appropriate substrate tension. Namely, the substrate tension plays a key role in the device peeling under the condition of large peel radius. The variation trend of the mode-mixity angle with the substrate tension is shown in Fig. 7.24b. As can be observed, the mode II peeling plays a major role under conditions of large substrate tension, which means an uncontrollable device peeling.

7.5 Experimental Determination of Adhesive Fracture Energy

The peeling-off of the device layer from the substrate results from the crack propagation along the adhesive layer. The basic principle of the peeling process is equivalent to the fracture between the adhesive layer and the substrate layer from the fracture mechanics point of view. The adhesive fracture energy between the adhesive layer and the substrate layer is not only related to the physical properties of the viscoelastic adhesive layer and the substrate layer, but also related to the peel speed [23–25].

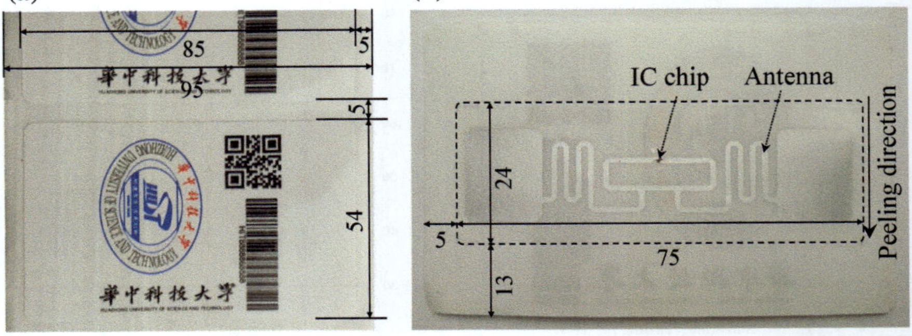

Fig. 7.25 RFID tag-silicone paper used in the experiment: **a** the front side, **b** the back side

If the effect of temperature is neglected, the adhesive fracture energy can be written as

$$G(v_{\text{peel}}) = G_0\left[1 + f(v_{\text{peel}})\right] \tag{7.45}$$

where G_0 is the adhesive fracture energy with the peel speed $v_{\text{peel}} = 0$. $G(v_{\text{peel}})$ is the adhesive fracture energy at a certain peel speed. $f(v_{\text{peel}})$ is a function about the peel speed.

Generally, there exists a nonlinear relationship between the adhesive fracture energy and the peel speed [26]. An experimental approach is adopted here to explore the exact relationship between the adhesive fracture energy and the peel speed. The size of the RFID tag used in the experiment is 85 mm × 54 mm, and there is a 5 mm gap between two tags, as shown in Fig. 7.25. The thickness of the RFID paper is 0.25 mm, the thickness of the antenna is 0.02 mm, the thickness of the silicone paper is 0.05 mm, and the width of the silicone paper is 95 mm. During the peeling experiment, the effect of the adhesive at the antenna part is considered. It is obviously infeasible that the RFID tag is too short to be used as the peel arm. The RFID tag is pasted onto the glass by the double-faced adhesive tape, and the silicone paper is used as the peel arm. Experiments show that the adhesive fracture energy between the double-faced adhesive tape and the glass is 50 times higher than that between the RFID tag and the silicone paper, this method is supposed reliable for the measurement of the adhesive fracture energy.

A homemade experimental platform is adopted to measure the adhesive fracture energy in the RFID system, as shown in Fig. 7.26. The parameters that can be controlled including the peel displacement from 0 to 350 mm, the peel angle from 0° to 180°, and the peel speed from 50 μm/s to 100 mm/s. In addition, the peel force can be obtained by using the 222-N LCM300 load cell (FUTEK, USA), the accuracy of which is 1 mN. When the peel angle is 90°, the test result of the peeling experiment

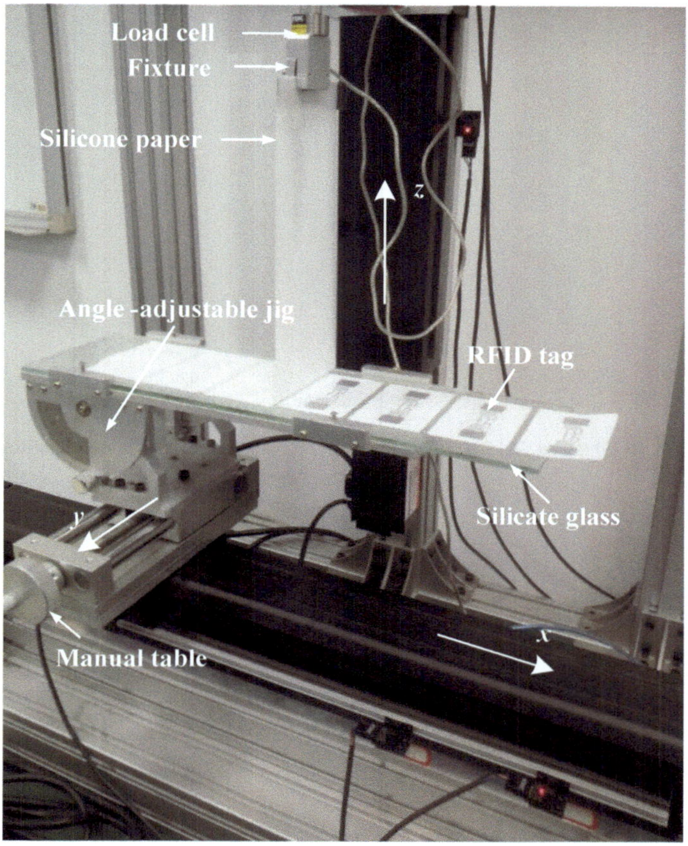

Fig. 7.26 Experimental peeling apparatus with an angle-adjustable jig. © 2018 IEEE. Reprinted, with permission, from Ref. [10]

in peeling the RFID tag from the silicone paper is shown in Fig. 7.27. When the peel speed is 4 mm/s, the real-time value of the peel force in peeling a single RFID tag is shown in Fig. 7.28. There is mutation when the crack propagates at the edge of the device layer. The real-time value of the peel force during a sequential peeling-off of five adjacent RFID tags is shown in Fig. 7.29.

The peel direction is along the short edge of the RFID tag. Ideally, the peel force should not be fluctuated when the peel speed is constant. However, the fluctuation of the peel force can be observed in Figs. 7.28 and 7.29 due to the gaps between RFID tags. In addition, there is antenna and chip at the middle of the RFID tag, which leads to the nonuniformity of the device layer. The orange line in Fig. 7.28 shows the maximum value of the peel force, and the blue one is the real-time value

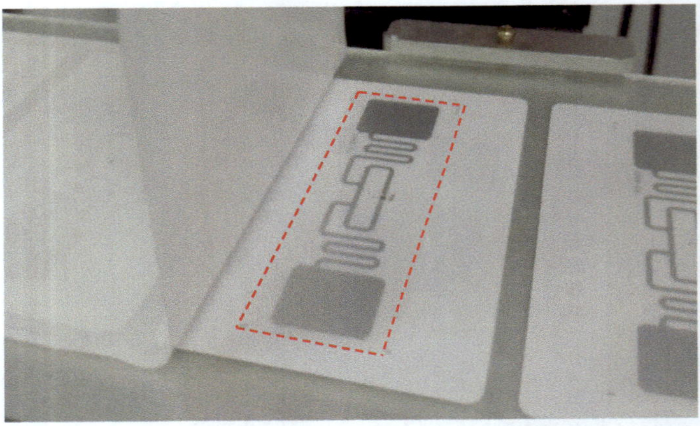

Fig. 7.27 The RFID tag peeled from the silicone paper. © 2018 IEEE. Reprinted, with permission, from Ref. [10]

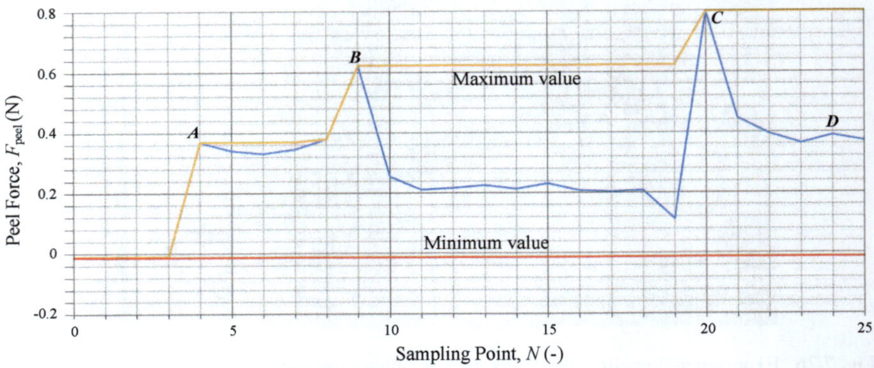

Fig. 7.28 Real-time peel force in peeling a single tag (peel speed: 4 mm/s)

of the peel force. The abscissa represents the sampling point. The interfacial peeling begins at Point *A*, and reaches the edge of the Inlay part at Point *B*. Because of the uneven distribution of the adhesive layer, the peel force fluctuates drastically. When the peeling process reaches another edge of the Inlay part, there is a force fluctuation at Point *C*, and the peeling process of the single tag is finished at Point *D*. Besides, a large fluctuation of the peel force can also be observed in Fig. 7.29.

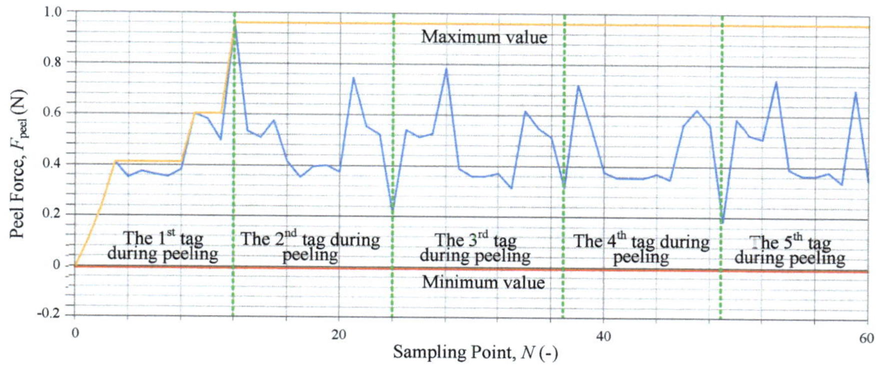

Fig. 7.29 Real-time peel force in sequentially peeling five adjacent RFID tags (peel speed: 4 mm/s). © 2018 IEEE. Reprinted, with permission, from Ref. [10]

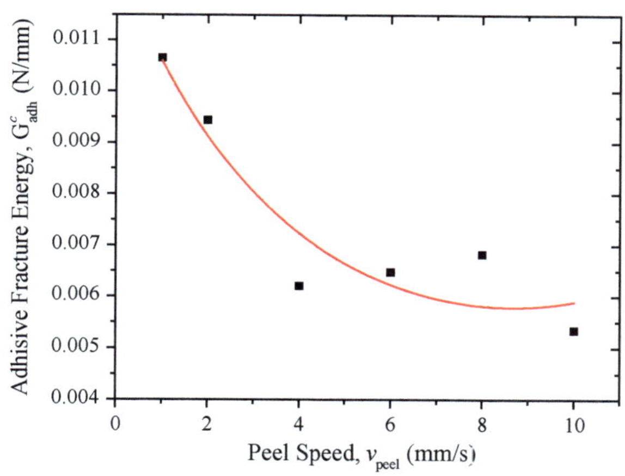

Fig. 7.30 The effect of the peel speed on the adhesive fracture energy. © 2018 IEEE. Reprinted, with permission, from Ref. [10]

Generally speaking, the planimeter method should be adopted in the peeling test. While whether the RFID tag can be peeled from the substrate layer is determined by the maximum peel strength of the adhesive layer. The peak value method will be used in the peel strength experiment. In order to avoid the effect of random factors in the experiment, the mean value of multiple experimental results is adopted. When the peel angle is 90°, the adhesive fracture energy in the RFID tag adhesive layer is shown in Fig. 7.30. It can be seen that the adhesive fracture energy decreases with the increase of the peel speed. The adhesive fracture energy tends to be stabilized at about 0.005 N/mm with further increase of the peel speed.

7.6 Design of Peel Blade and Its Application

7.6.1 Optimal Design of Peel Blade

As for the design of the peel blade, the most important parameters are the peel radius and the conformal angle. Based on the previous analysis in Fig. 7.22, the effect of the conformal angle can be neglected. The effect of the peel radius on the maximum tensile stress of the device layer and the ERR of interfacial peeling is shown in Fig. 7.31. It can be seen from Fig. 7.31a that the maximum tensile stress of the device layer is proportional to the reciprocal of the peel radius. To ensure that the peeled device layer is still intact after being peeled-off, the maximum tensile stress of the device layer should be smaller than σ_{dev}^{c}. The peel radius $r_{\text{blade}} > r_{\text{blade}}^{T_{\text{sub}}=6\text{N}}|\sigma_{\text{dev}}^{c}$, $r_{\text{blade}} > r_{\text{blade}}^{T_{\text{sub}}=10\text{N}}|\sigma_{\text{dev}}^{c}$, $r_{\text{blade}} > r_{\text{blade}}^{T_{\text{sub}}=15\text{N}}|\sigma_{\text{dev}}^{c}$ should thus be satisfied under conditions of $T_{\text{sub}} = 6$ N, 10 N, and 15 N, respectively. To ensure that the device layer can be peeled from the substrate layer, the ERR of interfacial peeling should be larger than the adhesive fracture energy, i.e., the peel radius should be smaller than $r_{\text{blade}}^{T_{\text{sub}}=6\text{N}}|G_{\text{adh}}^{c}$ under the condition of $T_{\text{sub}}= 6$ N. When $T_{\text{sub}} = 10$ N and $T_{\text{sub}} = 15$ N, the device layer can be peeled from the substrate layer even the peel radius is larger than 20 mm. Based on the above analysis, the proper range of the peel radius under the condition of $T_{\text{sub}}= 6$ N can be determined as $r_{\text{blade}}^{T_{\text{sub}}=6\text{N}}|\sigma_{\text{dev}}^{c} < r_{\text{blade}} < r_{\text{blade}}^{T_{\text{sub}}=6\text{N}}|G_{\text{adh}}^{c}$.

In order to validate the above theoretical results, a peel blade for the RFID tag is designed. The proper peel radius can be determined based on the results of Fig. 7.31. Considering that there are many random factors in the actual working condition, the peel radius is determined here as 2 mm, 4 mm, and 6 mm.

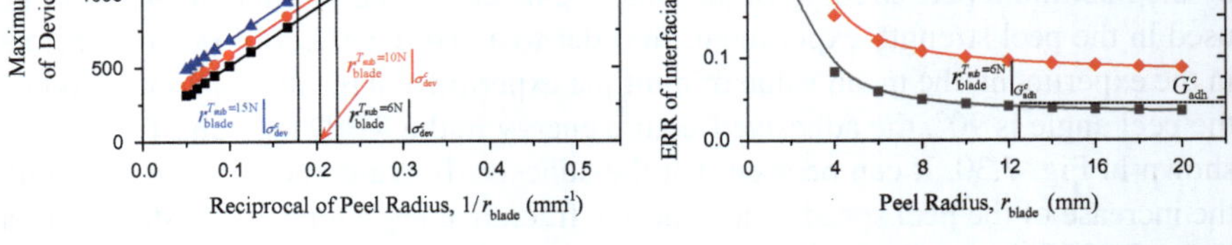

Fig. 7.31 Selection of proper peel radius. © 2018 IEEE. Reprinted, with permission, from Ref. [10]

7.6.2 Experimental Validation

The material properties used in experiment is listed in Table 7.2. The ERR of interfacial peeling is calculated as shown in Fig. 7.32. Combined with the adhesive fracture energy, one can get the maximum peel radius under different substrate tension conditions. It can be seen that when the peel radius is larger than 20 mm, the ERR of interfacial peeling tends to be constant, which means that increasing the peel radius has little effect on the ERR of interfacial peeling. Considering that the material used in the RFID tag is not isotropic, one can only get an approximate peel radius from Fig. 7.32.

Two different types of RFID tags are used to validate the design of the peel blade. To ensure that the RFID tag can be transferred to the target substrate successfully, the structure of the acceptor blade is designed close to the peel blade. During the peeling process, the working state of the peel blade is like a cantilever. Coupled with gravity, the axis of the peel blade will be sag. In order to solve this problem, the working plane for the acceptor blade is lower than that of the peel blade. The peeling experimental platform used to transfer RFID tags is shown in Fig. 7.33. Results of the peeling experiment under different conditions of peel radius are shown in Fig. 7.34. When the peel radius is 2 mm, the two types of RFID tags can be peeled-off successfully,

Table 7.2 Material properties used for the peeling experiment

Layer (–)	Thickness (mm)	Young's modulus (MPa)	Poisson's ratio (–)
Device	0.19	4800	0.30
Adhesive	0.01	3600	0.35
Substrate	0.05	8600	0.30

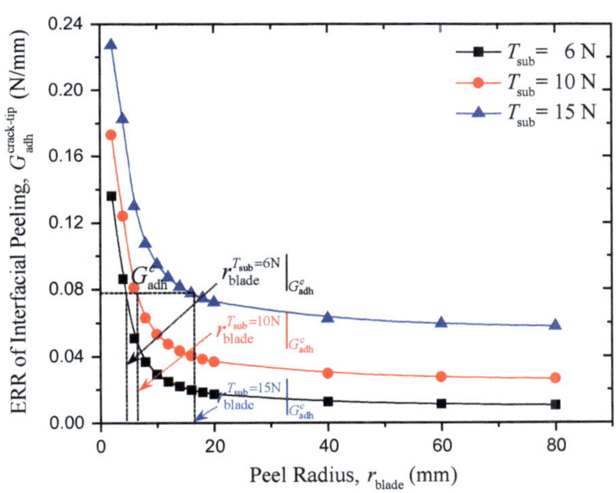

Fig. 7.32 Selection of the peel radius for RFID tags

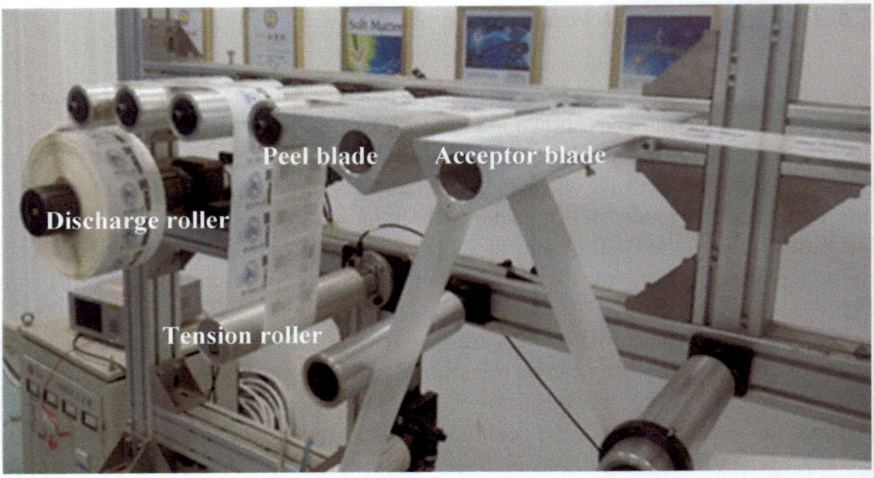

Fig. 7.33 RFID tag peeling experimental platform. © 2018 IEEE. Reprinted, with permission, from Ref. [10]

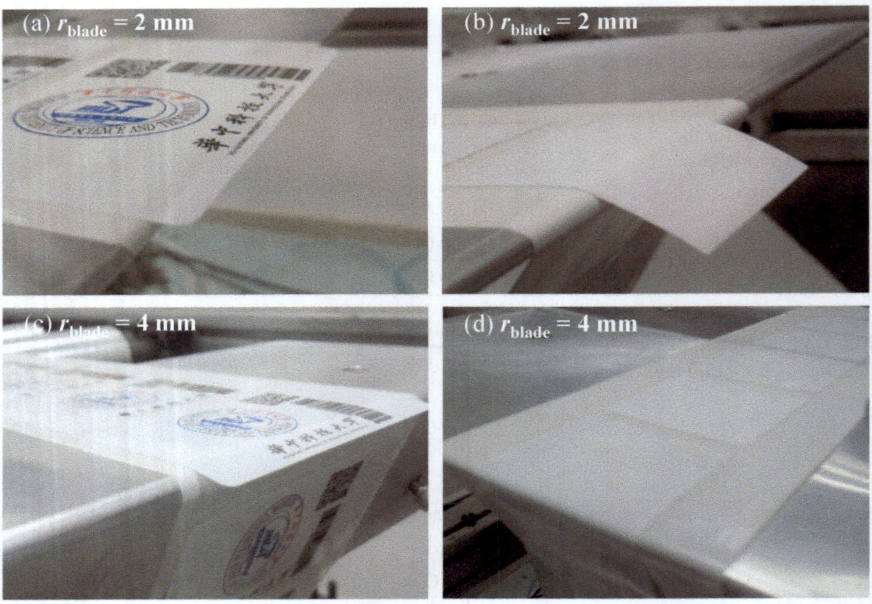

Fig. 7.34 Results of peeling RFID tags under different conditions of peel radius

as depicted in Fig. 7.34a, b. When the peel radius is 4 mm, only one type of the RFID tag can be peeled-off, as shown in Fig. 7.34c, d.

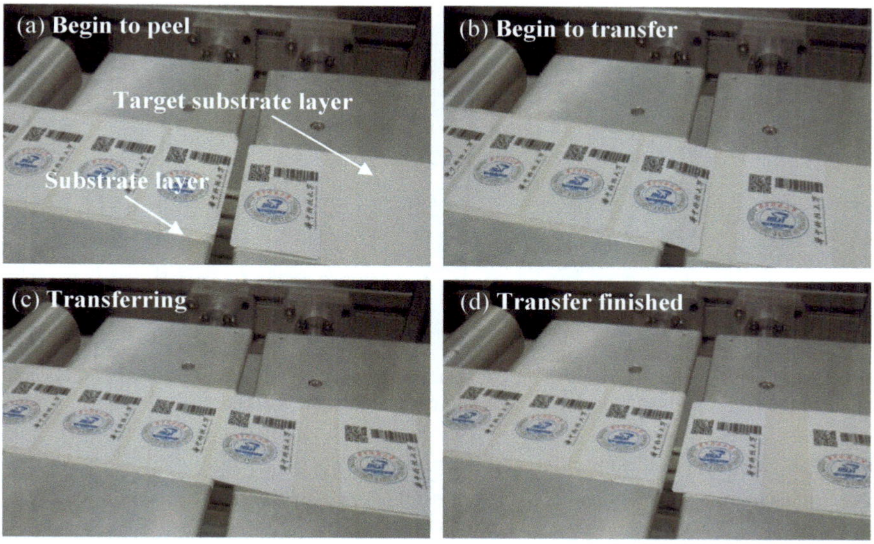

Fig. 7.35 Steps of the peel and transfer of RFID tags. © 2018 IEEE. Reprinted, with permission, from Ref. [10]

The peel and transfer of the RFID tag are shown in Fig. 7.35. There are four steps: (1) begin to peel, the RFID tag begins to be peeled from the substrate layer; (2) begin to transfer, one part of the RFID tag is on the substrate layer, while the other part is on the target substrate layer; (3) transferring; (4) transfer finished.

Experimental validation of the design results is shown in Fig. 7.36. The peel blade and the acceptor blade are put onto the label composite device, as depicted in Fig. 7.36a. The feeding speed is 0–30 m/min, and the control system is based on the industrial personal computer. The motor is controlled by 16-channel programmable multi-axis controller (PMAC) motion computer. The unwinding rollers are controlled by stepping motors, which are controlled by 6-channel PCI motion control card. The servo torque control mode is used to control the substrate tension. The label composite device is used to validate the peel blade and the acceptor blade, as shown in Fig. 7.36b, c. It proves that the peel blade and the acceptor blade can peel and transfer the flexible labels reliably and efficiently.

Different peel radii are used to validate the peel and transfer module, and the statistical results are listed in Table 7.3. It can be seen that when the peel radius is small at 2 mm, the three labels can be peeled with high peeling success rate. When the peel radius increases to 4 mm, the peeling success rate for one of the labels decreases significantly, which shows that different peel radii should be adopted if the peeling object changes.

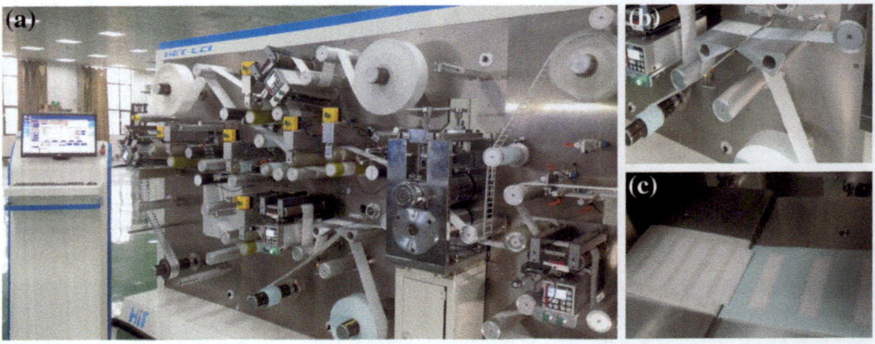

Fig. 7.36 Experimental validation of the peel blade: **a** application of the peel blade and the acceptor blade on the label composite device, **b** peel, and **c** transfer of the RFID tag

Table 7.3 Statistical results for different peeling objects

Peel radius (mm)	Sample number (–)	Successful number (–)	Success rate (%)	Tension (N)
2	137	137	100	15
4	148	143	96.6	10
2	183	162	88.5	10
4	23	2	8.7	6
2	276	258	93.4	6

7.7 Summary

The R2R manufacturing system with high efficiency and low cost is compatible with flexible substrates, which makes the R2R system an ideal manufacturing route for the flexible electronics. Therefore, it is of significance for the conformal peeling to be integrated with the R2R manufacturing system to improve the production efficiency and reduce the cost. Here, the conformal peeling and the R2R transferring of the RFID tags were studied from aspects of mechanical modeling, mechanism analysis, and experimental test. The multilayer device–adhesive–substrate model was established to solve the adhesive stresses and the ERR of interfacial peeling. Effects of material properties, geometric sizes, and process parameters were analyzed to reveal the mechanism of the conformal peeling process. The proper selection of the device thickness, substrate thickness, and the design of the peel blade and the substrate tension was achieved considering the competing fracture behavior between the device peeling and the device cracking. By using the peeling experimental test, the adhesive fracture energy was measured under different conditions of peel speed. The experimental validation on the peel and transfer of flexible RFID tags was finally conducted through the multilayer R2R manufacturing system. It was proved that the optimal design had a better practical application.

References

1. Jain K, Klosner M, Zemel M, Raghunandan S (2005) Flexible electronics and displays: high-resolution, roll-to-roll, projection lithography and photoablation processing technologies for high-throughput production. Proc IEEE 93(8):1500–1510
2. Lo C-Y, Hiitola-Keinänen J, Huttunen O-H, Petäjä J, Hast J, Maaninen A, Kopola H, Fujita H, Toshiyoshi H (2009) Novel roll-to-roll lift-off patterned active-matrix display on flexible polymer substrate. Microelectron Eng 86(4):979–983
3. Huang Y, Chen J, Yin Z, Xiong Y (2011) Roll-to-roll processing of flexible heterogeneous electronics with low interfacial residual stress. IEEE Trans Compon Packag Manuf Technol 1(9):1368–1377
4. Chen J, Liu H, Huang Y, Yin Z (2016) High-rate roll-to-roll stack and lamination of multilayer structured membrane electrode assembly. J Manuf Process 23:175–182
5. Eremeyev VA, Naumenko K (2015) A relationship between effective work of adhesion and peel force for thin hyperelastic films undergoing large deformation. Mech Res Commun 69:24–26
6. Williams JA, Kauzlarich JJ (2005) The influence of peel angle on the mechanics of peeling flexible adherends with arbitrary load-extension characteristics. Tribol Int 38:951–958
7. Dalbe M-J, Santucci S, Vanel L, Cortet P-P (2014) Peeling-angle dependence of the stick-slip instability during adhesive tape peeling. Soft Matter 10(48):9637–9643
8. Wang Y (2014) The effect of peeling rate and peeling angle on the peeling strength. The University of Akron
9. Kaelble DH (1992) Theory and analysis of peel adhesion: adhesive thickness effects. J Adhes 37(1–3):205–214
10. Huang Y, Liu H, Xu Z, Chen J, Yin Z (2018) Conformal peeling of device-on-substrate system in flexible electronic assembly. IEEE Trans Compon Packag Manuf Technol 8(8):1496–1506
11. Wong EH, Lim TB (2008) A more comprehensive solution for tri-material layers subjected to thermal stress. IEEE Trans Compon Packag Technol 31(1):54–64
12. Nicoleta W, Harald R, Tu H, Pawel G, Mathias S, Evangelos AA, Mahadi-Ul H, Joachim NB (2014) Stress analysis of ultra-thin silicon chip-on-foil electronic assembly under bending. Semicond Sci Technol 29(9):095007
13. Meitl MA, Zhu Z-T, Kumar V, Lee KJ, Feng X, Huang YY, Adesida I, Nuzzo RG, Rogers JA (2006) Transfer printing by kinetic control of adhesion to an elastomeric stamp. Nat Mater 5(1):33–38
14. Feng X, Meitl MA, Bowen AM, Huang Y, Nuzzo RG, Rogers JA (2007) Competing fracture in kinetically controlled transfer printing. Langmuir 23:12555–12560
15. Park S-I, Xiong Y, Kim R-H, Elvikis P, Meitl M, Kim D-H, Wu J, Yoon J, Yu C-J, Liu Z, Huang Y, K-c Hwang, Ferreira P, Li X, Choquette K, Rogers JA (2009) Printed assemblies of inorganic light-emitting diodes for deformable and semitransparent displays. Science 325:977–981
16. Saiki N, Inaba K, Kishimoto K, Seno H, Ebe K (2010) Study on peeling behavior in pick-up process of IC chip with adhesive tapes. J Solid Mech Mater Eng 4(7):1051–1060
17. Eitner U, Rendler LC (2014) The mechanical theory behind the peel test. Energy Procedia 55:331–335
18. Gent AN, Hamed GR (1975) Peel mechanics. J Adhes 7(2):91–95
19. Gent AN, Hamed GR (1977) Peel mechanics for an elastic-plastic adherend. J Appl Polym Sci 21:2817–2831
20. Chen J, Yin Z, Xiong Y, Quan J (2009) A hybrid control method of tension and position for a discontinuous web transport system. In: 2009 IEEE international conference on information and automation, Zhuhai/Macau, China, 2009, pp 265–270
21. Lee C-W, Shin K-H (2010) A study on taper-tension control considering telescoping in the winding system. IEEE Trans Ind Appl 46(2):687–692
22. Lee JS, Seong JW, Shin KH (2011) A study on the decision of operating tension by analysis of friction coefficient change in rewinding process. In: Paper presented at the 8th international conference on ubiquitous robots and ambient intelligence, Incheon, Korea

23. Bock K (2005) Polymer electronics systems-polytronics. Proc IEEE 93(8):1400–1406
24. Allen K (2005) Reel to real: Prospects for flexible displays. Proc IEEE 93(8):1394–1399
25. Choi J, Kim Y-J, Lee S, Son SU, Ko HS, Nguyen VD, Byun D (2008) Drop-on-demand printing of conductive ink by electrostatic field induced inkjet head. Appl Phys Lett 93(19):193508
26. Masuda R, Inoue H, Kishimoto K (2008) Peeling rate dependency of delamination behavior of adhesives. Adv Mater Res 33–37:339–344

Chapter 8
Laser Lift-off

8.1 Introduction

In flexible electronics manufacturing, it is usually demanded to transfer a device prepared on a rigid carrier to a flexible substrate. Typical process implementations include the transfer printing [1–3], needle ejecting [4–6], roll-to-roll [7–10] (R2R) conformal peeling, etc. However, wide applications of these processes are still limited by respective limitations, such as the small surface coverage, high cost, and short life span of the transfer printing process; limited chip size processing capability of the needle ejecting process, although improvement of which can be achieved by the adoption of multiple ejector needles [6, 11, 12]; the need of improved control over the peelability, accurate transfer, and collaborative control of the conformal peeling process [10, 13]. Innovative processes are required to achieve the mass transfer of large-area and ultra-thin devices from rigid carrier with high efficiency, reliability, and stability.

As a key technology in driving new developments in microelectronics, laser processing provides potential solutions to the above issues. Laser processing has unique advantages including high material-selectivity in multilayer structure, controllable processing areas, non-contact interaction, highly localized affected area, high processing speed, and versatility for various materials and geometries. Furthermore, lasers in industrial production often eliminate the use of hazardous, wet-chemical substances for etching or post processing. One of the most commonly applied laser processing techniques are based on the irradiation of a buried absorbing interface after transmitting through a transparent substrate. The interfacial reaction caused by the laser irradiation dominates the device peeling-off. A variety of related techniques have been developed for different flexible electronics applications, including the flexible energy system [14, 15], human health monitoring [16–21], flexible display [22, 23], and flexible solar cell [24, 25], etc.

In a typical LLO process, a thin film substrate is first coated onto a transparent rigid carrier. The thin film is then separated from the rigid carrier after irradiating the carrier/film interface through the LLO [26, 27]. The LLO process comes into play

Y. Huang et al., *Modeling and Application of Flexible Electronics Packaging*,
https://doi.org/10.1007/978-981-13-3627-0_8

as soon as the manufacturing industry demands for a gentle separation of thin layer system with various areas or sizes, and transfer of microelectronic functional layer onto lighter and thinner substrates to improve the mechanical flexibility. The LLO separation opens a novel path and becomes increasingly attractive to the microelectronics manufacturing considering that the feature size and film thickness in which are continuously being downscaled [28].

With the rapid development of flexible electronics, it has become increasingly important to study how to peel a large-area and ultra-thin flexible substrate like the PI film from a rigidly prepared carrier like the glass wafer. By now, the underlying mechanism of the LLO separation of the ultra-thin PI film from the glass carrier is still lacked. In this chapter, the related mechanism was systematically studied from aspects of the gas products, the microstructure at the PI/glass interface and the glass carrier, the residual interface adhesion strength, and deformation of the PI film. Process parameters were optimized to achieve a reliable LLO of ultra-thin PI film (thinner than 5 μm). The research results are beneficial for the preparation of large-area and ultra-thin flexible electronics.

8.2 Laser-Induced Interfacial Reaction (LIIR)

The LIIR occurs as the laser transmits through the transparent substrate and irradiates at the laser-reactive layer, as shown in Fig. 8.1. The interfacial reactions are mainly based on the absorption of the laser irradiation, which results in various phenomena like the material heating, melting, vaporization, ablation, and decomposition. Distinct phenomena depend primarily on the characteristics of laser parameters like the laser wavelength, laser fluence, and laser dwell time, etc., as well as the optical and physical/chemical properties of the irradiated material.

Three kinds of the LIIR techniques based on different interfacial phenomena are summarized as follows: (a) Using a shaped laser beam to scan over the whole interface after transmitting through the transparent substrate, leading to the release of the thin film of interest from the rigid carrier. This kind of process is generally termed as the LLO [28–30]. (b) Using a laser to irradiate through the transparent donor substrate to make the material of affected area transfer onto the receiver substrate, which is similar to the traditional printing process. This kind of laser-assisted printing processes [11, 31, 32] implement deposition and pattern of functional materials incompatible with traditional printing techniques in one single step. (c) Using a laser pulse to adjust the interfacial adhesion and delamination behavior between the transparent substrate and the transferred object. This kind of laser-assisted-releasing processes [33–35] make the transfer printing independent from the properties of the acceptor substrate available and have better performance in controllability, versatility, and scalability. The desired LIIR generally needs the consideration of an appropriate laser system and process. The optical, thermal, chemical, and physical properties of the reactive layer and the transparent substrate should be considered carefully to select appropriate laser parameters.

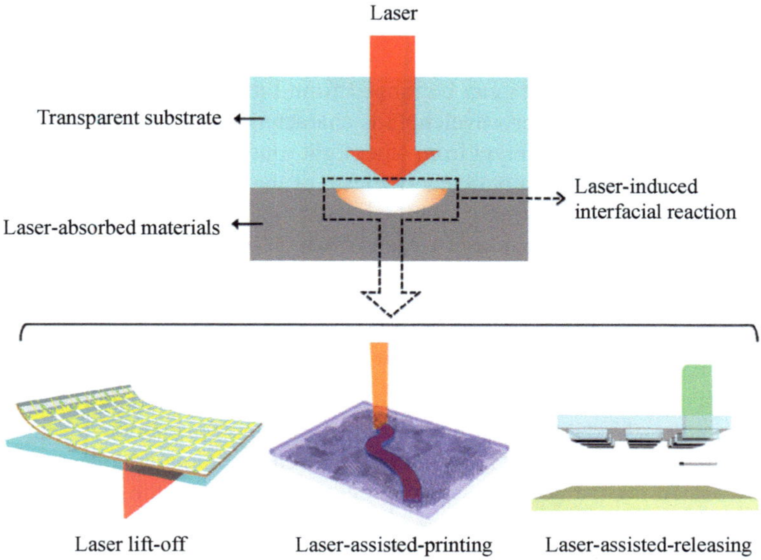

Fig. 8.1 Schematic diagrams of the LIIR

8.2.1 Band Gap

The appropriate laser source with a certain fluence at the interface is essential to a reliable LIIR process. The optical properties of both the substrate layer and the reactive layer are important factors that should be considered. The substrate that is transparent to a specific laser wavelength should be first prepared before the LIIR. The mechanism of the adsorption of the photon energy E_{laser} depends on the electronic structure of the specific material [36]. The band gap E_{band} is a critical factor determining optical properties for most materials applied in the LIIR process [37]. If $E_{\mathrm{laser}} < E_{\mathrm{band}}$, the phonon excitation and absorption are relatively weak. If $E_{\mathrm{laser}} > E_{\mathrm{band}}$, strong phonon absorption occurs along with resonant excitation involving mainly interband electronic transition. The laser transparency of the substrate layer and the strong laser adsorption of the reactive interface layer should be enabled. To make the laser-reactive layer accommodated with specific laser wavelength and transparent substrate, innovative approaches including the development of amorphous sacrificial layers [38], deposition of superlattice defects [39], localized laser adsorption through spatial control of the band gap [40], and dope of local absorption centers [41] have been presented.

8.2.2 *Laser Wavelength*

The wavelength of the lasers can be ranged from the deep ultraviolet (UV) to the far infrared region. The laser wavelength is characterized by the monochromaticity and coherence, which is different from other light sources. The wavelength of a laser beam is determined by the atomic energy level of the lasing medium. There can be excimer lasers, solid-state lasers, liquid-dye lasers, gas lasers, semiconductor lasers, fiber lasers, and free-electron lasers, each of which has its own specific ranges of wavelength.

Most LIIR techniques adopt short wavelength UV lasers, as the reactive layer typically exhibit much higher absorbance in the UV range than in the visual and infrared ranges. The optical penetration depth is then limited to the range of few hundreds of nanometers, thus the thickness of the reactive layer can be reduced to just few micrometers. Under the same circumstance, the long wavelength infrared laser is fully transparent to the reactive layer, which may cause thermal damage of adjacent layers. Considering the visible laser with semitransparency to the reactive layer, the laser energy is mainly absorbed by the intermediate region of the reactive layer, and the adjacent layer may also be heated due to the thermal diffusion. Therefore, the visible laser and infrared laser may be adequate for the LIIR of relatively thick reactive layer. In conclusion, the selection of the laser wavelength for the LIIR depends greatly on the optical absorbance and geometric dimensions of device structures.

8.2.3 *Laser Fluence*

The laser average power can be adjusted between mW and kW with precisely controlled spot size and spatial distribution. There exists a threshold laser fluence for almost all kinds of the LIIR-based techniques. If the laser fluence is lower than the threshold value, a desired LIIR may not be occurred. In contrast, when the laser fluence is much higher than the threshold value, the excessive laser energy will result in low processing efficiency and potential material damage.

The threshold laser fluence mainly arises because the laser-induced heating plays a dominant role in most kinds of LIIR. The high temperature generated at the interface can trigger a variety of temperature-dependent phenomena. Local deformation at the irradiated interface is controlled by the laser-induced thermal deformation behavior of the specific material [33]. Moreover, the high temperatures generated can induce phase and microstructural modifications. When the laser fluence is higher than the threshold value for material melting, the laser-induced local melting can make disconnected interface fuses in together with the help of applied pressure [11]. Large temperature gradient can be introduced by the localized laser heating. The induced material quenching, transient thermomechanical stress, decomposition, melting, and vaporization of interface material will result in the interface delamination [42]. In conclusion, different temperature distributions can be regulated based on the laser fluence employed, which provides the possibility of distinct phenomena in the LIIR.

8.2.4 Laser Irradiation Time

Lasers can be operated in either continuous or pulsed form, characterizing by whether the output energy is continuous over time or in the form of pulses over a certain time scale. The thermal diffusion depth, which is greatly determined by the laser irradiation time, indicates the effect of the energy spread during the laser irradiation time. The heat generated by the irradiation of the continuous laser or long pulse laser can lead to larger thermal diffusion depth than that of the short-pulse laser due to the difference in irradiate energy. However, the long-time irradiation is not preferred for the LIIR that demands highly localized heat-affected region around the interface, and may introduce thermal damage into heat-sensitive materials.

The short-pulse laser is popular in the LIIR that needs high heating/cooling rate and well-defined localized energy around the interface. When the laser pulse length is in the order of 10 ns, the thermal diffusion depth will be varied between 100 nm and 1000 nm for most reactive materials, and the laser fluence of around 500 mJ/cm^2 will be sufficient to induce a temperature rise in the order of 1000 °C in just few nanoseconds [43–45]. After the laser irradiation, high self-quenching rate can be achieved by the drastic heat dissipation to surrounding regions in tens of nanoseconds. The quenching-assisted LIIR is an important route to achieve the interface delamination. Besides, ultra-short pulsed lasers (picosecond, femtosecond) are especially useful for precise processing, considering that the delivered energy can only be adsorbed by the material through evaporating or flying apart [46]. The rapid material ejection can be used in direct-writing of various materials as laser-assisted-printing techniques. During the laser irradiation, the laser energy can hardly be transmitted from the irradiated region to surrounding regions before being ejected, which greatly reduces the heat-affected zone to achieve ultra-high processing resolution.

8.2.5 Laser Scanning Overlap

The laser scanning overlap can generally be characterized by the laser overlap rate and the scanning repetition rate that the irradiated area is experienced during the laser scanning process. An ideal LIIR requires suitable laser irradiations per site to produce desired physical/chemical changes of interface materials. There generally exists a threshold times of accumulated laser pulses per site for certain LIIR [43].

During typical laser scanning process, the laser overlap rate is indicated by the sum of spot overlap rate and scan line overlap rate. These overlap rates depend on the beam spot size, the pulse frequency, the scanning speed, and line scan spacing, etc. The spot overlap rate is proportional to the beam width (the width along the scanning direction) and the pulse frequency but inversely proportional to the scanning speed. The scan line overlap rate is proportional to the length along the perpendicular direction, but inversely proportional to the line scan spacing. Besides, the repetition rate of laser scanning should also be carefully chosen. If the repetition rate is too high, the over accumulated energy will cause undesirable material degradation or even severe thermal damage. If the repetition rate is too low, the efficiency of the laser irradiation will be reduced due to the increased heat dissipation.

8.2.6 Physical and Chemical Properties of Material

One of the most important factors determining the phenomena like the thermal strain, phase change, and microstructure change during the LIIR is the heat dissipation through the material thickness [29]. Materials with higher thermal diffusivity conduct heat more quickly and a sharper temperature gradient in the thickness direction can be detected. In practice, the relative values of the thermal diffusivity of the laser-irradiated layer, adjacent layers, and transparent substrate decide the temperature distribution and heat spread rate after the laser irradiation. The heat is likely to be accumulated at the interface if the materials near the interface have low thermal diffusivity. The rate and direction of the heat transfer to surrounding regions determine the rate and location of recrystallization of the molten zone, which ultimately determine the composition and structure of the laser-irradiated interface.

Other physical properties like the melting point, the coefficient of the thermal expansion, and mechanical property significantly affect the laser-induced phenomena. Moreover, chemical properties like the chemical composition and photochemical stability will affect the types and products of the chemical reaction. As a consequence, distinct phenomena rely on the specific combination of laser processing parameters and material properties.

8.3 Experimental Study of LLO Mechanism of Ultra-thin PI Film from Glass Carrier

8.3.1 Experimental Procedure

PI films were prepared by spin-coating and curing of the ZKPI-305IID non-photosensitive PI precursor on transparent 0.5 mm-thick glass carrier. The thickness of the PI film was controlled at 2–25 μm by changing the rotation speed of the spin-coating machine and the number of spin-coated layers. Here, the adhesively bonded PI film was first divided into multiple rectangular regions.

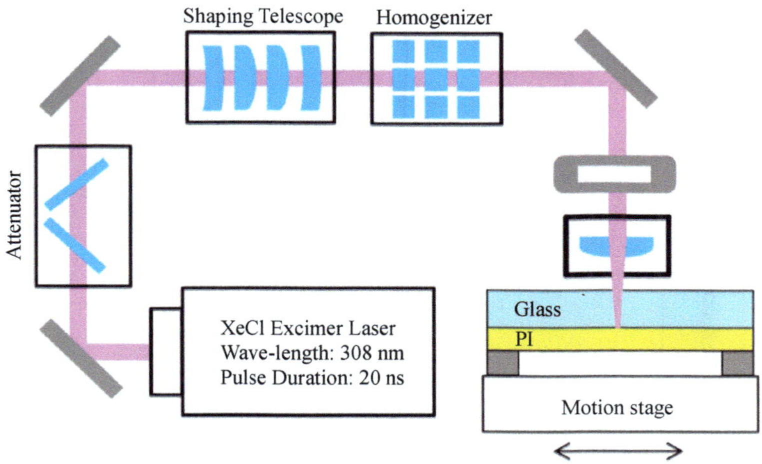

Fig. 8.2 Schematic illustration of the LLO experimental platform

A schematic illustration of the LLO experimental platform is shown in Fig. 8.2. The XeCl excimer laser (OptoSystems CL 7020) with the wavelength of 308 nm was used as the light source of the laser system. The laser penetration through the glass carrier is about 33%. The output energy of the laser beam was adjusted by the attenuator. The laser beam was shaped into narrow rectangular by the shaping telescope. The ununiformity of the raw beam was controlled less than ±5% using the homogenizer optics. A line-shaped laser beam profile (20 mm × 0.62 mm) was obtained on the focal plane by using the light bar and the focusing lens. The focal position of the laser spot can be located at the PI/glass interface by adjusting the position of the motion stage. The PI-on-glass structure was supported on both ends. The LLO process can be achieved in combination with the movement of the motion stage at a certain speed. The practical LLO experimental setup is shown in Fig. 8.3. Only the process parameters of the laser fluence and the laser scanning speed (v_{scan}) were varied in the LLO experiment. Combining the parameters of the line-shaped laser beam width (w_{beam}), the laser scanning speed (v_{scan}), and the laser pulse frequency (f_{pulse}), a new parameter characterizing the accumulated irradiation times per site (ART) was defined as $w_{beam} \times f_{pulse}/v_{scan}$ [47]. In all LLO experiments, the beam width and the pulse frequency of the line-shaped laser were fixed at 0.62 mm and 10 Hz, respectively. As a consequence, the ART was only related to the laser scanning speed.

In order to measure the adhesion strength of the PI/glass interface, a homemade adhesion strength measurement system was used [48]. The PI strip with the width of 5 mm and thickness of 22 μm was partially separated from the glass carrier and bonded to the end of the load cell, and the rest of the PI strip remained adhered to the glass carrier after the laser irradiation. The z axis stage moved upward from the glass carrier at a speed of 0.1 mm/s to maintain a stable debonding. The real-time

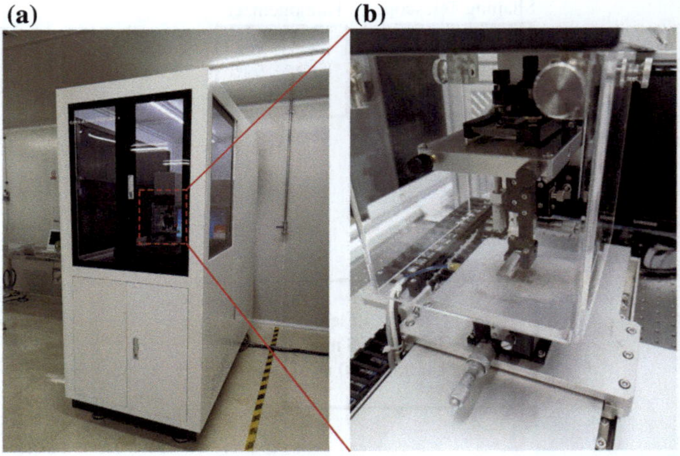

Fig. 8.3 Practical LLO experimental equipment (Laser Peeler α): **a** an overview, **b** local magnification

peel force can be obtained from the load sensor (222-N LCM300N, resolution of 1.0 mN). The interface adhesion strength can be obtained referring to the calculation method proposed by Liu et al. [48].

A confocal laser scanning microscope (EYENCE VK-X200K) was used to detect the micromorphology of the PI film after the laser irradiation. By using the profile measurement module of the microscope, the maximum height of the raised bubble (bubble height, h_{bubble}) formed by the separated PI film in the direction perpendicular to the plane of the glass carrier surface was measured after the laser irradiation. The PI-on-glass structure was cross-sectioned at the middle position of the irradiated region along the laser scanning direction. Microstructure of the cross sections near the PI/glass interface was observed by using the scanning electron microscopy (SEM, Helios NanoLab G3CX). The microstructure of the residues on the glass carrier after the interface adhesion strength test was also detected by using the SEM.

8.3.2 *Optical Phenomena Observed After Laser Irradiation*

The minimum laser fluence that is required to separate the PI film completely from the glass carrier is defined here as the threshold fluence for the PI film separation. If the laser fluence is lower than the threshold fluence, the film cannot be completely separated. In contrast, if the laser fluence is much higher than the threshold fluence, the damage risk of the separated PI film increases. Different optical phenomena can be observed at the irradiated region after the laser irradiation by using the laser scanning confocal microscope, which can be sequentially classified as the modes

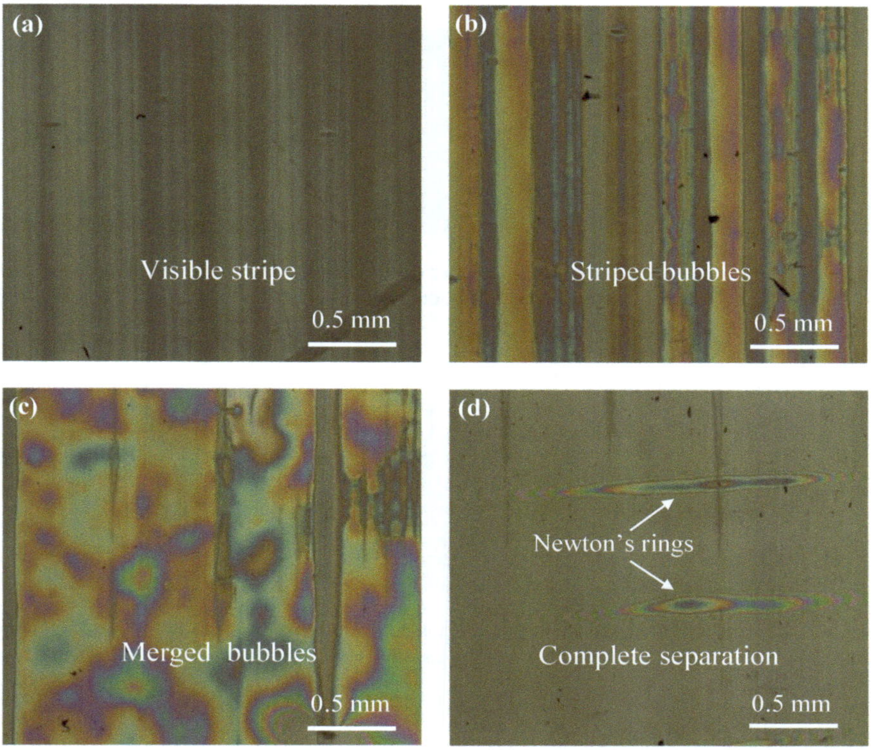

Fig. 8.4 Different optical phenomena observed from the top side of the PI film by using the laser scanning confocal microscope after the laser irradiation: **a** visible stripe, **b** striped bubbles, **c** merged bubbles, **d** complete separation. Reprinted by permission from Springer Nature, Ref. [27], Copyright 2018

of the visible stripe, striped bubbles, merged bubbles, and complete separation, as shown in Fig. 8.4a, b, c, d, respectively. The PI film is believed to be just separated when light fringes corresponding to the phenomena of Newton's rings are presented [49].

The effect of the laser scanning number on the optical phenomena observed by the laser scanning confocal microscope is shown in Fig. 8.5, in which the thickness of the PI film is constant at 2 μm. The laser scanning speed is set as 0 mm/s. The laser fluence is selected as 50 mJ/cm^2. As can be seen, the completely separated region enlarges with the increase of the laser scanning number. This is due to the sustained reaction of the PI with the excimer laser, leading to the continuous increase of the clearance at the PI/glass interface. The above results indicate that there exists an accumulated effect of the laser irradiation. The complete separation of the PI film occurs in case the cumulative irradiation energy is sufficient.

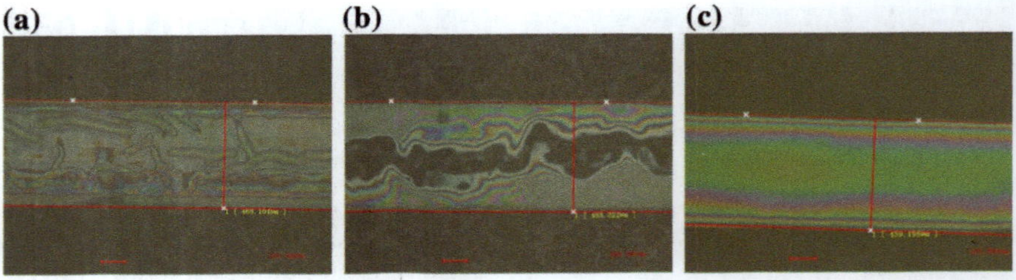

Fig. 8.5 The effect of the laser scanning number on the optical phenomena observed by the laser scanning confocal microscope under the condition of the laser scanning number of **a** 1, **b** 5, and **c** 10

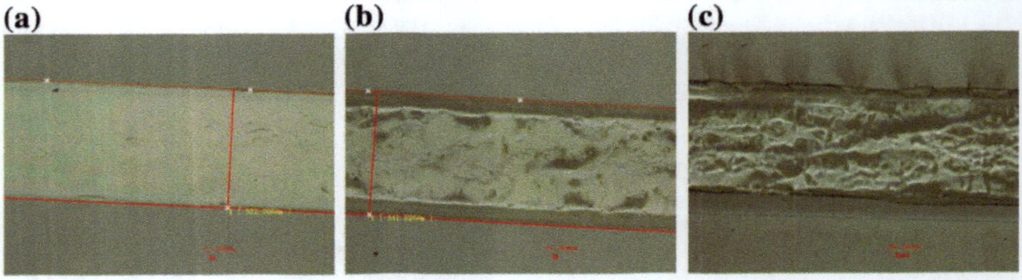

Fig. 8.6 The effect of the laser fluence on the optical phenomena observed by the laser scanning confocal microscope under the condition of the laser fluence of **a** 150 mJ/cm^2, **b** 200 mJ/cm^2, and **c** 250 mJ/cm^2

The effect of the laser fluence on the optical phenomena observed by the laser scanning confocal microscope is shown in Fig. 8.6, in which the PI film thickness is constant at 2 μm. The laser scanning speed is set as 0 mm/s. The laser scanning number is selected as 1. When the laser fluence is relatively high, wrinkles of the separated PI film or even damage at the central region can be observed, as shown in Fig. 8.6b. It is presumed that the amount of the irradiation-induced gas products increases as the laser fluence enlarges. As a consequence, irregular wrinkles are formed after the laser irradiation due to the impact effect of gas products. When the laser fluence is very high, too strong impact effect of the gas products will result in the initiation of the film crack, as indicated in Fig. 8.6c. In practical application, the above adverse phenomena in Fig. 8.6c, d should be avoided as much as possible.

8.3.3 *Effects of ART and PI Film Thickness on Threshold Fluence*

The effect of the ART on the threshold fluence is shown in Fig. 8.7. It can be observed that the threshold fluence decreases with the increase of the ART when the film thickness is constant. With the ART increased from 1 to 10, the reduction of the

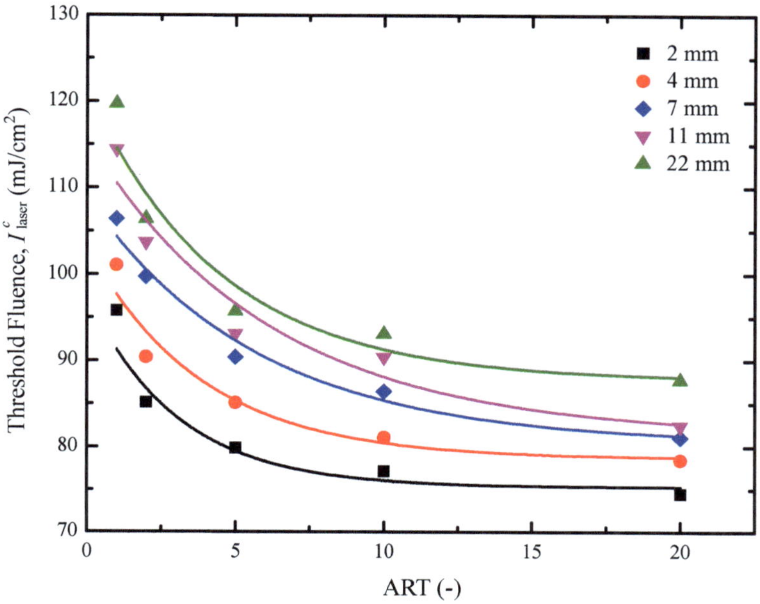

Fig. 8.7 The effect of the ART on the threshold fluence. Reprinted by permission from Springer Nature, Ref. [27], Copyright 2018

threshold fluence is about 30%. When the ART is further increased from 10 to 20, the downward trend slows down.

The effect of the PI film thickness on the threshold fluence is shown in Fig. 8.8. When the laser scanning speed is maintained constant, an increase in the threshold fluence can be observed with the increase of the film thickness from 2 to 22 μm. It can also be found that the threshold fluence gradually converges when the PI film thickness is larger than 15 μm. It can be concluded that a higher laser fluence is essential to separate thicker PI films from the glass carrier compared with thinner ones.

The above results are different from previous researches on the polymer-based LLO processes [28, 49–51]. It was previously indicated that there existed a fixed threshold fluence for a specific PI material, and the laser fluence should exceed the fixed threshold fluence to ablate enough interfacial material for a complete separation [51]. The explanation for this difference may be that the corresponding PI film was relatively thick, under which condition the threshold fluence was less susceptible to the film thickness, as derived from Fig. 8.8.

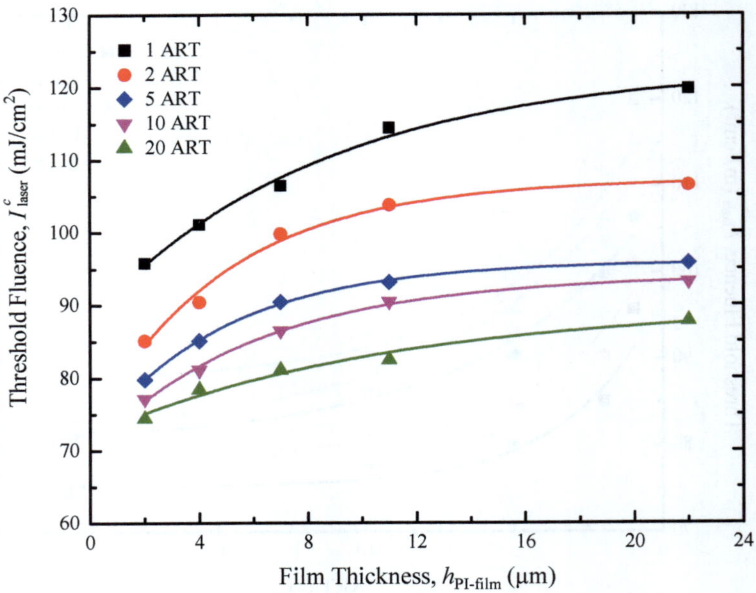

Fig. 8.8 The effect of the PI film thickness on the threshold fluence. Reprinted by permission from Springer Nature, Ref. [27], Copyright 2018

8.3.4 Typical SEM Images of PI/Glass Interface

Detection of the microstructure evolution at the PI/glass interface is beneficial for the understanding of the LLO mechanism. Typical SEM images of the PI/glass interface before and after the laser irradiation are shown in Fig. 8.9. The upper part is the PI film and the lower part is the glass carrier. The initial state of the PI/glass interface before the laser irradiation is depicted in Fig. 8.9a. The PI/glass interface can be found well bonded before the laser irradiation. The laser fluence was selected as 80 mJ/cm^2, 87 mJ/cm^2, and 93 mJ/cm^2 in Fig. 8.9b, c, d, respectively. The laser scanning speed was kept constant at 6 mm/s. It can be seen that the microstructure of the PI/glass interface changes significantly with the increase of the laser fluence.

After the laser irradiation under the condition of the laser fluence of 80 mJ/cm^2, many small voids are presented at the PI/glass interface, as depicted in Fig. 8.9b. In fact, the laser fluence of 80 mJ/cm^2 is much higher than the ablation threshold (the minimum laser fluence required for the initial ablation of the PI material) about 40 mJ/cm^2 for the excimer laser with the wavelength of 308 nm [52–54]. The mechanism of the separation of the PI film from the glass carrier has been supposed as the vaporization of interface material [49, 50]. Therefore, the formation of small voids should be related to the gas products generated by the laser irradiation. The solidification of the PI after the laser irradiation can be influenced by the gas products, which finally leads to the generation of small voids near the PI/glass interface.

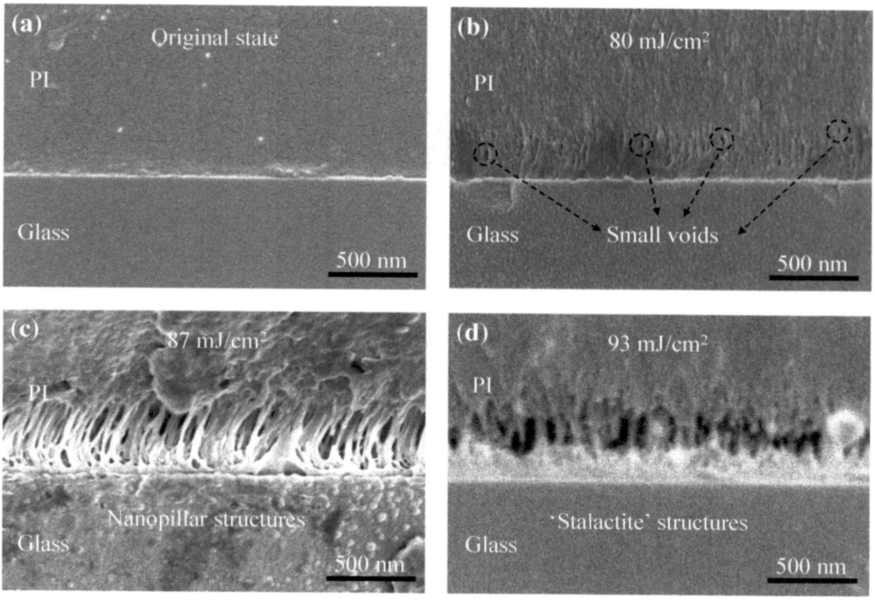

Fig. 8.9 SEM images at the PI/glass interface: **a** before laser irradiation, and after the laser irradiation under conditions of the laser fluence of **b** 80 mJ/cm^2, **c** 87 mJ/cm^2, and **d** 93 mJ/cm^2. Reprinted by permission from Springer Nature, Ref. [27], Copyright 2018

It can be speculated that higher laser energy yields more gas products, resulting in a higher impact effect of gas products. With the increase of the laser fluence (87 mJ/cm^2), the microstructure of the PI/glass interface is transformed into the nanoscale columnar connection structure after the solidification of the ablated material, as shown in Fig. 8.9c. With a further increase of the laser fluence (93 mJ/cm^2), fracture of the nano-columnar structure promotes the formation of stalactite-like structure at the PI/glass interface, as depicted in Fig. 8.9d. The interface clearance increases gradually from the initial state in Fig. 8.9a to the nearly complete separation state in Fig. 8.9d, indicating the increase of the extent of the laser irradiation. It can be expected that by further increasing the laser fluence, the interfacial nano-columnar connections will further decrease and eventually disappear, leading to a clear gap at the PI/glass interface and a complete separation of the PI film, as shown in Fig. 8.10.

The above results indicate that the impact effect of the gas products acts as the driving force for the PI film separation. Sufficient gas products are recommended to promote the PI film separation from the glass carrier by reducing the nano-columnar connections. If the gas products are insufficient, a part of the nano-columnar connections still exists at the PI/glass interface after the laser irradiation, resulting in a certain residual interface adhesion. Experimental results also show that the impact effect of the gas products can be accumulated through multiple laser irradiations,

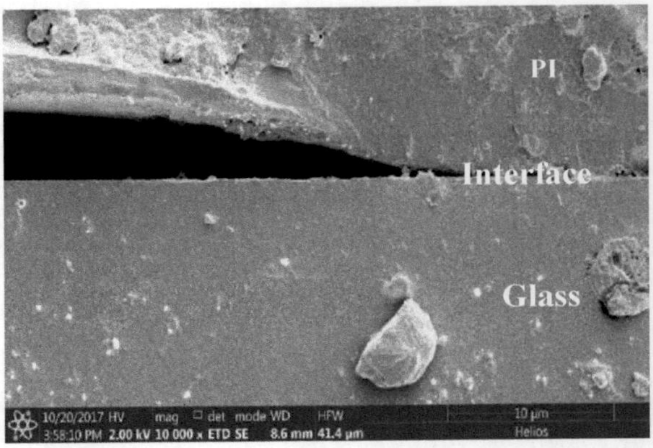

Fig. 8.10 Typical SEM image at the PI/glass interface with a clear interface gap

by which continuous changes in the microstructure at the PI/glass interface can be observed.

8.3.5 Effects of Laser Fluence and ART on Interface Adhesion Strength

The effect of the laser fluence on the measured residual interface adhesion strength is shown in Fig. 8.11. The measured interface adhesion strength of the common adhesive tape (Scotch Magic Tape) is around 0.08 N/mm. The thickness of the PI film was chosen as 22 μm. The ART was kept constant at 5. When the laser fluence is 82.5 mJ/cm^2, the residual interface adhesion strength is speculated over 0.688 N/mm because that the fracture of corresponding PI strip tends to occur during loading. A stable debonding process can be achieved once the laser fluence reaches 85 mJ/cm^2. It can be seen that the measured residual interface adhesion strength decreases from 0.428 N/mm to 0.0163 N/mm with the increase of the laser fluence from 85 mJ/cm^2 to 98.5 mJ/cm^2.

The effect of the ART on the residual interface adhesion strength is shown in Fig. 8.12, in which the measured interface adhesion strength of the common adhesive tape is also depicted. The thickness of the PI film was chosen as 22 μm. The laser fluence was fixed at 90.5 mJ/cm^2. When the ART is smaller than 3, the stable debonding cannot be achieved. When the ART is larger than 3, the reduced residual interface adhesion strength can be obtained experimentally. It can be found that the residual interface adhesion strength decreases from 0.390 N/mm to 0.0472 N/mm with the increase of the ART from 4 to 15. The above experimental results clearly demonstrate that the residual interface adhesion strength and the non-destructive interfacial debonding can be achieved finally by increasing the laser fluence and ART.

Microstructure analysis of the residues on the glass carrier after the above adhesion strength test is shown in Fig. 8.13. It can be observed that the micromorphology

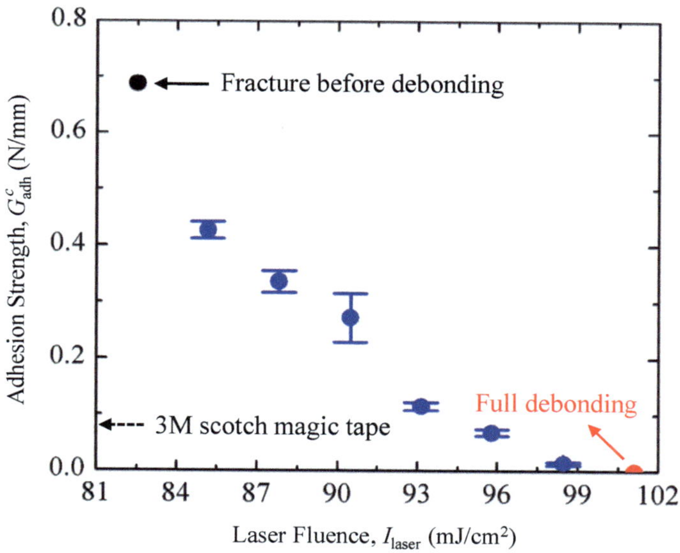

Fig. 8.11 The effect of the laser fluence on the residual interface adhesion strength (ART = 5). Reprinted by permission from Springer Nature, Ref. [27], Copyright 2018

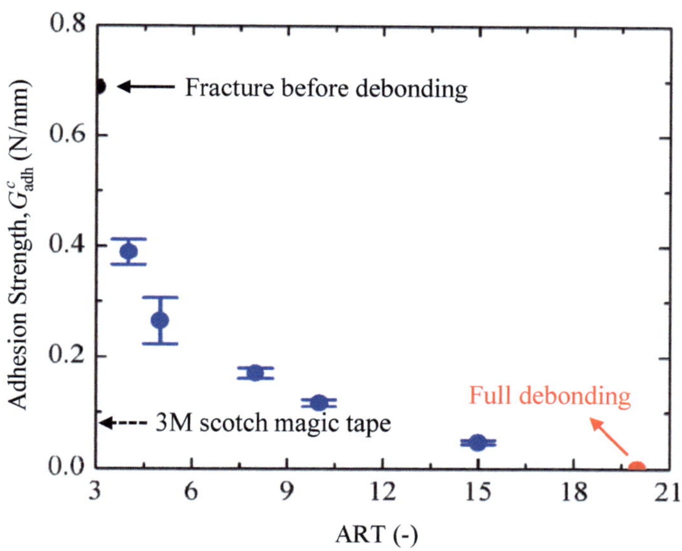

Fig. 8.12 The effect of the ART on the residual interface adhesion strength (laser fluence = 90.5 mJ/cm^2). Reprinted by permission from Springer Nature, Ref. [27], Copyright 2018

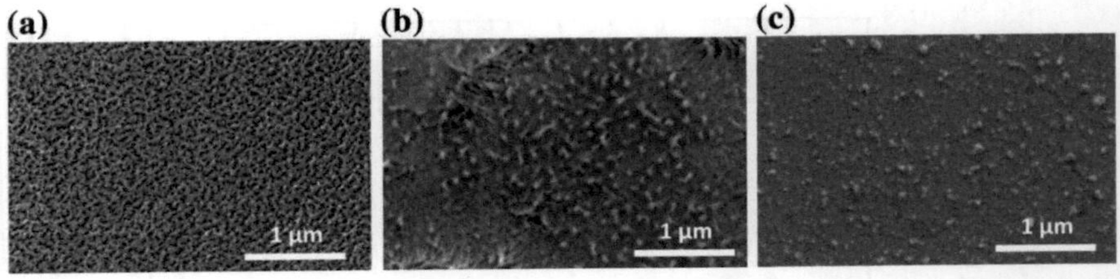

Fig. 8.13 The microstructure of residues on the glass carrier after the peeling test under conditions of the laser fluence of: **a** 85 mJ/cm^2, **b** 93 mJ/cm^2, and **c** 98 mJ/cm^2. Reprinted by permission from Springer Nature, Ref. [27], Copyright 2018

changes significantly with the increase of the laser fluence. When the laser fluence is relatively low (85 mJ/cm^2) along with high residual interface adhesion strength of 0.428 N/mm, high-density textured nano-structures are observed, as shown in Fig. 8.13a. This micromorphology is likely to be formed after the nano-columnar structure in Fig. 8.9c being tore off. With the increase of the laser fluence (93 mJ/cm^2), a significant drop in the residual interface adhesion strength (0.117 N/mm) can be detected and the density of textured nano-structures is clearly reduced, as depicted in Fig. 8.13b. This micromorphology seems to be formed relating to the interfacial microstructure in the case of Fig. 8.9d. When the laser fluence is close to the threshold fluence, the textured nano-structures nearly disappear as shown in Fig. 8.13c. Meanwhile, the measured interface adhesion strength approaches to zero (0.0145 N/mm). The observations are in accordance with results of the interface microstructure and the residual interface adhesion strength in Figs. 8.9, 8.10, 8.11 and 8.12.

8.3.6 Analysis of Separation Mode

In the LLO, it is not only required a complete separation of the PI film from the transparent glass carrier, but also the film should be maintained intact and flat without serious wrinkles. The wrinkled PI film will increase the damage risk of devices and difficulty of conforming to a complex surface. Four typical separation modes of the PI film after the laser irradiation are observed and depicted in Fig. 8.14. The film separations are typically classified as the flat, slightly wrinkled, severely wrinkled, and breakage modes, as shown in Fig. 8.14a, b, c, d, respectively. The flat mode is mostly demanded in the practical application. Experimental results show that the ultra-thin films (<5 μm) are prone to be wrinkled. The formation of wrinkles is attributed to the plastic deformation introduced by the impact effect of gas products. The slightly wrinkled mode with visible folds can generally be flattened. The severely wrinkled mode should be avoided as much as possible considering that it can hardly be flattened and conformed to complex surfaces.

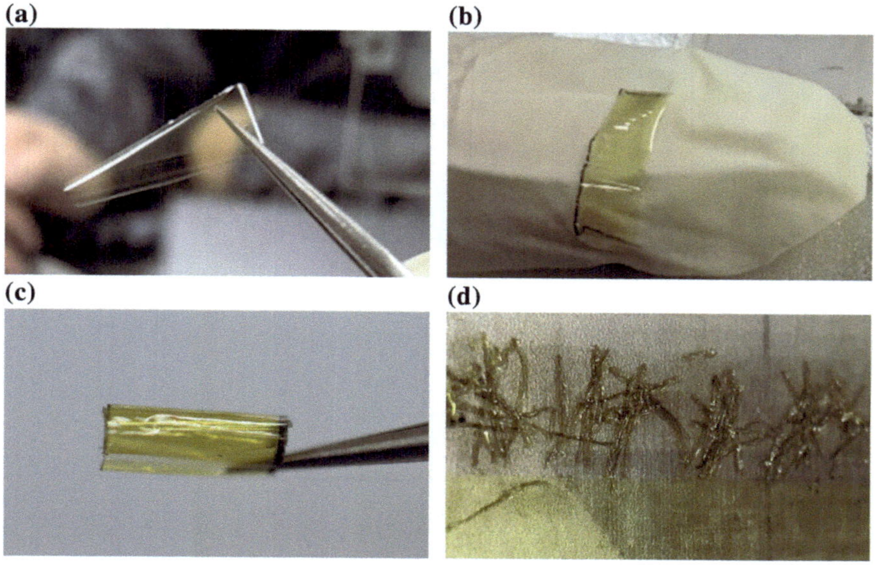

Fig. 8.14 Typical separation modes of the PI film after the LLO: **a** flat mode ($h_{\text{PI-film}} = 4\ \mu\text{m}$, ART = 5, $I_{\text{laser}} = 100\ \text{mJ/cm}^2$), **b** slightly wrinkled mode ($h_{\text{PI-film}} = 4\ \mu\text{m}$, ART = 5, $I_{\text{laser}} = 120\ \text{mJ/cm}^2$), **c** severely wrinkled mode ($h_{\text{PI-film}} = 4\ \mu\text{m}$, ART = 5, $I_{\text{laser}} = 146\ \text{mJ/cm}^2$), and **d** breakage mode ($h_{\text{PI-film}} = 2\ \mu\text{m}$, ART = 1, $I_{\text{laser}} = 160\ \text{mJ/cm}^2$). Reprinted by permission from Springer Nature, Ref. [27], Copyright 2018

Analysis of the experimental results of separation modes for PI films with the thickness of 2 μm and 4 μm is depicted in Fig. 8.15. When the PI film thickness is 2 μm, there is even no available laser fluence under smaller ARTs to obtain the flat mode although the laser energy is near the threshold fluence, as shown in Fig. 8.15a. The instantaneous high impact effect of the gas products during the LLO process may even lead to the breakage of the ultra-thin PI film. The flat mode of the ultra-thin PI film (2 μm) after the LLO separation can only be achieved by adopting both low laser fluence and large ART. When the film thickness increases up to 4 μm, the available range of the laser fluence that keeps the film flat extends, as shown in Fig. 8.15b. A broader process window can also be obtained with the increase of the ART. When the ART is set as 5, the separated PI film can still be flattened even when the laser fluence is 50% higher than the threshold fluence. When the ART is set as 10, serious folds of the separated PI film cannot be detected even when the laser fluence is 180% of the threshold fluence. The thicker film can be found more effective to resist the film deformation caused by the impact effect of the gas products.

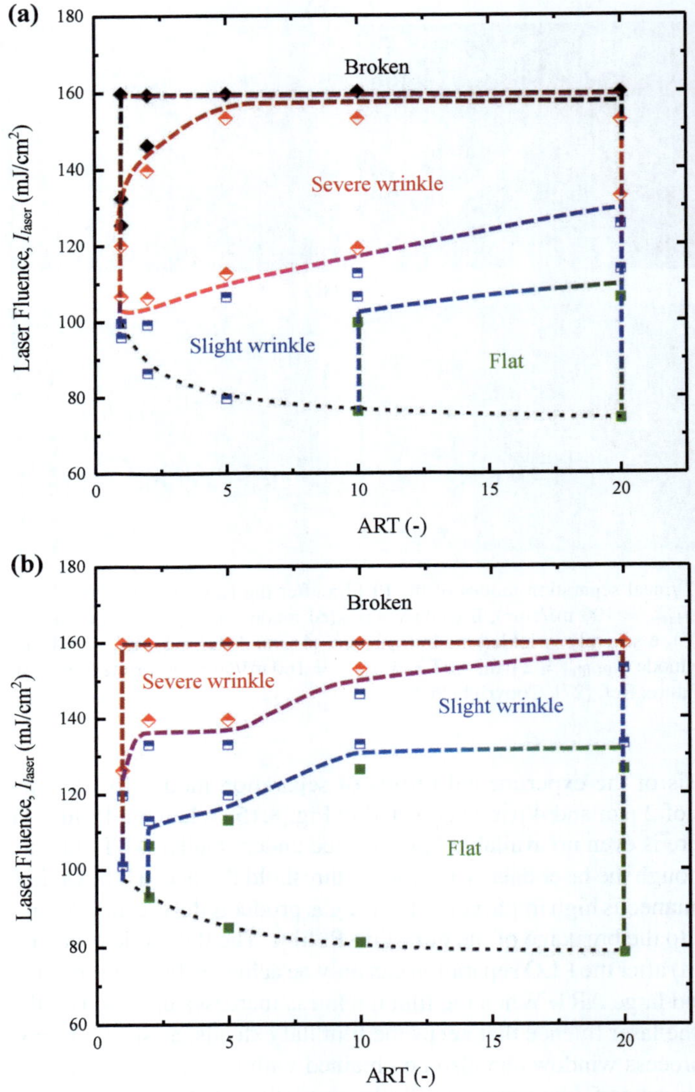

Fig. 8.15 Experimental results of the separation modes for PI films with thicknesses of **a** 2 μm and **b** 4 μm

8.3.7 Analysis of Bubble Height

The PI film bubble can be observed in experiment after the laser irradiation. The deformation degree of the PI film after the LLO is characterized here by the previously defined bubble height. Experimental results of the bubble height under different conditions of the laser irradiation and the PI film thickness are shown in Fig. 8.16. As can be seen, with the increase of the laser fluence or cumulative irradiation times, the bubble height increases because of the accumulated impact effect of gas products. The bubble height of thin PI films is found larger than that of thick PI films under the same condition of process parameter combination. The PI films with the thickness of 7 μm are easily to be broken under the condition of the laser fluence of 120 mJ/cm^2, as can be derived from Fig. 8.16a. The 7 μm-thick PI film cannot even withstand a single laser pulse with the laser fluence of 133 mJ/cm^2. In contrast, the thick PI films can withstand more pulses of high-energy irradiations, as shown in Fig. 8.16b, c. However, it is hard to form the bubble for the thick PI films under the condition of low-energy laser irradiation, which can be interpreted by the high mechanical stiffness of thick PI films. Compared with thick PI films, thin PI films are easily deformed and even broken under the same impact effect of gas products.

The deformation degree of the PI film after the laser irradiation can be further characterized by dividing the bubble height (h_{bubble}) by corresponding film thickness ($h_{\mathrm{PI\text{-}film}}$). The effect of the ART on the normalized bubble height ($h_{\mathrm{bubble}}/h_{\mathrm{PI\text{-}film}}$) is shown in Fig. 8.17. As can be seen, the deformation degree of ultra-thin films (<5 μm) is much higher than that of relatively thick ones. The normalized bubble height can even be greater than 1.5 for ultra-thin films, and serious wrinkles or film breakages can then be observed. When low laser fluence is adopted, the normalized bubble height can still be greater than 0.5 for ultra-thin films, and wrinkles of the film are easy to occur in these situations. For the PI films with thickness larger than 5 μm, the normalized bubble height is mostly smaller than 0.5, which maintains the flatness of the separated PI film.

In conclusion, the formation mechanism of the film bubble can be interpreted as follows: The decomposition of the PI material leads to the generation of gas products and the formation of interface clearance at the originally bonded region. A certain laser energy is converted into the impact energy of generated gas products. The greater impact energy and the thinner PI film promote the larger deformation of the PI film. In the range of the elastic deformation, the film deformation tends to revert back to the original state after the laser irradiation despite the internal gas pressure. Moreover, the occurrence of the plastic deformation will intensify the film deformation after the laser irradiation. As the film thickness increases, the impact effect of gas products decreases because of the enhancement in the film stiffness, which leads to the reduce in the bubble height.

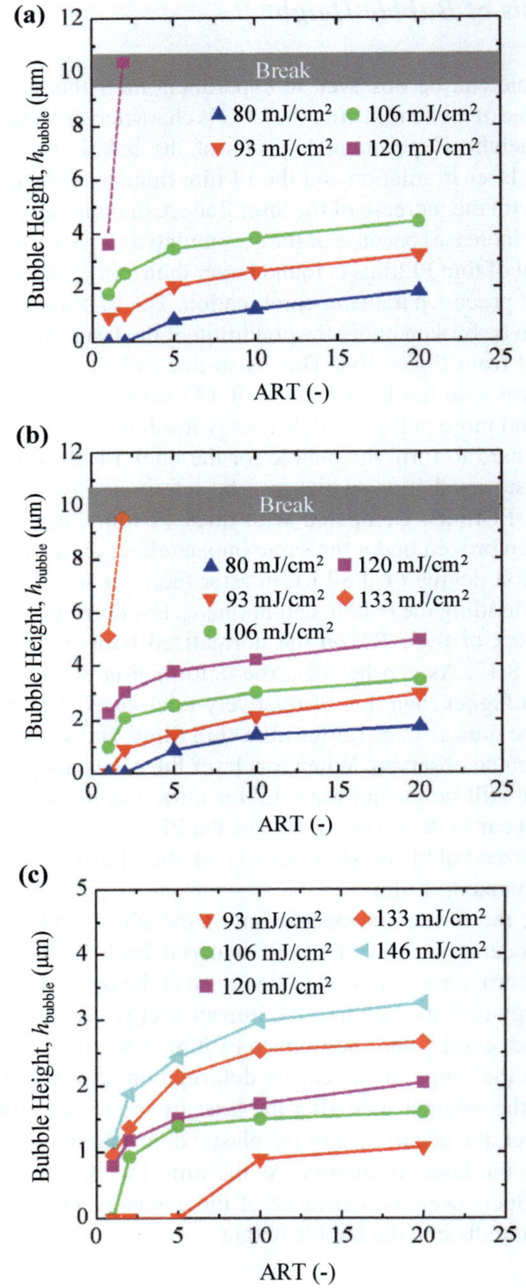

Fig. 8.16 Experimental results of the bubble height under conditions of the PI film thicknesses of: **a** 7 μm, **b** 11 μm, and **c** 22 μm. Reprinted by permission from Springer Nature, Ref. [27], Copyright 2018

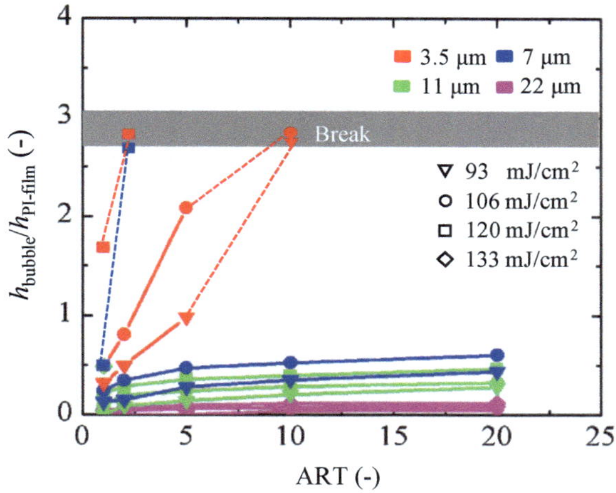

Fig. 8.17 The effect of the ART on the normalized bubble height ($h_{bubble}/h_{PI\text{-}film}$). Reprinted by permission from Springer Nature, Ref. [27], Copyright 2018

8.3.8 *Analysis of Process Window*

Based on the above analysis, process control of the LLO of the ultra-thin PI film is more crucial compared with thick films. A schematic illustration of the process window for the LLO of the PI film with different thicknesses is shown in Fig. 8.18. It can be seen that the process window is gradually reduced with the decrease of the PI film thickness. Besides, the increase of the ART not only reduces the threshold fluence but also results in a broader process window. The wrinkled degree of the PI film can be reduced by combining a large ART and low laser fluence, which is supposed as the optimized strategy for the LLO of ultra-thin PI films.

The practical significance of the proposed LLO is that the large-area and ultra-thin electronic device fabricated on the PI film can still be attached to the glass carrier after the laser irradiation despite the greatly decreased interface adhesion strength. The electronic device can then be easily transferred with rigid glass carrier for further fabrication without the use of extra rigid wafer. A reliable and complete separation of the PI film can be finally achieved after gently mechanical peeling.

8.4 Summary

LLO is an emerging technique for micro-/nano-fabrication, which involves the laser-induced impulsive effect to initiate separation at the interface between the substrate and the functional layer. The mechanism of the LLO of the ultra-thin PI film from the glass carrier was studied in this chapter. The disclosed mechanism showed that the impact effect of gas products generated by the laser irradiation promoted the microstructure change at the PI/glass interface, resulting in the decrease of the inter-

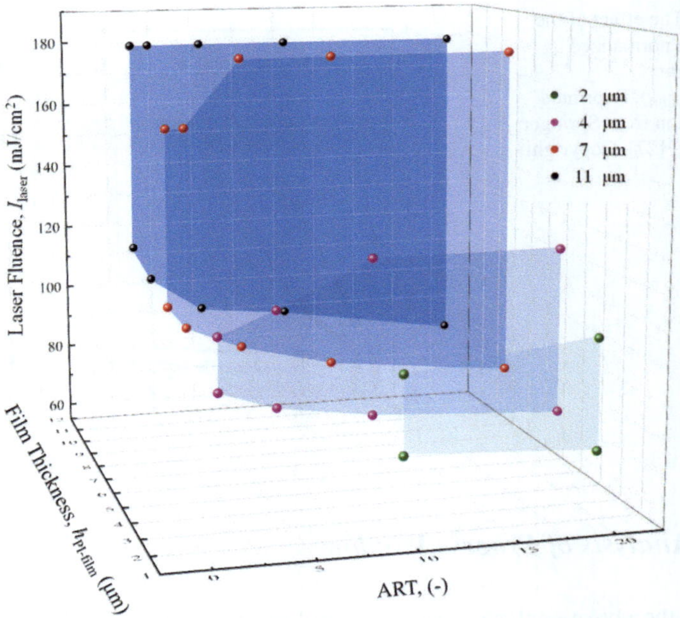

Fig. 8.18 A schematic illustration of the process window for the LLO of the PI film. Reprinted by permission from Springer Nature, Ref. [27], Copyright 2018

face adhesion strength. The nano-columnar structures were observed at the PI/glass interface after the laser irradiation due to the generation of gas products. A certain residual interface adhesion strength was retained as a consequence. A reliable LLO separation of the ultra-thin PI film was achieved by using the strategy of multiple irradiations of laser pulses with low laser fluence. The process window of the LLO of ultra-thin PI films was found much smaller than that of traditional thick films. This study provides an attractive route to achieve the fabrication of flexible devices on complex surfaces.

References

1. Meitl MA, Zhu Z-T, Kumar V, Lee KJ, Feng X, Huang YY, Adesida I, Nuzzo RG, Rogers JA (2006) Transfer printing by kinetic control of adhesion to an elastomeric stamp. Nat Mater 5(1):33–38
2. Feng X, Meitl MA, Bowen AM, Huang Y, Nuzzo RG, Rogers JA (2007) Competing fracture in kinetically controlled transfer printing. Langmuir 23(25):12555–12560
3. Kim T-H, Carlson A, Ahn J-H, Won SM, Wang S, Huang Y, Rogers JA (2009) Kinetically controlled, adhesiveless transfer printing using microstructured stamps. Appl Phys Lett 94(11):113502

4. Peng B, Huang Y, Yin Z, Xiong Y (2011) Analysis of interfacial peeling in IC chip pick-up process. J Appl Phys 110(7):073508
5. Liu Z, Huang Y, Chen J, Yin Z (2014) Tunable peeling technique and mechanism of thin chip from compliant adhesive tapes. IEEE Trans Compon Packag Manuf Technol 4(4):560–568
6. Liu Z, Huang YA, Liu H, Chen J, Yin Z (2014) Reliable peeling of ultrathin die with multineedle ejector. IEEE Trans Compon Packag Manuf Technol 4(9):1545–1554
7. Jain K, Klosner M, Zemel M, Raghunandan S (2005) Flexible electronics and displays: High-resolution, roll-to-roll, projection lithography and photoablation processing technologies for high-throughput production. Proc IEEE 93(8):1500–1510
8. Lo C-Y, Hiitola-Keinänen J, Huttunen O-H, Petäjä J, Hast J, Maaninen A, Kopola H, Fujita H, Toshiyoshi H (2009) Novel roll-to-roll lift-off patterned active-matrix display on flexible polymer substrate. Microelectron Eng 86(4):979–983
9. Chen J, Liu H, Huang Y, Yin Z (2016) High-rate roll-to-roll stack and lamination of multilayer structured membrane electrode assembly. J Manuf Process 23:175–182
10. Huang Y, Chen J, Yin Z, Xiong Y (2011) Roll-to-roll processing of flexible heterogeneous electronics with low interfacial residual stress. IEEE Trans Compon Packag Manuf Technol 1(9):1368–1377
11. Peng B, Huang Y, Yin Z, Xiong Y (2012) Competing fracture modeling of thin chip pick-up process. IEEE Trans Compon Packag Manuf Technol 2(7):1217–1225
12. Liu Z, Wan X, Huang Y, Chen J, Yin Z (2018) Theoretical and experimental studies of competing fracture for flexible chip-adhesive-substrate composite structure. IEEE Trans Compon Packag Manuf Technol 8(1):57–64
13. Huang Y, Liu H, Xu Z, Chen J, Yin Z (2018) Conformal peeling of device-on-substrate system in flexible electronic assembly. IEEE Trans Compon Packag Manuf Technol 8(8):1496–1506
14. Park KI, Son JH, Hwang GT, Jeong CK, Ryu J, Koo M, Choi I, Lee SH, Byun M, Wang ZL (2014) Highly-efficient, flexible piezoelectric PZT thin film nanogenerator on plastic substrates. Adv Mater 26(16):2514–2520
15. Do YH, Jung WS, Kang MG, Kang CY, Yoon SJ (2013) Preparation on transparent flexible piezoelectric energy harvester based on PZT films by laser lift-off process. Sens Actuators A 200:51–55
16. Jeong J-W, McCall Jordan G, Shin G, Zhang Y, Al-Hasani R, Kim M, Li S, Sim Joo Y, Jang K-I, Shi Y, Hong Daniel Y, Liu Y, Schmitz Gavin P, Xia L, He Z, Gamble P, Ray Wilson Z, Huang Y, Bruchas Michael R, Rogers John A (2015) Wireless optofluidic systems for programmable in vivo pharmacology and optogenetics. Cell 162(3):662–674
17. McCall JG, T-i Kim, Shin G, Huang X, Jung YH, Al-Hasani R, Omenetto FG, Bruchas MR, Rogers JA (2013) Fabrication and application of flexible, multimodal light-emitting devices for wireless optogenetics. Nat Protoc 8:2413
18. T-i Kim, McCall JG, Jung YH, Huang X, Siuda ER, Li Y, Song J, Song YM, Pao HA, Kim R-H, Lu C, Lee SD, Song I-S, Shin G, Al-Hasani R, Kim S, Tan MP, Huang Y, Omenetto FG, Rogers JA, Bruchas MR (2013) Injectable, cellular-scale optoelectronics with applications for wireless optogenetics. Science 340(6129):211–216
19. Lee HS, Chung J, Hwang GT, Jeong CK, Jung Y, Kwak JH, Kang H, Byun M, Kim WD, Hur S, Oh SH, Lee KJ (2014) Flexible inorganic piezoelectric acoustic nanosensors for biomimetic artificial hair cells. Adv Func Mater 24(44):6914–6921
20. Hwang GT, Annapureddy V, Han JH, Joe DJ, Baek C, Park DY, Kim DH, Park JH, Jeong CK, Park KI, Choi JJ, Kim DK, Ryu J, Lee KJ (2016) Self-powered wireless sensor node enabled by an aerosol-deposited pzt flexible energy harvester. Adv Energy Mater 6(13):1600237
21. Park DY, Joe DJ, Kim DH, Park H, Han JH, Jeong CK, Park H, Park JG, Joung B, Lee KJ (2017) Self-powered real-time arterial pulse monitoring using ultrathin epidermal piezoelectric sensors. Adv Mater 29(37):1702308
22. Shaw-Stewart JRH, Lippert TK, Nagel M, Nüesch FA, Wokaun A (2012) Sequential printing by laser-induced forward transfer to fabricate a polymer light-emitting diode pixel. ACS Appl Mater Interfaces 4(7):3535–3541

23. Shaw-Stewart JRH, Mattle T, Lippert TK, Nagel M, Nüesch FA, Wokaun A (2013) The fabrication of small molecule organic light-emitting diode pixels by laser-induced forward transfer. J Appl Phys 113(4):043104
24. Sanchez-Aniorte MI, Mouhamadou B, Alloncle AP, Sarnet T, Delaporte P (2016) Laser-induced forward transfer for improving fine-line metallization in photovoltaic applications. Appl Phys A 122(6):595
25. Chen Y, Munoz-Martin D, Morales M, Molpeceres C, Sánchez-Cortezon E, Murillo-Gutierrez J (2016) Laser induced forward transfer of high viscosity silver paste for new metallization methods in photovoltaic and flexible electronics industry. Phys Procedia 83:204–210
26. Bohandy J, Kim B, Adrian F (1986) Metal deposition from a supported metal film using an excimer laser. J Appl Phys 60(4):1538–1539
27. Bian J, Zhou L, Wan X, Liu M, Zhu C, Huang Y, Yin Z (2019) Experimental study of laser lift-off of ultra-thin polyimide film for flexible electronics. Sci China Technol Sci 62(2):233–242
28. Delmdahl R, Pätzel R, Brune J (2013) Large-area laser-lift-off processing in microelectronics. Phys Procedia 41:241–248
29. Lee Y, Kim T-S, Min S-Y, Xu W, Jeong S-H, Seo H-K, Lee T-W (2014) Individually position-addressable metal-nanofiber electrodes for large-area electronics. Adv Mater 26(47):8010–8016
30. Lee CH, Kim SJ, Oh Y, Kim MY, Yoon Y-J, Lee H-S (2010) Use of laser lift-off for flexible device applications. J Appl Phys 108(10):102814
31. Wan S, K. MB, Brandon L, Xiaowei Y, Joshua S, Xian H, Heng P (2017) Low-cost manufacturing of bioresorbable conductors by evaporation-condensation-mediated laser printing and sintering of Zn nanoparticles. Adv Mater 29(26):1700172
32. Constantinescu C, Diallo AK, Rapp L, Cremillieu P, Mazurczyk R, Serein-Spirau F, Lère-Porte JP, Delaporte P, Alloncle AP, Videlot-Ackermann C (2015) Laser-induced forward transfer of multi-layered structures for OTFT applications. Appl Surf Sci 336:11–15
33. Eisenhaure J, Kim S (2016) Laser-driven shape memory effect for transfer printing combining parallelism with individual object control. Adv Mater Technol 1(7):1600098
34. Marinov V, Swenson O, Miller R, Sarwar F, Atanasov Y, Semler M, Datta S (2012) Laser-enabled advanced packaging of ultrathin bare dice in flexible substrates. IEEE Trans Compon Packag Manuf Technol 2(4):569–577
35. Saeidpourazar R, Sangid MD, Rogers JA, Ferreira PM (2012) A prototype printer for laser driven micro-transfer printing. J Manuf Process 14(4):416–424
36. Bäuerle D (2011) Laser processing and chemistry. Springer-Verlag, Berlin, Heidelberg
37. Kittel C (2005) Introduction to solid state physics. Wiley, New York
38. Choi Y-Y, Choi K-H, Kim H-K (2011) Characteristics of a transparent GaON sacrificial layer for transfer of devices from sapphire to a flexible substrate. J Electrochem Soc 158(11):J349–J353
39. Tang L, Wang Y, Cheng G, Manfra MJ, Sands TD (2012) Free standing GaN nano membrane by laser lift-off method. In: MRS proceedings, vol 1432, mrss12-1432-g1403-1402
40. Arnold CB, Serra P, Piqué A (2007) Laser direct-write techniques for printing of complex materials. MRS Bull 32(1):23–31
41. Kim K, Kim SY, Lee J-L (2014) Flexible organic light-emitting diodes using a laser lift-off method. J Mater Chem C 2(12):2144–2149
42. Tsakalakos L, Sands T, Carleton E, Yu KM (2003) Modification of (Pb,La)(Zr,Ti)O_3 thin films during pulsed laser liftoff from MgO substrates. J Appl Phys 94(6):4047–4052
43. Kim SJ, Lee HE, Choi H, Kim Y, We JH, Shin JS, Lee KJ, Cho BJ (2016) High-performance flexible thermoelectric power generator using laser multiscanning lift-off process. ACS Nano 10(12):10851–10857
44. Lee HE, Kim S, Ko J, Yeom H-I, Byun C-W, Lee SH, Joe DJ, Im T-H, Park S-HK, Lee KJ (2016) Skin-like oxide thin-film transistors for transparent displays. Adv Func Mater 26(34):6170–6178
45. Seungjun K, Hwan SJ, Hyun LS, Kuk YB, Kwi-Il P, Keon LH, Myunghwan B, Jae LK (2014) Flexible crossbar-structured resistive memory arrays on plastic substrates via inorganic-based laser lift-off. Adv Mater 26(44):7480–7487

46. Phipps C (2007) Laser ablation and its applications, vol 129. Springer, US
47. Dang B, Andry P, Tsang C, Maria J, Polastre R, Trzcinski R, Prabhakar A, Knickerbocker J (2010) CMOS compatible thin wafer processing using temporary mechanical wafer, adhesive and laser release of thin chips/wafers for 3D integration. In: Proceedings 60th electronic components and technology conference (ECTC), 1–4 June 2010, pp 1393–1398
48. Liu Z, Tang P, Huang Y, Yin Z (2014) Experimental estimation of adhesive fracture energy of compliant adhesive tape. In: 15th international conference on electronic packaging technology, 12–15 Aug 2014, pp 842–846
49. MacCarthy N, Wood T, Ameri H, O'Connell D, Alderman J (2006) A laser release method for producing prototype flexible retinal implant devices. Sens Actuators A 132(1):296–301
50. Doany FE, Narayan C (1997) Laser release process to obtain freestanding multilayer metal-polyimide circuits. IBM J Res Dev 41(1–2):151–157
51. Delmdahl R, Fricke M, Fechner B (2014) Laser lift-off systems for flexible-display production. J Inform Disp 15(1):1–4
52. Küper S, Brannon J, Brannon K (1993) Threshold behavior in polyimide photoablation: Single-shot rate measurements and surface-temperature modeling. Appl Phys A 56(1):43–50
53. Babu S, D'Couto G, Egitto F (1992) Excimer laser induced ablation of polyetheretherketone, polyimide, and polytetrafluoroethylene. J Appl Phys 72(2):692–698
54. D'Couto GC, Babu SV (1994) Heat transfer and material removal in pulsed excimer-laser-induced ablation: Pulsewidth dependence. J Appl Phys 76(5):3052–3058

Chapter 9
Vacuum-Based Picking-up and Placing-on

9.1 Introduction

After partly being peeled off by the peeling-off processes discussed in previous chapters [1–3], the chip picking-up plays a critical role in flip-chip packaging to detach the chip intactly from the substrate. An illustration of the chip picking-up process through the vacuum adsorption is shown in Fig. 9.1. It is difficult to peel thin chip directly off from the substrate, which is restricted by a potential risk of the chip damage [4–7]. Whether the thin chip can be detached intactly from the remaining adhesive region or not is determined by the capability of the picking-up with the vacuum adsorption [8, 9]. With the increase of the ratio of the chip length to the chip thickness, the resulting incomplete detachment even leads to the chip failure. In order to determine a proper process window, it is necessary to explore effects of main process parameters on the chip picking-up process. Even though the chip peeling-off and the picking-up process are two consecutive operations, the corresponding models are independent of each other due to the change of external loads and boundary conditions [8]. It is technically desired to understand fundamentally the mechanism of the chip picking-up process.

After the chip picking-up, challenges are presented in the chip placing-on process for the multilayer structure system with different material properties concerning the thermal-mechanical integrity and residual stresses, especially when some layers are brittle inorganic materials and the remaining layers are compliant polymeric materials. The UTCOF assembly is susceptible to be warped by the mismatched residual stress. Some new issues related to the mechanical reliability should be resolved [11, 12]. Especially, serious warpage yielded during the thermal-compressing process has become an obstacle to achieve a reliable interconnection between the chip and thin film circuit, which may even give rise to the interfacial delamination of the bonding assembly [13, 14]. Aiming to reduce the process-induced warpage and achieve high-performance ultra-thin packaging, it is essentially needed to profoundly understand the mechanism and reveal the effects of the mismatch of material properties and structures [15, 16].

Y. Huang et al., *Modeling and Application of Flexible Electronics Packaging*,
https://doi.org/10.1007/978-981-13-3627-0_9

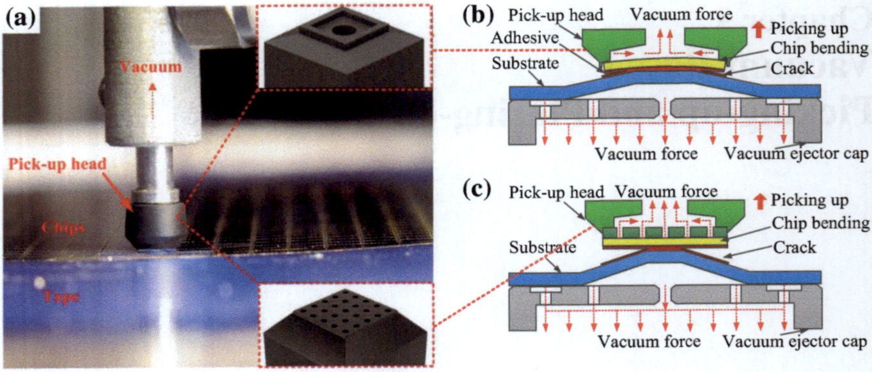

Fig. 9.1 Picking-up of thin chip from the substrate: **a** practical operation of thin-chip transferring with different vacuum picking-up heads, **b** schematic illustration of a large thin-chip bending under the vacuum adsorption through a single hole, **c** the chip is kept flat under the condition of the vacuum adsorption through distributed hole-matrix. It should be noted that the top surface of the vacuum ejector cap is flat to tightly hold the substrate, which can establish a flat contact between the substrate and the cap. © 2015 IEEE. Reprinted, with permission, from Ref. [10]

The outline of this chapter is as follows: A theoretical method concerning the effect of the vacuum adsorption on the chip picking-up process under the hypotheses of constant and variable lengths of the substrate was first proposed. A co-optimization approach was then put forward to reduce the interfacial stress in the R2R processing. An analytical model based on the layerwise theory (LWT) was finally proposed to investigate the thermally induced warpage of the UTCOF assembly.

9.2 Vacuum-Based Chip Picking-up Process

9.2.1 *Mechanical Model*

A mechanical model of the chip picking-up process is depicted in Fig. 9.2. The upward height and crack length at the end of the chip peeling-off process in Fig. 9.2a is termed as $h_{\text{peel-off}}$ and $a_{\text{peel-off}}$, respectively. The VAP below the substrate (p_{sub}) is used to hold the substrate. The chip gripped by the VAP above the chip (p_{chip}) moves up a certain distance Δh_{lift} in the picking-up process, generating a small length of the crack propagation $\Delta a_{\text{pick-up}}$. It is assumed that the crack propagates along the midplane of the adhesive layer [17], and the VAP is uniformly distributed. Obviously, the chip picking-up with the vacuum adsorption can be treated as a further detaching behavior of the chip with an upward height $h_{\text{pick-up}} = h_{\text{peel-off}} + \Delta h_{\text{pick-up}}$ and a crack length $a_{\text{pick-up}} = a_{\text{peel-off}} + \Delta a_{\text{pick-up}}$. Here, the picking-up height $h_{\text{pick-up}}$ is defined at the edges of the chip layer. The bonded region is represented by $l_{\text{adh}} =$

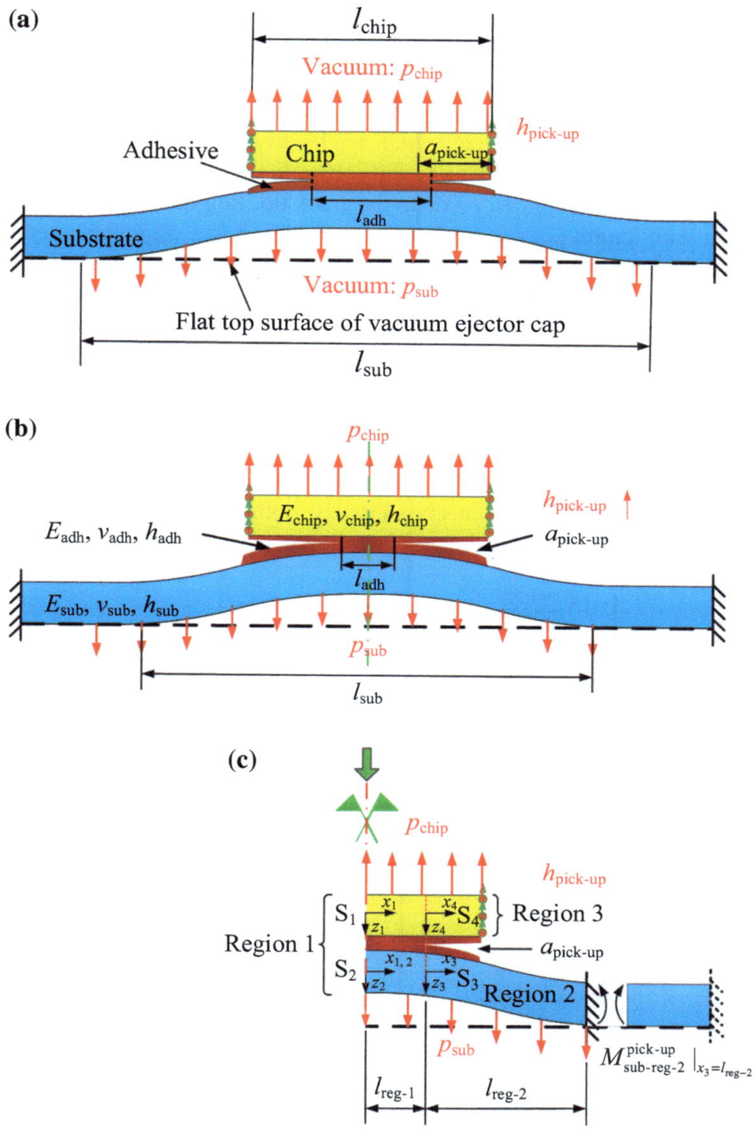

Fig. 9.2 Mechanical model of the chip picking-up process with the vacuum adsorption considering variable substrate length: **a** initial stage, **b** final stage, **c** right-half portion of the symmetric model of (**b**). Considering the flat contact between the substrate and vacuum ejector cap, the effective substrate length from (**a**) to (**b**) decreases with the crack propagation. © 2015 IEEE. Reprinted, with permission, from Ref. [10]

$l_{chip} - 2a_{pick\text{-}up}$. The chip thickness and substrate thickness are, respectively, denoted by h_{chip} and h_{sub}, both of which are much thicker than the adhesive thickness h_{adh}. The materials are taken as isotropic and linear elastic.

In view of the model symmetry, only the right-half portion of the model in Fig. 9.2b is considered, as shown in Fig. 9.2c, which can be divided into three regions. Region 1 is constituted by the chip segment S_1 and substrate segment S_2 with length $l_{reg\text{-}1} = l_{adh}/2$ starting from the symmetry axis to the crack tip. Region 2 is a single substrate segment S_3 with length $l_{reg\text{-}2} = l_{sub}/2 - l_{reg\text{-}1}$. Region 3 is a single chip segment S_4 with length a. Each segment has a local rectangular coordinate (x_i, y_i, z_i) with the origin at midpoint of the left end, where $i = 1, 2, 3, 4$. The stresses and strains in the thickness direction can be assumed to be constant in the model considering that the adhesive layer is much thinner and softer than the substrate and chip layers [18]. Here, each segment is modeled as an elastic beam based on Timoshenko theory. Detailed approach for the construction of the adhesive model can be referred to in Chap. 2.

Region 1 is composed of the chip and substrate segments. Considering that $x_1 = x_2 \in [0, l_{reg\text{-}1}]$, x_1 coordinate can thus be shared by both segments. The corresponding equilibrium equations can be expressed as follows:

$$\begin{cases} \dfrac{\mathrm{d}N_{chip\text{-}reg\text{-}1}^{pick\text{-}up}}{\mathrm{d}x_1} + \tau_{adh} = 0 \\ \dfrac{\mathrm{d}Q_{chip\text{-}reg\text{-}1}^{pick\text{-}up}}{\mathrm{d}x_1} + \left(\sigma_{adh} - p_{chip}\right) = 0 \\ \dfrac{\mathrm{d}M_{chip\text{-}reg\text{-}1}^{pick\text{-}up}}{\mathrm{d}x_1} + h_{chip\text{-}half}\tau_{adh} - Q_{chip\text{-}reg\text{-}1}^{pick\text{-}up} = 0 \end{cases} \tag{9.1}$$

$$\begin{cases} \dfrac{\mathrm{d}N_{sub\text{-}reg\text{-}1}^{pick\text{-}up}}{\mathrm{d}x_1} - \tau_{adh} = 0 \\ \dfrac{\mathrm{d}Q_{sub\text{-}reg\text{-}1}^{pick\text{-}up}}{\mathrm{d}x_1} - (\sigma_{adh} - p_{sub}) = 0 \\ \dfrac{\mathrm{d}M_{sub\text{-}reg\text{-}1}^{pick\text{-}up}}{\mathrm{d}x_1} + h_{sub\text{-}half}\tau_{adh} - Q_{sub\text{-}reg\text{-}1}^{pick\text{-}up} = 0 \end{cases} \tag{9.2}$$

In the analytical model, the peel stress and shear stress of the adhesive layer are treated as the main unknowns. Detailed solution strategy can refer to Chap. 2 [18]. By substituting the internal forces $N_i(x)$, $Q_i(x)$, and $M_i(x)$ into Eqs. (9.1) and (9.2), the following sixth-order constant coefficient nonhomogeneous differential equations for the adhesive stresses can be obtained:

$$\frac{\mathrm{d}^6\sigma_{adh}(x_1)}{\mathrm{d}x_1^6} + \eta_1\frac{\mathrm{d}^4\sigma_{adh}(x_1)}{\mathrm{d}x_1^4} + \eta_2\frac{\mathrm{d}^2\sigma_{adh}(x_1)}{\mathrm{d}x_1^2} + \eta_3\sigma_{adh}(x_1) = f_1 \tag{9.3}$$

$$\frac{\mathrm{d}^7\tau_{adh}(x_1)}{\mathrm{d}x_1^7} + \eta_1\frac{\mathrm{d}^5\tau_{adh}(x_1)}{\mathrm{d}x_1^5} + \eta_2\frac{\mathrm{d}^3\tau_{adh}(x_1)}{\mathrm{d}x_1^3} + \eta_3\frac{\mathrm{d}\tau_{adh}(x_1)}{\mathrm{d}x_1} = f_2 \tag{9.4}$$

where $f_1 = -k_\sigma k_\tau[(a_{\text{chip}} + a_{\text{sub}})(d_{\text{chip}}p_{\text{chip}} + d_{\text{sub}}p_{\text{sub}}) + d_{\text{chip}}d_{\text{sub}}(h_{\text{chip-half}} + h_{\text{sub-half}})(h_{\text{chip-half}}p_{\text{sub}} + h_{\text{sub-half}}p_{\text{chip}})]$, $f_2 = k_\sigma k_\tau d_{\text{chip}}d_{\text{sub}}(h_{\text{chip-half}} + h_{\text{sub-half}})(p_{\text{sub}} - p_{\text{chip}})$, $\eta_1 = -k_\sigma(c_{\text{chip}} + c_{\text{sub}}) - k_\tau(a_{\text{chip}} + a_{\text{sub}} + d_{\text{chip}}h^2_{\text{chip-half}} + d_{\text{sub}}h^2_{\text{sub-half}})$, $\eta_2 = k_\sigma k_\tau(c_{\text{chip}} + c_{\text{sub}})(a_{\text{chip}} + a_{\text{sub}} + d_{\text{chip}}h^2_{\text{chip-half}} + d_{\text{sub}}h^2_{\text{sub-half}}) + k_\sigma(d_{\text{chip}} + d_{\text{sub}})$, $\eta_3 = -k_\sigma k_\tau[d_{\text{chip}}d_{\text{sub}}(h_{\text{chip-half}} + h_{\text{sub-half}})^2 + (a_{\text{chip}} + a_{\text{sub}})(d_{\text{chip}} + d_{\text{sub}})]$.

After solving the characteristic roots of Eqs. (9.3) and (9.4), the expressions for the adhesive peel stress and shear stress can be given by

$$\sigma_{\text{adh}}(x_1) = \sum_{n=1}^{6} F_n \exp(\lambda_n x_1) + \frac{f_1}{\eta_3} \tag{9.5}$$

$$\tau_{\text{adh}}(x_1) = -\frac{\sum_{n=1}^{6} F_n\left(\frac{\lambda_n^3}{k_\sigma} - (c_{\text{chip}} + c_{\text{sub}})\lambda_n + \frac{(d_{\text{chip}} + d_{\text{sub}})}{\lambda_n}\right)\exp(\lambda_n x_1)}{(d_{\text{chip}}h_{\text{chip-half}} - d_{\text{sub}}h_{\text{sub-half}})} + \frac{f_2}{\eta_3}x_1 + F_7 \tag{9.6}$$

where F_1, F_2, F_3, F_4, F_5, F_6, and F_7 are integration constants to be determined. λ_1, λ_2, λ_3, λ_4, λ_5, and λ_6 are the characteristic roots of Eq. (9.3).

The internal forces of the chip and substrate layers in Region 1 can then be obtained by integrating Eqs. (9.5) and (9.6), introducing six more integration constants, F_8, F_9, F_{10}, F_{11}, F_{12}, and F_{13}. The displacements can be deduced by further integrating internal forces, introducing another six more integration constants, F_{14}, F_{15}, F_{16}, F_{17}, F_{18}, and F_{19}. Therefore, there exist 21 integration constants that should be determined in Region 1.

As to Regions 2 and 3, the corresponding equilibrium equations can be expressed by Eqs. (9.7) and (9.8), respectively

$$\frac{dN^{\text{pick-up}}_{\text{sub-reg-2}}}{dx_3} = 0,\quad \frac{dQ^{\text{pick-up}}_{\text{sub-reg-2}}}{dx_3} + p_{\text{sub}} = 0,\quad \frac{dM^{\text{pick-up}}_{\text{sub-reg-2}}}{dx_3} - Q^{\text{pick-up}}_{\text{sub-reg-2}} = 0 \tag{9.7}$$

$$\frac{dN^{\text{pick-up}}_{\text{chip-reg-3}}}{dx_4} = 0,\quad \frac{dQ^{\text{pick-up}}_{\text{chip-reg-3}}}{dx_4} - p_{\text{chip}} = 0,\quad \frac{dM^{\text{pick-up}}_{\text{chip-reg-3}}}{dx_4} - Q^{\text{pick-up}}_{\text{chip-reg-3}} = 0 \tag{9.8}$$

Analytical solutions can be deduced directly by integrating with respect to x_3 and x_4, separately. The internal forces and displacements in each region introduce seven new integration constants, which are termed as F_{20}, F_{21}, F_{22}, F_{23}, F_{24}, F_{25}, F_{26} for Region 2, and F_{27}, F_{28}, F_{29}, F_{30}, F_{31}, F_{32}, F_{33} for Region 3.

A practical chip picking-up process with the vacuum adsorption is depicted in Fig. 9.3. With the increase in the picking-up displacement and the crack length, the increasing contact region between the substrate and the vacuum adsorption cap gives rise to a decreasing effective substrate length, as seen in Fig. 9.2a, b.

An initial substrate length is given in the beginning. In fact, if the flat contact is established under the condition of the vacuum adsorption, there is no bending moment at ends of contact region, i.e., $M^{\text{pick-up}}_{\text{sub-reg-2}}(l_{\text{reg-2}}) = 0$, as shown in Fig. 9.2c.

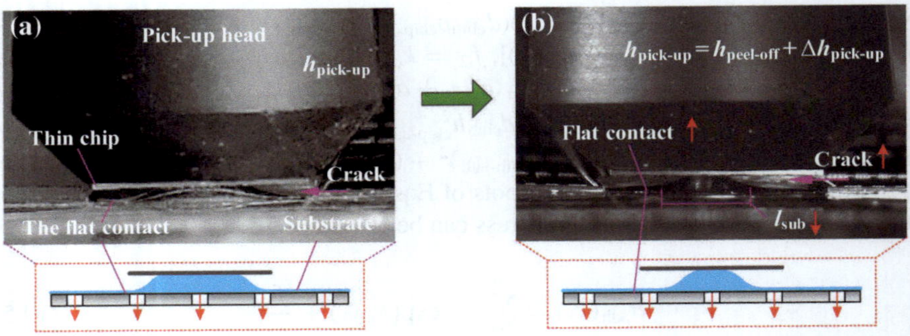

Fig. 9.3 Practical chip picking-up process with the vacuum adsorption: **a** initial stage, **b** final stage, where a flat contact between the bottom surface of the substrate and the top flat surface of the vacuum ejector cap is formed. Since the contact region increases as the crack propagates from (**a**) to (**b**), the effective substrate length that contributes to the crack propagation decreases meanwhile. © 2015 IEEE. Reprinted, with permission, from Ref. [10]

That is, if $M^{\text{pick-up}}_{\text{sub-reg-2}}(l_{\text{reg-2}}) \neq 0$, i.e., $M^{\text{pick-up}}_{\text{sub-reg-2}}(l_{\text{reg-2}}) > 0$ or $M^{\text{pick-up}}_{\text{sub-reg-2}}(l_{\text{reg-2}}) < 0$, the unbalance of bending moment will lead to the reduction or enlargement of the contact region, respectively. Therefore, $M^{\text{pick-up}}_{\text{sub-reg-2}}(l_{\text{reg-2}}) = 0$ can be treated as the criterion to search out the edges of the effective substrate length. $M^{\text{pick-up}}_{\text{sub-reg-2}}(l_{\text{reg-2}})$ can be obtained from Eq. (9.9), which is a function of the peel stress and shear stress in the adhesive layer. A simple bisection algorithm can be used to search out the effective substrate length, as described by the following flow chart in Fig. 9.4.

$$\begin{aligned} M^{\text{pick-up}}_{\text{sub-reg-2}}(x_3)\big|_{x_3=l_{\text{reg-2}}} &= l_{\text{reg-2}} \int_0^{l_{\text{reg-1}}} [\sigma_{\text{adh}}(x_1) - p_{\text{sub}}]\,\mathrm{d}x_1 \\ &+ \int_0^{l_{\text{reg-1}}} \left\{ h_{\text{sub}} \tau_{\text{adh}}(x_1) - \int_0^{x_1} [\sigma_{\text{adh}}(x_1) - p_{\text{sub}}]\mathrm{d}x_1 \right\} \mathrm{d}x_1 \\ &- \frac{1}{2} p_{\text{sub}} l^2_{\text{reg-2}} \end{aligned} \tag{9.9}$$

9.2.2 Boundary and Continuity Conditions

(a) Symmetric conditions at the left-side end of segments S_1 and S_2

$$u^{\text{pick-up}}_{\text{chip-reg-1}}\big|_{x_1=0} = 0, \quad \phi^{\text{pick-up}}_{\text{chip-reg-1}}\big|_{x_1=0} = 0, \quad Q^{\text{pick-up}}_{\text{chip-reg-1}}\big|_{x_1=0} = 0 \tag{9.10}$$

$$u^{\text{pick-up}}_{\text{sub-reg-1}}\big|_{x_1=0} = 0, \quad \phi^{\text{pick-up}}_{\text{sub-reg-1}}\big|_{x_1=0} = 0, \quad Q^{\text{pick-up}}_{\text{sub-reg-1}}\big|_{x_1=0} = 0 \tag{9.11}$$

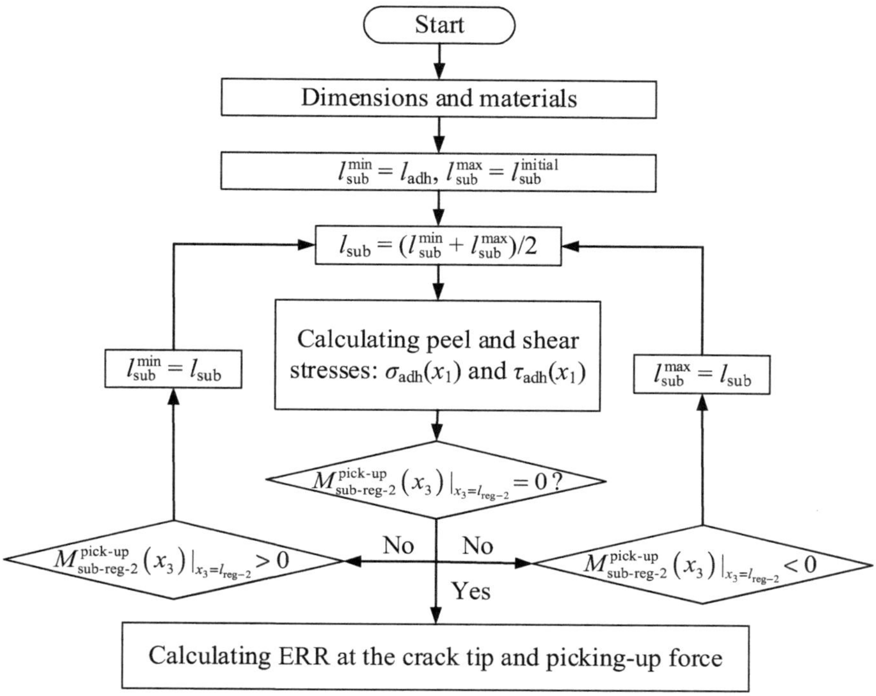

Fig. 9.4 Flow chart of the bisection algorithm used to search out the effective substrate length with the criterion $M_{sub\text{-}reg\text{-}2}^{pick\text{-}up}(l_{reg\text{-}2}) = 0$. Here, l_{sub}^{min} and l_{sub}^{max} standing for the minimum and maximum substrate length are, respectively, set as l_{adh} and $l_{sub}^{initial}$ in the beginning, where the latter represents the initial substrate length satisfying $l_{sub}^{initial} > l_{adh}$. If there exists $M_{sub\text{-}reg\text{-}2}^{pick\text{-}up} > 0$ or $M_{sub\text{-}reg\text{-}2}^{pick\text{-}up} < 0$, l_{sub}^{min} or l_{sub}^{max} should be replaced by the current $l_{sub} = (l_{sub}^{min} + l_{sub}^{max})/2$ and return to the loop to search a new l_{sub}. The loop is skipped out when $M_{sub\text{-}reg\text{-}2}^{pick\text{-}up} = 0$, which indicates that the current substrate length is the desired effective substrate length, and corresponding stresses can be used to calculate the interfacial total ERR at the crack tip and the picking-up force.

(b) Continuity conditions of forces and displacements at the intersection between segments S_2 and S_3

$$\begin{cases} u_{sub\text{-}reg\text{-}1}^{pick\text{-}up}\big|_{x_1=l_{reg\text{-}1}} = u_{sub\text{-}reg\text{-}2}^{pick\text{-}up}\big|_{x_3=0} \\ \phi_{sub\text{-}reg\text{-}1}^{pick\text{-}up}\big|_{x_1=l_{reg\text{-}1}} = \phi_{sub\text{-}reg\text{-}2}^{pick\text{-}up}\big|_{x_3=0} \\ w_{sub\text{-}reg\text{-}1}^{pick\text{-}up}\big|_{x_1=l_{reg\text{-}1}} = w_{sub\text{-}reg\text{-}2}^{pick\text{-}up}\big|_{x_3=0} \end{cases} \quad (9.12)$$

$$\begin{cases} N_{sub\text{-}reg\text{-}1}^{pick\text{-}up}\big|_{x_1=l_{reg\text{-}1}} = N_{sub\text{-}reg\text{-}2}^{pick\text{-}up}\big|_{x_3=0} \\ Q_{sub\text{-}reg\text{-}1}^{pick\text{-}up}\big|_{x_1=l_{reg\text{-}1}} = Q_{sub\text{-}reg\text{-}2}^{pick\text{-}up}\big|_{x_3=0} \\ M_{sub\text{-}reg\text{-}1}^{pick\text{-}up}\big|_{x_1=l_{reg\text{-}1}} = M_{sub\text{-}reg\text{-}2}^{pick\text{-}up}\big|_{x_3=0} \end{cases} \quad (9.13)$$

(c) Continuity conditions of forces and displacements at the intersection between segments S_1 and S_4

$$\begin{cases} u_{\text{chip-reg-1}}^{\text{pick-up}}\big|_{x_1=l_{\text{reg-1}}} = u_{\text{chip-reg-3}}^{\text{pick-up}}\big|_{x_4=0} \\ \phi_{\text{chip-reg-1}}^{\text{pick-up}}\big|_{x_1=l_{\text{reg-1}}} = \phi_{\text{chip-reg-3}}^{\text{pick-up}}\big|_{x_4=0} \\ w_{\text{chip-reg-1}}^{\text{pick-up}}\big|_{x_1=l_{\text{reg-1}}} = w_{\text{chip-reg-3}}^{\text{pick-up}}\big|_{x_4=0} \end{cases} \tag{9.14}$$

$$\begin{cases} N_{\text{chip-reg-1}}^{\text{pick-up}}\big|_{x_1=l_{\text{reg-1}}} = N_{\text{chip-reg-3}}^{\text{pick-up}}\big|_{x_4=0} \\ Q_{\text{chip-reg-1}}^{\text{pick-up}}\big|_{x_1=l_{\text{reg-1}}} = Q_{\text{chip-reg-3}}^{\text{pick-up}}\big|_{x_4=0} \\ M_{\text{chip-reg-1}}^{\text{pick-up}}\big|_{x_1=l_{\text{reg-1}}} = M_{\text{chip-reg-3}}^{\text{pick-up}}\big|_{x_4=0} \end{cases} \tag{9.15}$$

(d) Clamped boundary conditions at the right-side end of segment S_3

$$u_{\text{sub-reg-2}}^{\text{pick-up}}\big|_{x_3=l_{\text{reg-2}}} = \phi_{\text{sub-reg-2}}^{\text{pick-up}}\big|_{x_3=l_{\text{reg-2}}} = w_{\text{sub-reg-2}}^{\text{pick-up}}\big|_{x_3=l_{\text{reg-2}}} = 0 \tag{9.16}$$

(e) Upward displacement applied at the right-side end of the chip layer

$$w_{\text{chip-reg-3}}^{\text{pick-up}}\big|_{x_4=a_{\text{pick-up}}} = h_{\text{pick-up}} \tag{9.17}$$

(f) Free traction along x axis and free bending moment at the right-side end of segment S_4

$$N_{\text{chip-reg-3}}^{\text{pick-up}}\big|_{x_4=a_{\text{pick-up}}} = M_{\text{chip-reg-3}}^{\text{pick-up}}\big|_{x_4=a_{\text{pick-up}}} = 0 \tag{9.18}$$

9.2.3 Determination of Integration Constants

Extra nine equations can be used to describe relationships among the nonindependent integration constants. Combining the 24 boundary conditions, there are 33 equations to solve the 33 integration constants. As to the adhesive stresses, the first six integration constants F_1, F_2, F_3, F_4, F_5, and F_6, can be obtained by using MATLAB software to solve the following linear equation set

$$\sum_{n=1}^{6} \frac{F_n}{\lambda_n} = 0 \tag{9.19}$$

$$\sum_{n=1}^{6} \frac{F_n}{\lambda_n^3} = 0 \tag{9.20}$$

$$\sum_{n=1}^{6} F_n \lambda_n = 0 \tag{9.21}$$

$$\sum_{n=1}^{6} F_n \left[\begin{array}{l} \alpha_1 \left(\frac{\lambda_n^2}{k_\sigma} - c_{\text{chip}} - c_{\text{sub}} \right) \\ + \frac{\alpha_2}{k_\sigma} \lambda_n + \frac{\alpha_3}{\lambda_n} + \frac{\alpha_4}{\lambda_n^3} - l_{\text{reg-1}} \frac{\alpha_4}{\lambda_n^2} \end{array} \right] \exp(\lambda_n l_{\text{reg-1}}) = \alpha_5 + \alpha_6 + \alpha_7 \quad (9.22)$$

$$\sum_{n=1}^{6} F_n \left[\begin{array}{l} \beta_2 \left(\frac{\lambda_n^2}{k_\sigma} - c_{\text{chip}} - c_{\text{sub}} \right) \\ - \frac{2a_{\text{sub}} d_{\text{sub}}}{l_{\text{sub}}} \frac{1}{\lambda_n^3} + \frac{\beta_3}{\lambda_n^2} - \frac{\beta_3}{\lambda_n} \end{array} \right] \exp(\lambda_n l_{\text{reg-1}}) = \beta_5 + \beta_6 + \beta_7 \quad (9.23)$$

$$\sum_{n=1}^{6} F_n \left[\begin{array}{l} \frac{1}{k_\sigma} (\gamma_1 \lambda_n^2 - a_{\text{pick-up}} \lambda_n - 1) \\ -c_{\text{chip}} - c_{\text{sub}} + \frac{\gamma_2}{\lambda_n} \end{array} \right] \exp(\lambda_n l_{\text{reg-1}}) = \gamma_3 + \gamma_4 + \gamma_5 + \gamma_6 + h_{\text{pick-up}} \quad (9.24)$$

where $\alpha_1 = -(d_{\text{chip}} h_{\text{chip-half}} l_{\text{sub-half}} - d_{\text{sub}} h_{\text{sub-half}} \, a_{\text{pick-up}})/(d_{\text{chip}} h_{\text{chip-half}} - d_{\text{sub}} h_{\text{sub-half}}) l_{\text{sub-half}}$, $\alpha_2 = d_{\text{sub}} h_{\text{sub-half}}/(d_{\text{chip}} h_{\text{chip-half}} - d_{\text{sub}} h_{\text{sub-half}}) l_{\text{sub-half}}$, $\alpha_3 = [2 l_{\text{sub-half}} a_{\text{pick-up}} d_{\text{chip}} + l_{\text{leg-2}}^2 d_{\text{sub}} - 2\alpha_2 l_{\text{sub-half}} (c_{\text{chip}} + c_{\text{sub}})]/2 l_{\text{sub-half}}$, $\alpha_4 = d_{\text{chip}} d_{\text{sub}} (h_{\text{chip-half}} + h_{\text{sub-half}})/(d_{\text{chip}} h_{\text{chip-half}} - d_{\text{sub}} h_{\text{sub-half}}) l_{\text{sub-half}}$, $\alpha_5 = -f_1 [6 l_{\text{sub-haf}} (c_{\text{chip}} + c_{\text{sub}}) + 3 l_{\text{sub-haf}} d_{\text{chip}} l_{\text{reg-1}} (l_{\text{chip-half}} + a_{\text{pick-up}}) + d_{\text{sub}} (l_{\text{sub}}^3 - l_{\text{leg-2}}^3]/6 l_{\text{sub-half}} \eta_3$, $\alpha_6 = f_2 (l_{\text{sub-half}} l_{\text{leg-1}}^2 d_{\text{sub}} h_{\text{sub-half}} + 2 l_{\text{reg-2}} l_{\text{leg-1}}^2 d_{\text{sub}} h_{\text{sub-half}} - 3 l_{\text{sub-half}} l_{\text{leg-1}}^2 d_{\text{chip}} h_{\text{chip-half}})/6 l_{\text{sub-half}} \eta_3$, $\alpha_7 = (c_{\text{sub}} + d_{\text{sub}} h_{\text{sub-half}}^2/6) p_{\text{sub}} + (c_{\text{chip}} + d_{\text{chip}} l_{\text{chip-half}}^2/2) p_{\text{chip}}$, $\beta_1 = d_{\text{chip}} (a_{\text{sub}} + d_{\text{sub}} h_{\text{chip-half}} h_{\text{sub-half}} + d_{\text{sub}} h_{\text{sub-half}}^2)$, $\beta_2 = (a_{\text{chip}} d_{\text{sub}} h_{\text{sub-half}} + \beta_1 h_{\text{chip-half}})/(d_{\text{chip}} h_{\text{chip-half}} - d_{\text{sub}} h_{\text{sub-half}})$, $\beta_3 = a_{\text{sub}} d_{\text{sub}} l_{\text{reg-1}}/l_{\text{sub-half}} - d_{\text{sub}} h_{\text{sub-half}} \eta_3/k_\sigma k_\tau (d_{\text{chip}} h_{\text{chip-half}} - d_{\text{sub}} h_{\text{sub-half}})$, $\beta_4 = a_{\text{pick-up}} \beta_1 + a_{\text{sub}} d_{\text{sub}} l_{\text{leg-2}}^2/2 l_{\text{sub-half}}$, $\beta_5 = [f_2 d_{\text{sub}} h_{\text{sub-half}} (2 + k_\tau a_{\text{chip}} l_{\text{leg-1}}^2) + k_\tau f_2 h_{\text{chip-half}} l_{\text{leg-1}}^2 \beta_1]/2 k_\tau \eta_3$, $\beta_6 = f_1 [6 l_{\text{sub-half}} (d_{\text{sub}} h_{\text{sub-half}}^2 + a_{\text{sub}})(c_{\text{chip}} + c_{\text{sub}}) + l_{\text{sub-half}} \beta_1 (3 l_{\text{leg-1}}^2 + 6 a_{\text{pick-up}} l_{\text{reg-1}}) + a_{\text{sub}} d_{\text{sub}} (l_{\text{sub-half}}^3 - l_{\text{reg-2}}^2)]/6 l_{\text{sub-half}} \eta_3$, $\beta_7 = -(d_{\text{sub}} h_{\text{sub-half}}^2 + a_{\text{sub}})(c_{\text{chip}} p_{\text{chip}} + c_{\text{sub}} p_{\text{sub}}) - \beta_1 l_{\text{chip-half}}^2 p_{\text{chip}}/2 - a_{\text{sub}} d_{\text{sub}} l_{\text{sub-half}}^2 p_{\text{sub}}/6$, $\gamma_1 = l_{\text{leg-2}}^2/2 - a_{\text{pick-up}} l_{\text{reg-2}}$, $\gamma_2 = a_{\text{pick-up}} l_{\text{leg-2}}^2 (d_{\text{sub}} - d_{\text{chip}})/2 - (d_{\text{chip}} a_{\text{pick-up}}^3 + d_{\text{sub}} l_{\text{leg-2}}^3)/3 + c_{\text{sub}} (a_{\text{pick-up}} - l_{\text{reg-2}}) + a_{\text{pick-up}}^2 d_{\text{chip}} l_{\text{reg-2}}$, $\gamma_3 = p_{\text{chip}} [d_{\text{chip}} l_{\text{chip-half}}^4 - 6 (l_{\text{reg-2}} - a_{\text{pick-up}})^2 d_{\text{chip}} l_{\text{chip-half}}^2 - 12 (l_{\text{reg-2}} - a_{\text{pick-up}})^2 c_{\text{chip}}]/24$, $\gamma_4 = p_{\text{sub}} [(l_{\text{chip-half}} - 3 l_{\text{reg-2}} + 3 a_{\text{pick-up}}) d_{\text{sub}} l_{\text{sub-half}}^3 - 24 l_{\text{sub-half}} (l_{\text{reg-2}} - a_{\text{pick-up}})^2 c_{\text{sub}}]/24$, $\gamma_5 = f_1 [2 - k_\sigma (2 l_{\text{chip-half}} l_{\text{reg-2}} + l_{\text{reg-1}}^2 - l_{\text{sub-half}}^2) c_{\text{sub}} + 2 k_\sigma (l_{\text{reg-1}} \gamma_2 - c_{\text{chip}} \gamma_1)]/2 k_\sigma \eta_3$, $\gamma_6 = -(l_{\text{reg-1}}^4 + 4 l_{\text{reg-1}}^3 a_{\text{pick-up}} - 12 l_{\text{reg-1}}^2 \gamma_1)[(d_{\text{chip}} + d_{\text{sub}}) f_1 + (d_{\text{chip}} h_{\text{chip-half}} - d_{\text{sub}} h_{\text{sub-half}}) f_2]/24 \eta_3$.

9.2.4 Determination of Picking-up Criterion

The crack propagation in the adhesive layer can be predicted by comparing the total ERR at the crack tip $G_{\text{adh}}^{\text{crack-tip}}$ with the adhesive fracture energy G_{adh}^c. Additionally, in order to grip the chip, the picking-up force $P_{\text{pick-up}}$ should be smaller than the vacuum adsorption force above the chip $p_{\text{chip}} l_{\text{chip}}$. Thus, the criteria for the chip picking-up with the vacuum adsorption can be determined as

$$G_{\text{adh}}^{\text{crack-tip}} > G_{\text{adh}}^{c}, \text{ and } P_{\text{pick-up}} < p_{\text{chip}} l_{\text{chip}} \tag{9.25}$$

Detailed approach on the determination of the total ERR at the crack tip can refer to Chap. 2. The picking-up force is related to the VAP and reaction force caused by the frame edge of the picking-up head, and is equal to the adhesion force based on the static equilibrium condition.

$$P_{\text{pick-up}} = 2\int_0^{l_{\text{reg-1}}} \sigma_{\text{adh}}(x)\text{d}x_1 = 2\left[\sum_{n=1}^{6} \frac{F_n}{\lambda_n} \exp\left(\lambda_n l_{\text{reg-1}}\right) + \frac{f_1}{\eta_3} l_{\text{reg-1}}\right] \tag{9.26}$$

In order to verify the accuracy of the analytical ERR of interfacial peeling, results of the finite element approach with the virtual crack closure technique (VCCT) are obtained by using Abaqus software. In view of the convergence of the ERR calculation, a small virtual crack propagation length of 1 μm is adopted.

9.2.5 Effect of Ratio of Adhesive Length to Chip Length

The effect of the ratio of the adhesive length to the chip length on the total ERR at the crack tip and the picking-up force under various conditions of chip thicknesses is shown in Fig. 9.5, where $h_{\text{pick-up}} = 0.5$ mm, $p_{\text{chip}} = 90$ kPa, and $p_{\text{sub}} = 50$ kPa. Here, the hypothesis of the constant substrate length is adopted. Good agreement can be found between the theoretical analysis and finite element approach. A fast decrease in the total ERR at the crack tip can be first observed in Fig. 9.5a with the decrease of $l_{\text{adh}}/l_{\text{chip}}$. The decreasing trend slows down obviously as $l_{\text{adh}}/l_{\text{chip}}$ further decreases. Results indicate that the substrate tension plays a dominant role in the chip delamination at first. As the crack propagates and the length of Region 2 increases, the tension under the same condition decreases gradually. However, a constant VAP always exists below the substrate, which generates a nearly flat variation trend of the total ERR at the crack tip. Further slow decrease of the total ERR at the crack tip is due to the occurrence of the convergent debonding [19]. The crack propagation will stop when $G_{\text{adh}}^{\text{crack-tip}} < G_{\text{adh}}^{c}$. If $G_{\text{adh}}^{c} = 0.0263$ N/mm, the crack will stop propagating when $l_{\text{adh}}/l_{\text{chip}} = 0.4$. If $G_{\text{adh}}^{c} < 0.02$ N/mm, the chip can be detached completely with the given picking-up displacement.

Moreover, the thinner chip corresponds to a smaller total ERR at the crack tip when $l_{\text{adh}}/l_{\text{chip}} > 0.25$, but the difference of ERRs between thinner and thicker chip gets smaller with the crack propagation, indicating that it is tougher to detach a thinner chip from the substrate to an anticipated extent, especially for the initial cracking. When $l_{\text{adh}}/l_{\text{chip}} < 0.25$, as shown in the partial enlargement in Fig. 9.5a, the total ERR at the crack tip of the thinner chip turns to be slightly larger than that of the

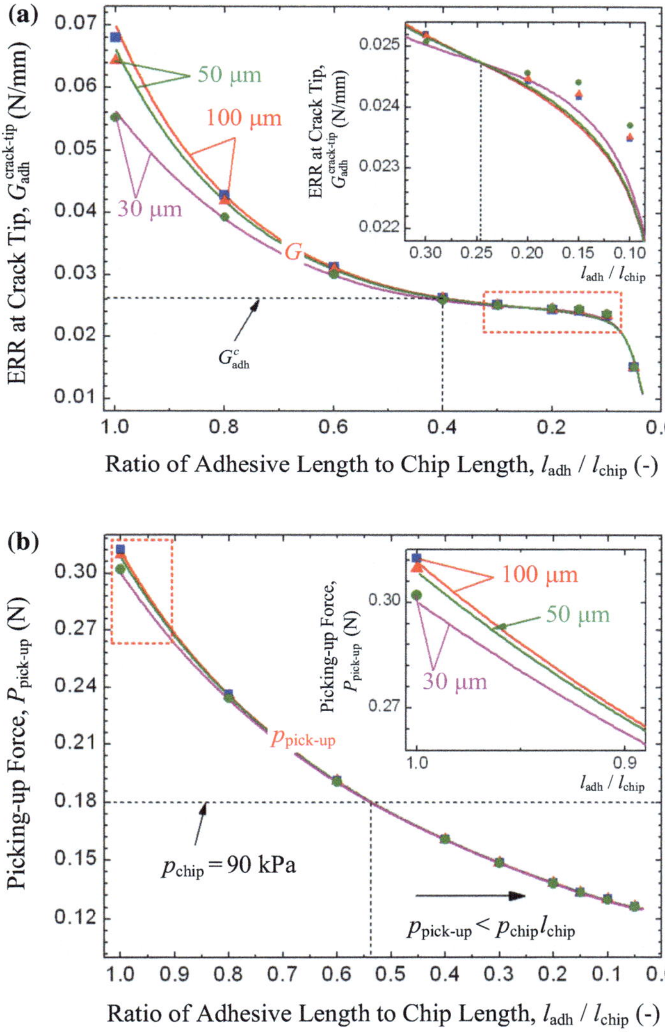

Fig. 9.5 The effect of the ratio of the length of the adhesive region to the chip length: **a** result of the interfacial total ERR at the crack tip, **b** result of the picking-up force. Theoretical predictions and numerical results are represented by solid lines and scatter plots, respectively. Reprinted from Ref. [20] by permission of Taylor & Francis Ltd.

thicker one, indicating that the delamination of the thinner chip is relatively easier. The trend is primarily due to that the thinner chip is more flexible, which suffers from the warpage more easily under conditions of the vacuum adsorption and the picking-up force. Besides, the trend also shows that the effect of the chip thickness is negligible when $l_{adh}/l_{chip} < 0.4$.

As shown in Fig. 9.5b, the picking-up force decreases with the decrease of l_{adh}/l_{chip}, and the trend is almost independent of the chip thickness. In practice, the smaller picking-up force is helpful to hold the chip with the vacuum adsorption reliably, indicating that the size of bonded region should be limited for a reliable chip picking-up process. For example, if $p_{chip} = 90$ kPa, it is suggested that the size of remaining bonded region after peeling-off should be controlled less than about 50% of the chip length. Besides, according to Eq. (9.25), the appropriate length of the bonded region can be determined by comparing both results of the total ERR at the crack tip and the picking-up force. In this case, the appropriate length should be controlled less than 40% of the chip length.

9.2.6 *Effect of Ratio of VAP Below Substrate*

In practice, the VAP is one of the critical parameters affecting the performance of the chip picking-up [8, 21]. The total ERR at the crack tip and the picking-up force versus the VAP below the substrate for various chip thicknesses and VAP above the chip is shown in Fig. 9.6, where $l_{adh}/l_{chip} = 0.2$, $p_{chip} = 70$ and 90 kPa, and $h_{pick\text{-}up} =$ 0.5 mm. Here, the hypothesis of the constant substrate length is adopted. As can be seen, the theoretical predictions agree well with results of the finite element analysis. In Fig. 9.6a, the total ERR at the crack tip for the given length of the adhesive layer enlarges with the increase of the VAP below the substrate. In addition, the thinner chip leads to a larger total ERR at the crack tip, especially for the higher VAP below the substrate. It can be concluded that the higher VAP below the substrate facilitates the chip delamination. Moreover, the total ERR at the crack tip for the 30-μm-thick chip decreases very slightly as the VAP above the chip increases from 70 to 90 kPa, which indicates that the VAP above the chip has little effect on the chip delamination.

As illustrated in Fig. 9.6b, the picking-up force increases with the increasing VAP below the substrate. It is evident that the variation of the chip thickness as well as the VAP above the chip does not significantly change the picking-up force. Results indicate that the VAP below the substrate can be adjusted to obtain the appropriate picking-up force. There is a range limitation of the VAP below the substrate to achieve the purpose of reliable chip picking-up. The criterion for holding the chip is $P_{pick\text{-}up} < p_{chip}l_{chip}$. If the VAP above the chip is fixed, the upper range of the VAP below the substrate can be determined from Fig. 9.6b. The criterion for holding the substrate throughout the process is $p_{sub}l_{sub} > P_{pick\text{-}up}$. The maximum value of the picking-up force cannot exceed the vacuum adsorption force above the chip, $p_{sub}l_{sub} > p_{chip}l_{chip}$ should thus be satisfied. As shown in Fig. 9.6b, if $p_{chip} = 70$ kPa, the tunable range of the VAP below the substrate is 35 kPa $< p_{sub} <$ 50 kPa. When $p_{chip} = 90$ kPa, the

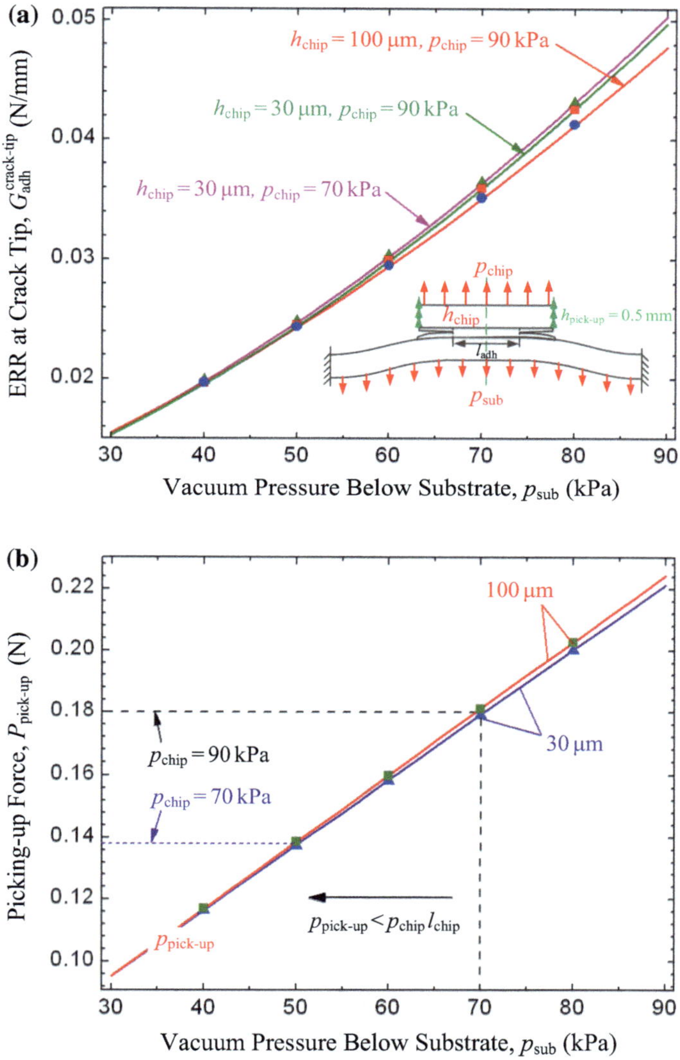

Fig. 9.6 The effect of the VAP below the substrate: **a** result of the interfacial total ERR at the crack tip, **b** result of the picking-up force. Theoretical predictions and numerical results are represented by solid lines and scatter plots, respectively. Reprinted from Ref. [20] by permission of Taylor & Francis Ltd.

corresponding range is 45 kPa $< p_{sub} <$ 70 kPa. Therefore, even though the higher VAP below the substrate facilitates the chip delamination, the VAP should still be controlled within a proper range, otherwise the incomplete chip detachment should occur.

9.2.7 Effect of Crack Propagation

The chip-adhesive-substrate structure characterized by initial geometric dimensions and material properties are set as: $E_{chip} = 129$ GPa, $\nu_{chip} = 0.28$, $h_{chip} = 50$ μm, $E_{adh} = 40$ MPa, $\nu_{adh} = 0.4$, $h_{adh} = 5$ μm, $E_{sub} = 160$ MPa, $\nu_{sub} = 0.45$, and $h_{sub} = 0.1$ mm. It is assumed that the thin chip can be kept flat during the picking-up process through the application of a picking-up head (flat plate with distributed hole-matrix). Here, the hypothesis of the variable substrate length is adopted. The mechanism of the chip picking-up considering flat contact between the substrate and the vacuum adsorption cap is investigated, where $l_{chip} = 5$ mm, and $h_{chip} = 50$ μm, and the initial length of the substrate is 8.5 and 7 mm separately for comparison. Additionally, the VAP below the substrate is given at 60 and 0 kPa to represent the condition with and without the vacuum adsorption.

The total ERR at the crack tip and the picking-up force versus the crack propagation is shown in Fig. 9.7, where $h_{pick\text{-}up} = 0.5$ mm. Theoretical predictions and numerical simulations are represented by solid lines and scatter plots, respectively. The theoretical results agree very well with the finite element simulations. The displacement contour in Fig. 9.7a describes the contact behavior between the substrate and the top surface of the vacuum adsorption cap. Under the condition of $p_{sub} =$ 60 kPa, the compliant substrate with an initial length of 8.5 mm is large enough to establish a flat contact at the initial stage of the crack propagation. The effective substrate length decreases as the crack propagates, while the length of the substrate segment S_3 is constant at about 1.72 mm, and the total ERR at the crack tip is almost constant. The constant length of $l_{reg\text{-}2}$ is termed as $l_{reg\text{-}2}^{constant}$, which is related to the substrate thickness, material properties of the substrate, the VAP below the substrate, and the picking-up displacement. While for the compliant substrate with an initial length of 7 mm, there is a rapid decrease of the total ERR at the crack tip at the initial stage of the crack propagation due to the extra substrate tension before establishing a flat contact. Once the contact is established, the variation trend of the total ERR at the crack tip changes back to the flat contact mode. Therefore, the length of ERR-decreased region can be calculated as $l_{reg\text{-}2}^{constant} - (l_{sub}^{initial} - l_{chip})/2$. Moreover, the decreasing trend is probably due to the convergent debonding, which has been described in Ref. [19].

It can be seen from Fig. 9.7b that the picking-up force decreases linearly with the crack propagation when $l_{sub}^{initial} = 8.5$ mm. As to $l_{sub}^{initial} = 7$ mm, the picking-up force also decreases from a larger value at the initial stage before forming the flat contact. It indicates that the largest picking-up force occurs at the initial cracking, and may give rise to the risk of the chip cracking. Moreover, the smaller initial

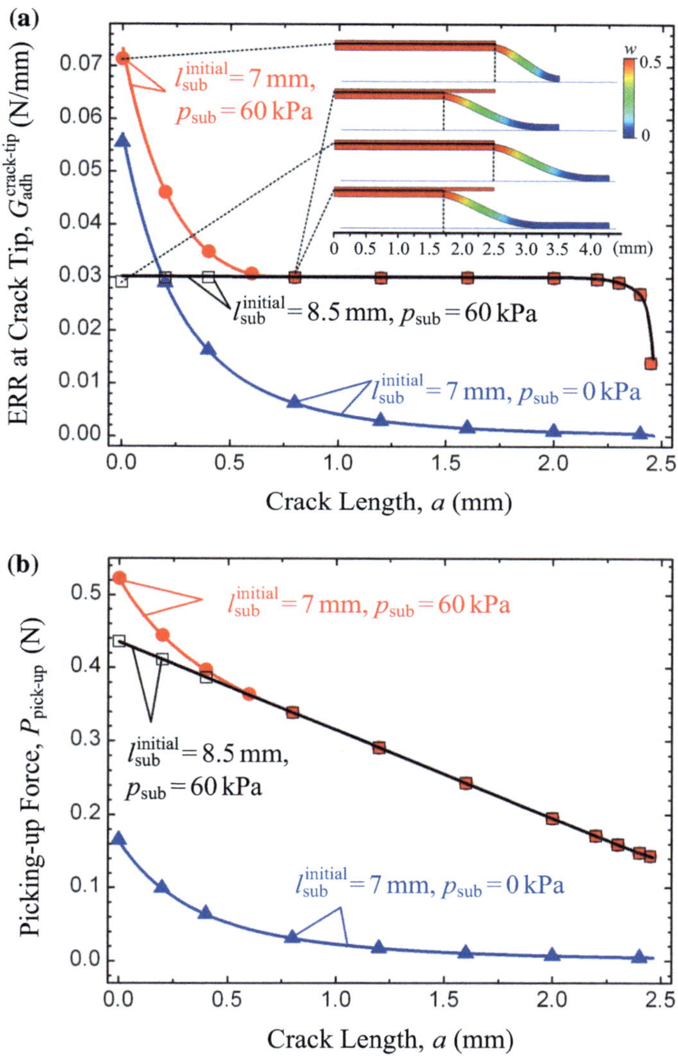

Fig. 9.7 Mechanism of the chip picking-up considering a flat contact between the substrate and vacuum ejector cap: **a** total ERR at the crack tip versus the crack length, **b** picking-up force versus the crack length. For comparison, different initial substrate lengths and VAP below the substrate are considered. Theoretical predictions and numerical results are represented by solid lines and scatter plots, respectively. © 2015 IEEE. Reprinted, with permission, from Ref. [10]

substrate length generates not only the larger ERR at the crack tip, but also the larger picking-up force. In fact, a clamped condition of the target chip may be naturally formed by adjacent chips [21]. Especially for the large ultra-thin chip, it is more likely to cause the asymmetric chip delamination and chip breakage. Besides, the total ERR at the crack tip and the picking-up force without the VAP below the substrate are much smaller than those with the vacuum adsorption. The significance of the vacuum adsorption is verified through the above analysis. It also suggests that the vacuum adsorption should be considered for a precise theoretical prediction of the chip picking-up process. It has been pointed out that a higher VAP below the substrate would lead to a larger total ERR at the crack tip and the picking-up force [20].

It is critical to keep the chip flat in the process. The total ERR at the crack tip and the picking-up force versus the crack propagation is shown in Fig. 9.8. The conditions are the same as before and $l_{\text{sub}}^{\text{initial}} = 8.5\,\text{mm}$. In the case of the chip bending, the chip is absorbed by a picking-up head with one vacuum hole under the condition of $p_{\text{chip}} = 80\,\text{kPa}$. The total ERR at the crack tip under the condition of the chip bending can be found increased with the crack propagation. It implies that once the crack initiates, it can propagate by itself. Nevertheless, compared with the case of the chip-keeping-flat, the total ERR at the crack tip is much smaller at the initial stage under the condition of the chip bending. Therefore, it can be concluded that the chip bending is extremely not helpful to the chip delamination from the substrate. As to the result of the picking-up force, there is almost no difference between both cases, which implies that the chip bending has little effect on the picking-up force. It should be noted that the maintenance of a completely flat chip throughout the picking-up process is generally impossible, although the adoption of a picking-up head with distributed hole-matrix can reduce the bending extent of a large and thin chip substantially.

9.2.8 Effects of Properties of Chip and Substrate Layers

Considering two cases of the chip-keeping-flat and chip bending, effects of the chip length, substrate thickness, and Young's modulus of the substrate on the total ERR at the crack tip, picking-up force, and $l_{\text{reg-2}}^{\text{constant}}$ are shown in Fig. 9.9. Here, the hypothesis of the variable substrate length is adopted. The initial substrate length is given to be large enough to establish the initial flat contact between the substrate and the vacuum adsorption cap. $p_{\text{chip}} = 80\,\text{kPa}$ and $p_{\text{sub}} = 60\,\text{kPa}$ are adopted. Theoretical predictions and numerical results represented respectively by lines and scatter plots agree well with each other. The ERR of initial cracking is independent of the changes of these factors in the case of the chip-keeping-flat, while the corresponding ERRs decrease with the increase of these factors in the case of the chip bending. It can thus be concluded that the chip picking-up process is insensitive and sensitive to the above parameters under conditions of the chip-keeping-flat and chip bending, respectively.

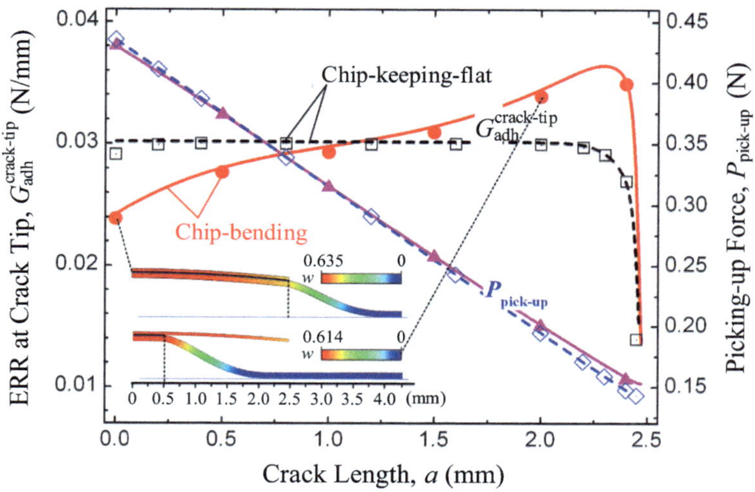

Fig. 9.8 Total ERR at the crack tip and the picking-up force versus the crack propagation. The case of the chip bending with $p_{chip} = 80$ kPa is also included for comparison, corresponding to the chip picking-up process by a picking-up head with one vacuum hole in practice. © 2015 IEEE. Reprinted, with permission, from Ref. [10]

In both cases of the chip bending and chip-keeping-flat, the picking-up force increases with the increase of the chip length, substrate thickness, and Young's modulus of the substrate. In detail, the picking-up force in the case of the chip bending is just slightly smaller than that in the case of the chip-keeping-flat, which may be due to that $l_{reg\text{-}2}^{constant}$ for the case of the chip bending is smaller. A larger picking-up force is obviously needed to detach a larger chip from the substrate, which demonstrates that it is much tougher to pick up a chip with larger length under the condition of the vacuum adsorption, especially for the case of the chip bending. For example, the total ERR at the crack tip for a chip with the length of 5 mm under the condition of the vacuum adsorption in the case of the chip bending reduced about 20% compared with the case of the chip-keeping-flat. However, results also indicate that either decreasing the substrate thickness from 100 μm to 50 μm or decreasing the Young's modulus of the substrate from 160 MPa to 20 MPa can increase about 15% of the total ERR at the crack tip and decrease about 10% of the picking-up force. Therefore, it can be concluded that the smaller thickness and Young's modulus of the substrate improve the reliability of the chip picking-up process. Nevertheless, it is worth noting that the increasing picking-up displacement is the main driving force to delaminate a chip from the substrate in practice.

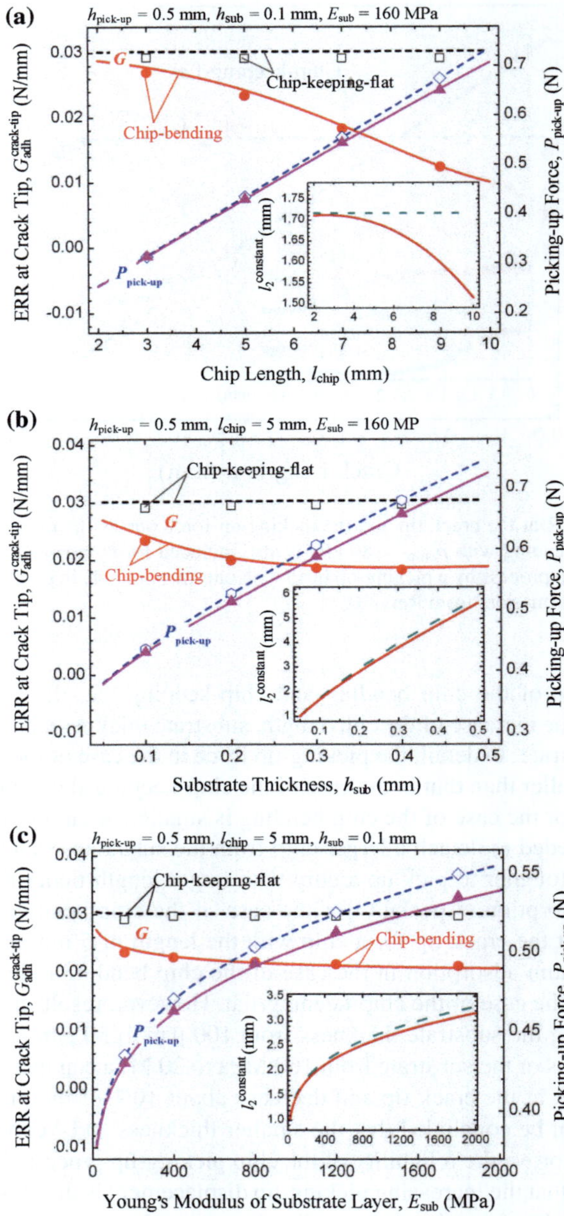

Fig. 9.9 Results of the total ERR at the crack tip, picking-up force, and $l_{reg-2}^{constant}$ considering the effects of: **a** the chip length, **b** the substrate thickness, and **c** the Young's modulus of the substrate. Theoretical predictions and numerical results are represented by lines and scatter plots, respectively. In order to understand effects of these factors on geometry of the chip-adhesive-substrate structure, results of $l_{reg-2}^{constant}$ is also given here, which indicates the change of the effective substrate length for a constant adhesive length. © 2015 IEEE. Reprinted, with permission, from Ref. [10]

9.2.9 Prediction of Chip Picking-up

Variation of the extent of the chip delamination from the substrate and the picking-up force with increasing upward displacement under conditions of the chip bending and chip-keeping-flat is shown in Fig. 9.10. Here, the hypothesis of the variable substrate length is adopted. The initial geometric dimensions and material properties are the same as those mentioned before, e.g., $l_{chip} = 5$ mm, $h_{chip} = 50$ μm, $p_{sub} = 60$ kPa, and $G^{c}_{adh} = 0.03$ N/mm. Here, $l_{adh}/l_{chip} = 0$ represents the complete detachment of the chip from the substrate. In Fig. 9.10a, when $l^{initial}_{sub} = 8.5$ mm, l_{adh}/l_{chip} in both cases changes like a step function with increasing picking-up displacement. This is because that once the initial cracking occurs, the crack can propagate spontaneously due to the

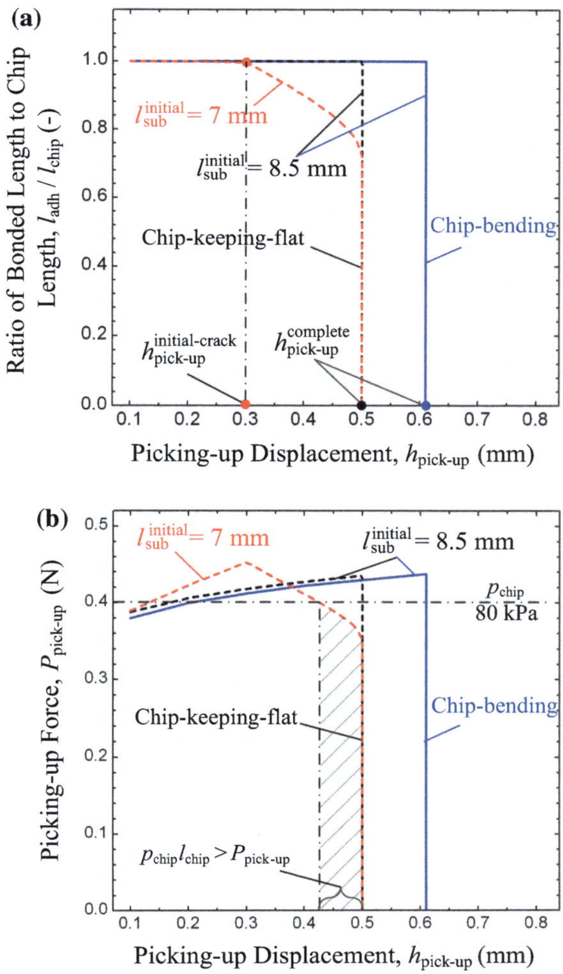

Fig. 9.10 Theoretical prediction of the picking-up process for the 50-μm-thick chip with increasing picking-up displacement: **a** result of the extent of the chip delamination, **b** result of the picking-up force. © 2015 IEEE. Reprinted, with permission, from Ref. [10]

trend of ERRs at the crack tip, as illustrated in Fig. 9.8. However, it is worth noting that in the case of the chip bending, a larger picking-up displacement is required to accomplish the process. When $l_{sub}^{initial} = 7$ mm, the additional substrate tension makes the initial cracking occur early, after which l_{adh}/l_{chip} decreases gradually with the increasing picking-up displacement, and finally jumps to zero. The critical picking-up displacement for generating the initial crack and accomplishing the chip detachment are defined as $h_{\text{pick-up}}^{\text{initial-crack}}$ and $h_{\text{pick-up}}^{\text{complete}}$, respectively. It is evident that $h_{\text{pick-up}}^{\text{initial-crack}} = h_{\text{pick-up}}^{\text{complete}}$ for a sufficiently large $l_{sub}^{initial}$ to establish the initial flat contact, whereas, $h_{\text{pick-up}}^{\text{initial-crack}} < h_{\text{pick-up}}^{\text{complete}}$ for much small $l_{sub}^{initial}$ to establish the initial flat contact.

As illustrated in Fig. 9.10b, with the increase of the picking-up displacement, the picking-up force increases to a peak value just before the occurrence of the initial cracking. Additionally, the smaller initial substrate length leads to a larger peak value of the picking-up force. Actually, if the vacuum adsorption force is smaller than the picking-up force, the chip cannot be picked up directly just by the vacuum adsorption. In this case, the occurrence of the interfacial crack, especially for the initial cracking, relies on the capability of the chip peeling-off process, where a multi-needle ejector can be used to lift the chip effectively [22]. Moreover, the analytical prediction can be used not only to evaluate the capability of the chip picking-up with the vacuum adsorption, but also to establish the process window through Eq. (9.25). In case of $l_{sub}^{initial} = 7$ mm, when the chip is lifted by a multi-needle ejector to an upward displacement exceeding 0.43 mm, it can be picked away directly by the picking-up head with distributed hole-matrix.

9.2.10 *Design of Chip Picking-up*

The ratio of the adhesive length to the chip length and the picking-up force versus the picking-up displacement is depicted in Fig. 9.11. Here, $h_{chip} = 100$ μm, $G_{adh}^{c} = 0.0263$ N/mm, $p_{sub} = 40$ kPa, 50 kPa, 60 kPa, and 70 kPa, $p_{chip} = 90$ kPa. The picking-up displacement applied at both edges of the chip is set to increase gradually. $l_{adh}/l_{chip} = 1$ indicates no crack initiation in the adhesive layer, as shown in Fig. 9.11a. When the picking-up displacement exceeds the critical value, l_{adh}/l_{chip} starts to decrease, which implies that the crack starts to propagate in the adhesive layer. Additionally, a smaller critical and total picking-up displacement are needed for higher VAP below the substrate to generate the initial cracking and accomplish the chip detachment, respectively. Results indicate that the higher VAP below the substrate can accelerate the chip detachment from the substrate and make the chip picking-up process more efficient.

As illustrated in Fig. 9.11b, with the increase of the picking-up displacement, the picking-up force increases to a peak value just before the crack initiation, and decreases subsequently as the crack propagates. The trend is consistent with that extracted from experiments [23]. In addition, the length of bonded region that can be delaminated is restricted due to the fixed VAP above the chip. The lower VAP

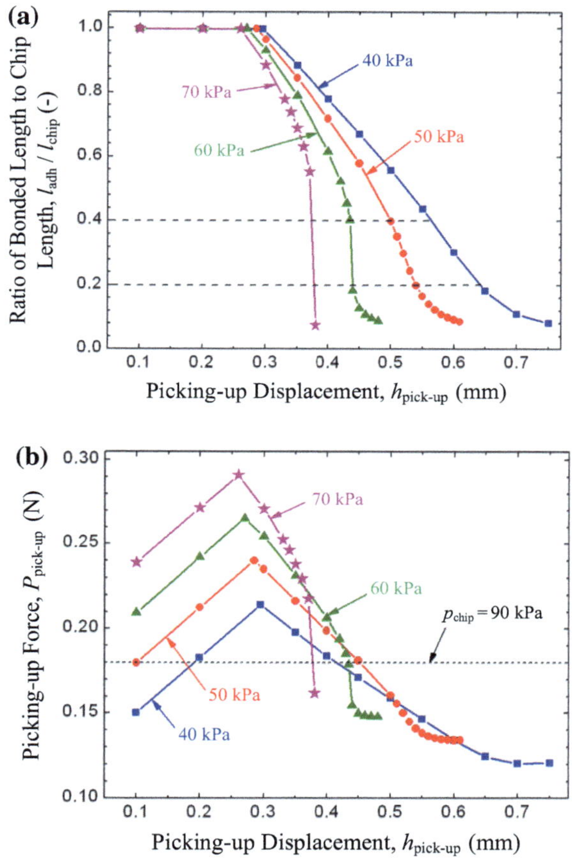

Fig. 9.11 The effect of the picking-up displacement on **a** the ratio of the bonded length to chip length, and **b** the picking-up force. Reprinted from Ref. [20] by permission of Taylor & Francis Ltd.

below the substrate gives rise to the smaller picking-up force, which indicates that the lower VAP below the substrate is beneficial to the chip picking-up from larger remaining bonded region. If the VAP below the substrate decreases from 70 kPa to 40 kPa, the length of the bonded region that can be delaminated increases from 20% to 70% of the chip length. Therefore, the theoretical model can be used to predict the capability of the chip picking-up and determine the corresponding critical picking-up displacement. In practice, the VAP is a controllable process parameter that can be used to design a proper process window.

Based on the above discussion, the picking-up displacement is the main driving force to detach the chip from the substrate, and the delamination process is greatly affected by the conditions of the vacuum adsorption and the chip thickness. In practice, the thick chip is easy to be picked up intactly from the substrate by the current technique. In order to prevent the chip damage in the picking-up process, the lower VAPs above the chip and below the substrate should be adopted, which are beneficial to the decrease in the picking-up force and chip cracking stress. If the length of the remaining bonded region after the peeling-off exceeds the anticipated length, the VAP below the substrate should be decreased and the VAP above the chip should be increased to enhance the reliability of the ultra-thin chip picking-up process.

9.3 R2R Process with Low Interfacial Residual Stress

9.3.1 Temperature and Force-Induced Misfits in R2R Processing

Residual stresses are inevitably generated in the flip-chip and film deposition, and will lead to curling, eliminating, and cracking. They mainly arise from high deposition temperature, mismatch of the coefficient of thermal expansion (CTE), and substrate tension in the R2R processing. Three essential steps, including deposition, patterning, and packaging, can be integrated into the R2R processing to achieve the purpose of high-throughput manufacturing. The substrate in the R2R processing is a patterned film-on-substrate structure composed of different materials [24]. The film is deposited at high-temperature $T_{\text{deposition}}$, patterned by photolithography at room temperature T_{room}, and the fabricated device will run at the working temperature T_{work}. Any change of the temperature will generally lead to the interfacial stresses. The substrate tension is inherently sensitive to external disturbances during operation, especially when the substrate thickness is less than 50 μm. In practice, the substrate tension may be affected by many uncertainties.

Thin patterned films are usually deposited onto polymeric substrate layer-by-layer at high temperature. The variation of strain can be calculated according to the film deposition process [25]. Variation of the strain during different fabrication steps is shown in Fig. 9.12, as follows: (a) polymeric substrate in freestanding state, (b) polymeric substrate stretched by the substrate tension to make unwind/rewind rollers release/accumulate material, (c) stretched substrate elongated as the temperature rises to the deposition temperature, (d) built-in strain $\varepsilon_{\text{built-in}}$ formed as the film deposits or the chip is mounted, which will force the substrate to shrink or expand in case the substrate is not constrained, (e) when films grown on a stretched substrate, the built-in strain leads to the contraction of the substrate, (f) when the tensile stress is released, the substrate will be further shrunk, and the film-on-substrate structure will roll into a cylinder, (g) occurrence of the strain mismatch between the film and substrate at the room temperature, the radius becomes smaller due to that the CTE

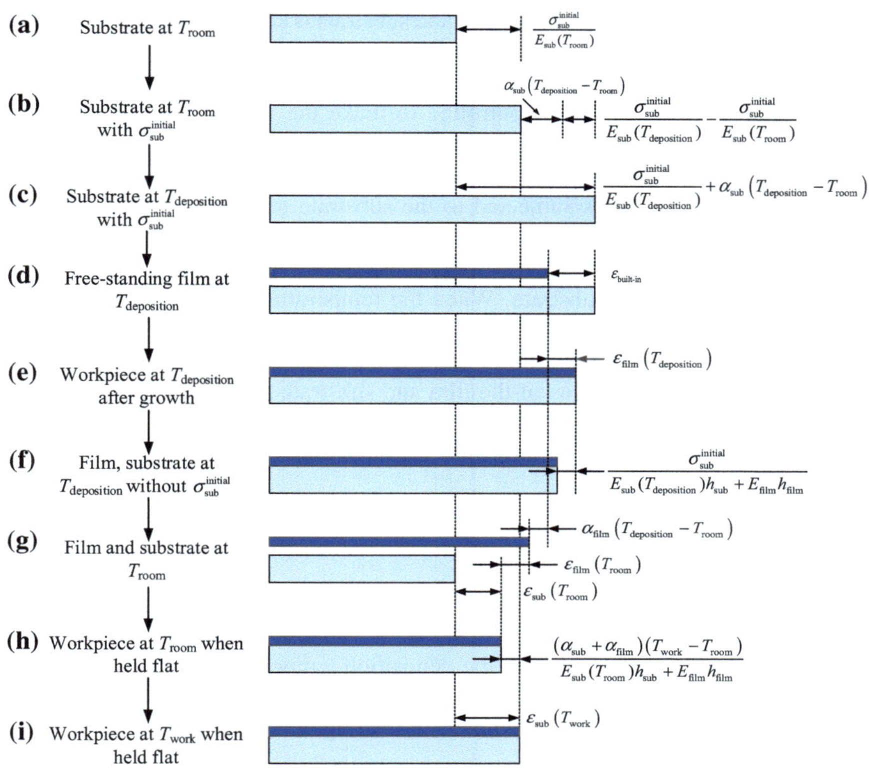

Fig. 9.12 Length of a film-on-substrate structure at different stages. © 2011 IEEE. Reprinted, with permission, from Ref. [26]

of the substrate, α_{sub}, is typically much larger than that of the film, α_{film}, (h) the state of the film bonded onto the substrate at room temperature, (i) the state of the film bonded onto the substrate at the working temperature.

The deformation behavior of the film-on-substrate structure depends strongly on the Young's modulus and thickness of the film (E_{film}, h_{film}) and substrate (E_{sub}, h_{sub}) [27]. When $E_{film}h_{film} \ll E_{sub}h_{sub}$, the deformation of the substrate plays a dominate role and the film complies with it. The stress in the substrate is small, and the film/substrate couple is only slightly curved, even when the film is highly stressed. When $E_{film}h_{film} \gg E_{sub}h_{sub}$, the deformation of film plays a dominant role and the deformation of the substrate can be neglected. When $E_{film}h_{film} \approx E_{sub}h_{sub}$, the film and substrate interact strongly. Such equal strength of the film and substrate usually appears in flexible electronic devices, which gives rise to a complicated mechanical behavior. Efforts should be taken to adjust process parameters to eliminate the residual stress occurred in the R2R processing.

The strain mismatch between the film and substrate is composed of three dominant components: (1) mismatch of the CTEs between the film and substrate, (2) built-in strain in deposited films, (3) substrate strain induced by the substrate tension. The substrate tension $\sigma_{sub}^{initial}$ is controlled to make the system run smoothly. The substrate is stress free at the room temperature before depositing thin films, where the room temperature is set as a reference state of the substrate, namely $\varepsilon_{sub}(T_{room}) = 0$. When the substrate is subjected to the substrate tension and elongated from $l_{sub}^{initial}$ to l_{sub}^{max}, there exists $\varepsilon_{sub}^{initial}(T_{room}) = (l_{sub}^{max} - l_{sub}^{initial})/l_{sub}^{initial}$ and $\sigma_{sub}^{initial}(T_{room}) = \overline{E}_{sub}\,\varepsilon_{sub}^{initial}$, where $\overline{E}_{sub} = E_{sub}/(1 - \nu_{sub}^2)$, $E_{sub}(T)$ is the temperature-dependent Young's modulus of the substrate. When the temperature raises to the deposition temperature, the substrate strain results from the combined effects of the thermal and mechanical loading. The film is bonded onto the substrate and cooled to the room temperature, then strains of the film and substrate are set as new values. The temperature changes during different fabrication steps, namely, $T_{room} \rightarrow T_{deposition} \rightarrow T_{room} \rightarrow T_{work}$, and the strain mismatch can be expressed as

$$\varepsilon_{sub}(T_i) = \alpha_{sub}(T_i - T_{room}) + \frac{\sigma_{sub}^{initial}}{\overline{E}_{sub}(T_i)} \tag{9.27}$$

where i = deposition, room, work.

The film experiences $T_{deposition} \rightarrow T_{room} \rightarrow T_{work}$, and the film is stress free at the deposition temperature. Hence, the deposition temperature is set as a reference state for the film, namely $\varepsilon_{film}(T_{deposition}) = 0$. The strain of films results from the combined effects of the thermal and built-in loading, namely,

$$\varepsilon_{film}(T_i) = \varepsilon_{built\text{-}in} + \alpha_{film}(T_i - T_{deposition}) \tag{9.28}$$

When the substrate tension is released, the formula of strain mismatch $\varepsilon_{mis}(T_i) = \varepsilon_{sub}(T_i) - \varepsilon_{film}(T_i)$ can be obtained. Here, a thermal-mechanical model of the film-on-substrate system is presented to discuss the misfit and interfacial stress. During working, the resistive layer dissipates heat in two ways: on the upside to the surrounding air and on the downside to the substrate. When the deposited film is cooled down to room temperature or heated up to higher temperature, new values of strain and stress will be generated. The strain of thin films considering the temperature can be obtained as [25]

$$\varepsilon_{film}(T_{work}, T_{deposition}, T_{room}) = \frac{\begin{bmatrix} \overline{E}_{sub}(T_{work})h_{sub}\alpha_{sub}(T_{work} - T_{room}) \\ -\overline{E}_{film}h_{film}\alpha_{film}(T_{depositiion} - T_{work}) \\ +\overline{E}_{sub}(T_{work})h_{sub}\sigma_{sub}^{initial}/\overline{E}_{sub}(T_{deposition}) - h_{film}\sigma_{bulit\text{-}in} \end{bmatrix}}{\overline{E}_{sub}(T_{work})h_{sub} + \overline{E}_{film}h_{film}} \tag{9.29}$$

where $\overline{E}_{film} = E_{film}/(1 - \nu_{film}^2)$, E_{film} is the Young's modulus of the film that usually considered to be constant, $\sigma_{built\text{-}in}$ is the built-in stress in the film. The strain of the film at the working temperature can be described by Eq. (9.29).

If a structure of two layers with similar Young's modulus such as silicon and steel, the neutral surface is the midplane, and the strain in the film is given by $\varepsilon_{film} = (h_{film} + h_{sub})/2R$. If a stiff film is deposited onto a compliant substrate, the neutral surface shifts from the midplane toward the thin film. The location of the bending axis, which is the line in the cross section where the bending strain component of the system is zero, can be treated as the distance from the neutral axis to the film-substrate interface. The deviation value is calculated by

$$h_{bend} = \frac{E_{sub}h_{sub}^2 - E_{film}h_{film}^2}{2E_{sub}h_{sub} - 2E_{film}h_{film}} \tag{9.30}$$

When the process temperatures are considered, the strain $\varepsilon_{film}(T_i)$ can be obtained as [28]

$$\varepsilon_{film}(T_i) = \frac{(h_{sub} + h_{film})}{2R(T_i)} \times \frac{(1 + 2h_{film}/h_{sub} + \overline{E}_{film}h_{film}^2/\overline{E}_{sub}h_{sub}^2)}{(1 + E_{film}/E_{sub})\left[(1 + \overline{E}_{film}h_{film}/\overline{E}_{sub}h_{sub}) - h_{film}(1 - \overline{E}_{film}/\overline{E}_{sub})/2R(T_i)\right]} \tag{9.31}$$

In case the silicon film is adopted, three kinds of substrates are adopted and compared here: steel ($E_{film}/E_{sub} \approx 1$), plastic ($E_{film}/E_{sub} \approx 100$), and rubber ($E_{film}/E_{sub} \approx 10{,}000$).

The effect of the thickness ratio of the film to substrate on the normalized strain in the film is shown in Fig. 9.13. A more compliant substrate will lead to a smaller strain. The strains of the film and substrate are related to the radius, which is further related to the misfit strain. If the working temperature is dynamic, the strain of the film can be represented by extending Eq. (9.31) where the temperature T_i increases accordingly from the room temperature to the working temperature.

9.3.2 Interfacial Stress Related to Structures and Processes

Free-body diagram of a film-on-substrate structure is shown in Fig. 9.14, where the length of the structure is $2l_{film\text{-}on\text{-}sub\text{-}half}$ and the origin of coordinates is located at the middle point. Due to that the stress state in the film is symmetric at $x = 0$, only the condition of $0 \leq x \leq l_{film\text{-}on\text{-}sub\text{-}half}$ is considered here. The film is subjected to axial, shear, and peel forces. The external compressive stress from the roller is not included.

The longitudinal deformation mainly arises from following sources: axial force, continuously distributed axial forces $\tau(x)\mathrm{d}x$ formed by the shear stress $\tau(x)$, continuously distributed couples $0.5h_{film}\tau(x)\mathrm{d}x$ or $0.5h_{sub}\tau(x)\mathrm{d}x$, peel stress $\sigma(x)$, and bending moment. The membrane force in the film induced from the contraction-elongation is $N_{film}(x) = \sigma_{film}(x)h_{film} = \int_x^{l_{sub\text{-}on\text{-}sub\text{-}half}} \tau(\lambda)\mathrm{d}\lambda$, which is assumed to be uniformly distributed within the thickness direction. In order to balance the bending

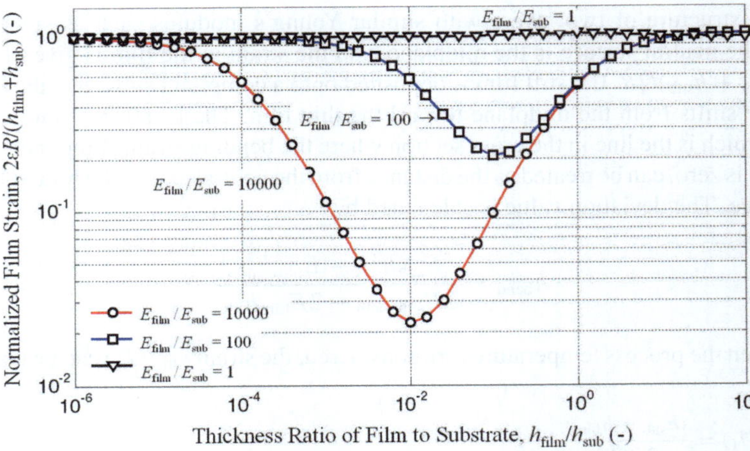

Fig. 9.13 Normalized strain in the film versus the ratio of film thickness to substrate thickness. © 2011 IEEE. Reprinted, with permission, from Ref. [26]

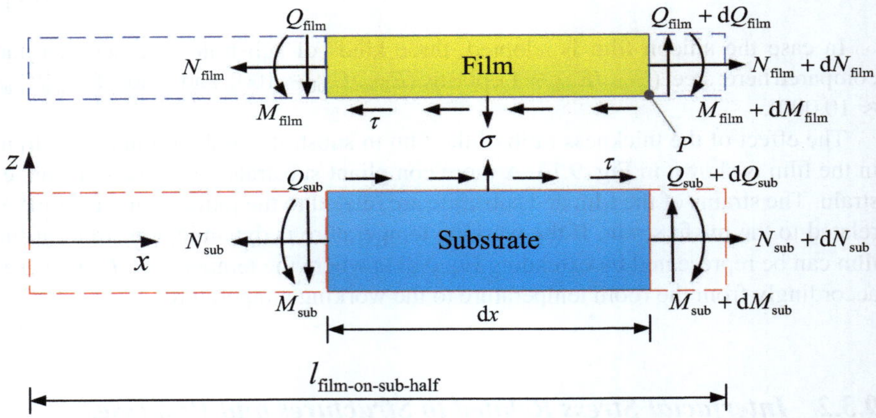

Fig. 9.14 Free-body diagram of the film-on-substrate structure. © 2011 IEEE. Reprinted, with permission, from Ref. [26]

moment induced by the axial force, there exists bending moment $M(x)$ acting on the x cross section. $M(x)$ can be decomposed into two parts. One acting on the film is $M_{\mathrm{film}}(x) = N_{\mathrm{film}}(x)(h_{\mathrm{film}}/2 + h_{\mathrm{bend}}) = \sigma_{\mathrm{film}}(x)h_{\mathrm{film}}(h_{\mathrm{film}}/2 + h_{\mathrm{bend}})$, and the other acting on the substrate is $M_{\mathrm{sub}}(x) = N_{\mathrm{sub}}(x)(h_{\mathrm{sub}}/2 - h_{\mathrm{bend}}) = \sigma_{\mathrm{sub}}(x)h_{\mathrm{sub}}(h_{\mathrm{sub}}/2 - h_{\mathrm{bend}})$. The axial force $N_i(0)$ and bending moment $M_i(0)$ appear at the section of $x = 0$. The axial forces in the film and substrate are converted from the shear stress and peel stress at the interface. The shear force can be decomposed into continuously distributed couples and axial forces with the magnitude of $0.5h_{\mathrm{film}}\tau(x)\mathrm{d}x$ and $\tau(x)\mathrm{d}x$, respectively. The action position of axial forces is at the centroid of the film. Similarly, the shear force on the substrate can be decomposed into couples $-0.5h_{\text{sub}}\tau(x)\text{d}x$ and axial forces $-\tau(x)\text{d}x$.

Equilibrium of the axial forces in the x-direction requires $N(x) = \int_x^{l_{\text{sub-on-sub-half}}} \tau(\lambda)\text{d}\lambda$, where $N(x)$ is a continuously distributed axial force. Relationship between the shear stress and peel stress can be derived as $h_{\text{bend}}\tau(x) - \int_0^x \sigma(\lambda)\text{d}\lambda = 0$. According to the self-equilibrium beam theory, equilibrium of each layer of the film-on-substrate structure gives [29, 30]

$$\frac{\text{d}M_{\text{film}}}{\text{d}x} = Q_{\text{film}} - \frac{1}{2}\tau h_{\text{film}}, \quad \frac{\text{d}Q_{\text{film}}}{\text{d}x} = \sigma, \quad \frac{\text{d}N_{\text{film}}}{\text{d}x} = \tau \tag{9.32}$$

$$\frac{\text{d}M_{\text{sub}}}{\text{d}x} = Q_{\text{sub}} - \frac{1}{2}\tau h_{\text{sub}}, \quad \frac{\text{d}Q_{\text{sub}}}{\text{d}x} = -\sigma, \quad \frac{\text{d}N_{\text{sub}}}{\text{d}x} = -\tau \tag{9.33}$$

Axial displacements at the corresponding central axes in the film u_{film} and substrate u_{sub} are given by

$$\frac{\text{d}u_{\text{film}}}{\text{d}x} = (1 + \nu_{\text{film}})\alpha_{\text{film}}(T_{\text{work}} - T_{\text{deposition}}) + \frac{1 - \nu_{\text{film}}^2}{E_{\text{film}}h_{\text{film}}}N_{\text{film}} - \frac{1 - \nu_{\text{film}}^2}{E_{\text{film}}h_{\text{film}}}x\tau \tag{9.34}$$

$$\frac{\text{d}u_{\text{sub}}}{\text{d}x} = (1 + \nu_{\text{sub}})\alpha_{\text{sub}}(T_{\text{work}} - T_{\text{room}}) - \frac{1 - \nu_{\text{sub}}^2}{E_{\text{sub}}h_{\text{sub}}}N_{\text{sub}} + \frac{1 - \nu_{\text{sub}}^2}{E_{\text{sub}}h_{\text{sub}}}x\tau + \frac{1 - \nu_{\text{sub}}^2}{E_{\text{sub}}h_{\text{sub}}}\sigma_{\text{sub}}^{\text{initial}} \tag{9.35}$$

where the first items of the right side of Eqs. (9.34) and (9.35) are, respectively the axial displacements associated with the thermal expansion at corresponding central axes in the film and substrate, the second items are associated with the tensile stress, the third items are associated with the shear stress, and the last item of Eq. (9.35) is associated with the initial substrate tension.

Transverse displacements at the middle surface in the film w_{film} and substrate w_{sub} associated with the bending moments are

$$\frac{\text{d}^2 w_{\text{film}}}{\text{d}x^2} = -\frac{2h_{\text{bend}}}{h_{\text{film}}}\frac{1 - \nu_{\text{film}}^2}{E_{\text{film}}I_{\text{film-2}}}M_{\text{film}} + \frac{1 - \nu_{\text{film}}^2}{E_{\text{film}}I_{\text{film}}}x\frac{h_{\text{film}}}{2}\tau - \frac{1 - \nu_{\text{film}}^2}{E_{\text{film}}I_{\text{film}}}\int_0^x (x - \lambda)\sigma(\lambda)\text{d}\lambda \tag{9.36}$$

$$\frac{\text{d}^2 w_{\text{sub}}}{\text{d}x^2} = -\frac{2h_{\text{bend}}}{h_{\text{sub}}}\frac{1 - \nu_{\text{sub}}^2}{E_{\text{sub}}I_{\text{sub-2}}}M_{\text{sub}} - \frac{1 - \nu_{\text{sub}}^2}{E_{\text{sub}}I_{\text{sub}}}x\frac{h_{\text{sub}}}{2}\tau + \frac{1 - \nu_{\text{sub}}^2}{E_{\text{sub}}I_{\text{sub}}}\int_0^x (x - \lambda)\sigma(\lambda)\text{d}\lambda \tag{9.37}$$

where $I_{\text{film-1}} = h_{\text{film}}^3/12$, $I_{\text{sub-1}} = h_{\text{sub}}^3/12$, $I_{\text{film-2}} = \int_{h_{\text{bend}}}^{h_{\text{film}}+h_{\text{bend}}} y^2\text{d}y$, $I_{\text{sub-2}} = \int_{h_{\text{bend}}-h_{\text{sub}}}^{h_{\text{bend}}} y^2\text{d}y$.

The first items of the right side of Eqs. (9.36) and (9.37) are, respectively, associated with the bending moments in the film and substrate, the second terms are associated with the shear stress, and the last terms are associated with the peel stress.

According to the condition of compatibility at the interface, the longitudinal and transverse deformations of the substrate at the interface should be equal to that of the film. Following ordinary differential equations for the interfacial shear stress and peel stress can be obtained:

$$\left(\frac{1-\nu_{\text{film}}^2}{E_{\text{film}}h_{\text{film}}}x+\frac{1-\nu_{\text{sub}}^2}{E_{\text{sub}}h_{\text{sub}}}x+\frac{1-\nu_{\text{film}}^2}{E_{\text{film}}I_{\text{film-1}}}x\frac{h_{\text{film}}}{2}\frac{h_{\text{film}}}{2}+\frac{1-\nu_{\text{sub}}^2}{E_{\text{sub}}I_{\text{sub-1}}}x\frac{h_{\text{sub}}}{2}\frac{h_{\text{sub}}}{2}\right)\frac{\mathrm{d}^2\tau}{\mathrm{d}x^2}$$
$$+\left(3\frac{1-\nu_{\text{film}}^2}{E_{\text{film}}h_{\text{film}}}+3\frac{1-\nu_{\text{sub}}^2}{E_{\text{sub}}h_{\text{sub}}}+\frac{1-\nu_{\text{film}}^2}{E_{\text{film}}I_{\text{film-1}}}\frac{3}{2}h_{\text{film}}\frac{h_{\text{film}}}{2}+\frac{1-\nu_{\text{sub}}^2}{E_{\text{sub}}I_{\text{sub-1}}}\frac{3}{2}h_{\text{sub}}\frac{h_{\text{sub}}}{2}\right)\frac{\mathrm{d}\tau}{\mathrm{d}x}$$
$$-\left[h_{\text{film}}\frac{(1-\nu_{\text{film}}^2)}{E_{\text{film}}I_{\text{film-1}}}+h_{\text{sub}}\frac{(1-\nu_{\text{sub}}^2)}{E_{\text{sub}}I_{\text{sub-1}}}\right]\sigma=0 \tag{9.38}$$

$$\left(\frac{h_{\text{sub}}}{2\overline{E}_{\text{sub}}I_{\text{sub-1}}}+\frac{h_{\text{film}}}{2\overline{E}_{\text{film}}I_{\text{film-1}}}\right)x\frac{\mathrm{d}^2\tau}{\mathrm{d}x^2}+\left(\frac{h_{\text{sub}}}{2\overline{E}_{\text{sub}}I_{\text{sub-1}}}+\frac{h_{\text{film}}}{2\overline{E}_{\text{film}}I_{\text{film-1}}}\right)\frac{\mathrm{d}\tau}{\mathrm{d}x}$$
$$-\left(\frac{1}{\overline{E}_{\text{film}}I_{\text{film-1}}}+\frac{1}{\overline{E}_{\text{sub}}I_{\text{sub-1}}}\right)\sigma=0 \tag{9.39}$$

By rearranging Eqs. (9.38) and (9.39), the peel stress can be eliminated, and the ordinary differential equations of the interfacial shear stress can be written as

$$3Ax\frac{\mathrm{d}\tau(x)}{\mathrm{d}x}+(2A-B)\tau(x)=0 \tag{9.40}$$

where $A = 1/E_{\text{film}}h_{\text{film}} + 1/E_{\text{sub}}h_{\text{sub}}$, $B = 6h_{\text{bend}}(1/E_{\text{film}}h_{\text{film}}^2 + 1/E_{\text{sub}}h_{\text{sub}}^2)$.

Then the interfacial shear stress can be obtained as

$$\tau(x) = CX^{\left(\frac{B-2A}{3A}\right)} \tag{9.41}$$

where C is an integration constant. The boundary conditions at $x = \pm l_{\text{film-on-sub-half}}$ are $M_{\text{film}} = 0$, $N_{\text{film}} = 0$, $Q_{\text{film}} = 0$, $M_{\text{sub}} = 0$, $N_{\text{sub}} = 0$, and $Q_{\text{sub}} = 0$. The interfacial shear stress and peel stress can be obtained as

$$\tau(x)=\frac{(1+\nu_{\text{film}})\alpha_{\text{film}}(T_{\text{work}}-T_{\text{deposition}})-(1+\nu_{\text{sub}})\alpha_{\text{sub}}(T_{\text{work}}-T_{\text{room}})-\frac{1}{\overline{E}_{\text{sub}}h_{\text{sub}}}\sigma_{\text{sub}}^{\text{initial}}}{3A^2l_{\text{film-on-sub-half}}}$$
$$\times\,(A+B)\left[\frac{x}{l_{\text{film-on-sub-half}}}\right]^{\left(\frac{B-2A}{3A}\right)} \tag{9.42}$$

$$\sigma(x)=\frac{(1+\nu_{\text{film}})\alpha_{\text{film}}(T_{\text{work}}-T_{\text{deposition}})-(1+\nu_{\text{sub}})\alpha_{\text{sub}}(T_{\text{work}}-T_{\text{room}})-\frac{1}{\overline{E}_{\text{sub}}h_{\text{sub}}}\sigma_{\text{sub}}^{\text{initial}}}{9A^3l_{\text{film-on-sub-half}}^2}$$
$$\times\,h_{\text{bend}}(A+B)(B-2A)\left[\frac{x}{l_{\text{film-on-sub-half}}}\right]^{\left(\frac{B-5A}{3A}\right)} \tag{9.43}$$

The distribution of the interfacial shear stress, interfacial peel stress, and normal stress in the film along the interface is shown in Fig. 9.15. The maximum value of the in-plane normal stress presents at $x = 0$. The normal stress decreases with the increase of x coordinate and then converges to zero at $x = l_{\text{film-on-sub-half}}$. The shear

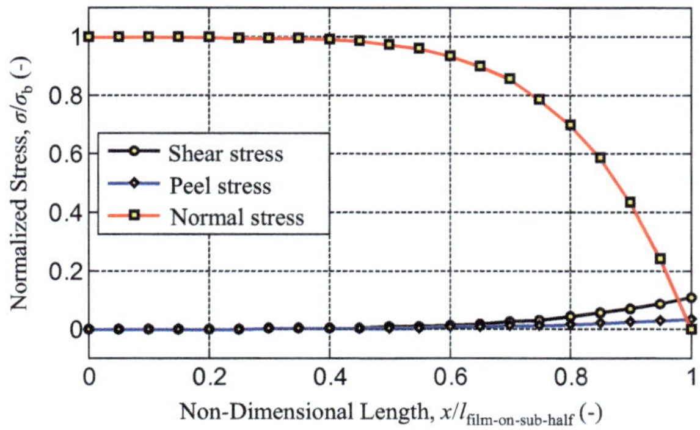

Fig. 9.15 Distribution of the interfacial shear stress, peel stress, and normal stress in the film along the interface. © 2011 IEEE. Reprinted, with permission, from Ref. [26]

stress and peel stress at $x = 0$ are equal to zero and monotonically varied with the increase of x coordinate. It is shown that the end effect results in the decrease of the in-plane normal stress and increase of the shear stress and peel stress.

9.3.3 Material Selection and Structural Optimization of Film-on-Substrate System

The interfacial stresses can be generally minimized by [27]: (1) choosing a substrate with the CTE close to that of the thin film; (2) decreasing the variation of process temperature; (3) utilizing the built-in strain in film growing; and (4) compensating the CTE mismatch between the temperature-dependent material properties of the substrate and the tension of the substrate. The first method is not suitable for flexible electronics application because the CTE of polymeric/rubber substrate is much larger than that of silicon/metal device films. Low deposition temperature should be adopted for the second approach, which will reduce the structural performance. Moreover, the dynamic working temperature is another important factor for products such as the solar cell. The built-in strain depends on the deposition condition and can be easily achieved, but the effectiveness is limited. The method based on temperature-dependent substrate and substrate tension is simple and practical for the R2R processing to keep a stable interfacial stress with the change of temperature [28].

The temperature-dependent Young's modulus is common in polymeric material, but very complex. Even a linear relationship with temperature for Young's modulus is substituted into Eq. (9.29), there exists obvious nonlinear behavior. The substrate strain versus the temperature with and without external stress is shown in Fig. 9.16. It indicates in Fig. 9.16a that the model based on temperature-independent and temperature-dependent Young's modulus show different results even without external stress. If the working temperature is equal to the room temperature, results

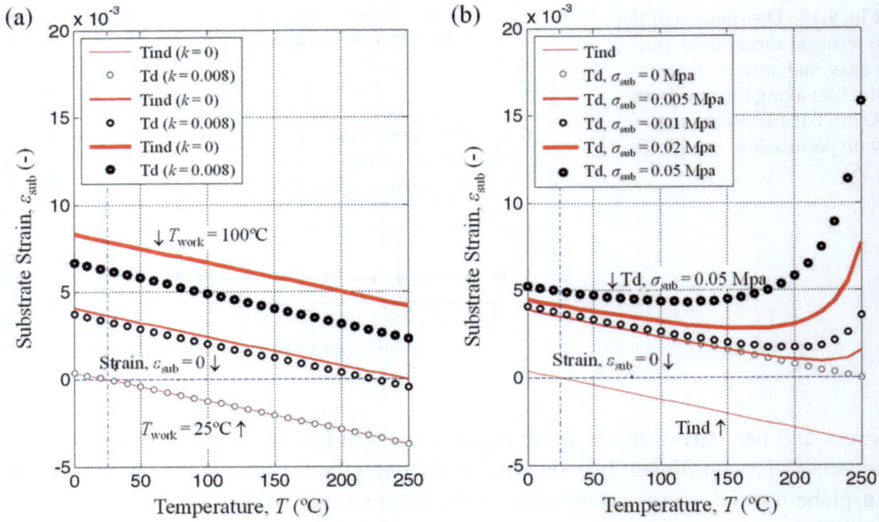

Fig. 9.16 Strain of the substrate versus the temperature: **a** without external stress, **b** with external stress. Tind and Td represent the temperature-independent and temperature-dependent conditions, respectively. © 2011 IEEE. Reprinted, with permission, from Ref. [26]

of the model based on two kinds of Young's modulus remain the same. Difference between these two kinds of the model depends on the working temperature.

A strong nonlinear relationship between the substrate strain and the temperature is shown in Fig. 9.16b. The degree of nonlinearity depends strongly on both the deposition temperature and substrate tension. However, the nonlinearity will disappear when the two process parameters act on the film-on-substrate structure alone. There exists an extreme point to achieve the lowest mismatch of strain. Sensitivity analysis of the three different temperatures is given as follows:

$$\partial \varepsilon_{\mathrm{film}}\left(T_{\mathrm{work}}, T_{\mathrm{deposition}}, T_{\mathrm{room}}\right)/\partial T_i = 0 \tag{9.44}$$

It can be noted from Fig. 9.16 that there exists a parameter k meeting $\partial \varepsilon_{\mathrm{film}}/\partial T_i$ ($i =$ deposition, room, work). There is a certain parameter to attain the smallest strain according to the three different temperatures, separately. By adopting Eq. (9.44), the parameter k can be calculated in combination with the deposition temperature

Table 9.1 Input parameters for numerical analysis

Item	Unit	Kapton (E)	SiN*x*
Young's modulus	GPa	5.3	210
Poisson's ratio	–	0.32	0.25
CTE	ppm/K	16	2.7
Thickness	m	0.0005	0.005
Width	m	0.005	0.005
Length	m	0.05	0.05

$$\frac{\partial \varepsilon_{\text{film}}}{\partial T_{\text{deposition}}} = \frac{kh_{\text{sub}}\alpha_{\text{sub}}(T_{\text{deposition}} - T_{\text{room}}) + \overline{E}_{\text{film}}h_{\text{film}}\alpha_{\text{film}} + \dfrac{kh_{\text{sub}}\sigma_{\text{sub}}^{\text{initial}}}{\left[\overline{E}_{\text{sub}}^{\text{initial}} + k(T_{\text{deposition}} - T_{\text{room}})\right]}}{\left[\overline{E}_{\text{sub}}^{\text{initial}} + k(T_{\text{work}} - T_{\text{room}})\right]h_{\text{sub}} + \overline{E}_{\text{film}}h_{\text{film}}}$$
$$- \frac{kh_{\text{sub}}\left\{\begin{array}{l}\left[\overline{E}_{\text{sub}}^{\text{initial}} + k(T_{\text{work}} - T_{\text{room}})\right]\left(\begin{array}{c}h_{\text{sub}}\alpha_{\text{sub}}(T_{\text{deposition}} - T_{\text{room}})\\ + \dfrac{h_{\text{sub}}\sigma_{\text{sub}}^{\text{initial}}}{\left[\overline{E}_{\text{sub}}^{\text{initial}} + k(T_{\text{deposition}} - T_{\text{room}})\right]}\end{array}\right)\\ -\overline{E}_{\text{film}}h_{\text{film}}\alpha_{\text{film}}(T_{\text{deposition}} - T_{\text{work}}) - h_{\text{film}}\sigma_{\text{built-in}}\end{array}\right\}}{\left\{\left[\overline{E}_{\text{sub}}^{\text{initial}} + k(T_{\text{work}} - T_{\text{room}})\right]h_{\text{sub}} + \overline{E}_{\text{film}}h_{\text{film}}\right\}^2} \tag{9.45}$$

The silicon and PDMS are usually adopted as the film and substrate, respectively. The mechanical properties are: $E_{\text{film}} = 130 \times 10^3$ MPa, $E_{\text{sub}} = 2$ MPa, $\nu_{\text{film}} = 0.27$, $\nu_{\text{sub}} = 0.48$, $\alpha_{\text{film}} = 2.7$ ppm/K, and $\alpha_{\text{sub}} = 16$ ppm/K. If denoting $T_{\text{room}} = 25$ °C, $T_{\text{work}} = 75$ °C, and $T_{\text{deposition}} = 200$ °C, by combining Eqs. (9.44) and (9.45), three values can be obtained, one complex number and two real numbers, $k = -0.000207$ and $k = -0.900640$. If the temperature increases to 225 °C, the Young's modulus varies about 2% when $k = -0.000207$, which is relatively reasonable and feasible.

The strain of film may be reduced if the film is placed in the neutral plane and sandwiched between the substrate and encapsulation layer. The traditional method to decrease the interfacial normal stress is to make the interface as neutral plane, namely $h_{\text{bend}} = 0$, which can be regarded as a neutral plane-based method. $E_{\text{sub}}h_{\text{sub}}^2 = E_{\text{film}}h_{\text{film}}^2$ can then be obtained. This method is just suitable for ideal but impractical manufacturing process, such as the room temperature deposition and R2R processing without any tension stress.

Only the shear stress is examined here in detail. If the shear stress is low and more or less evenly distributed along the interface, the peel stress should be small as well. Based on Eq. (9.42), effects of the film thickness, substrate thickness, temperature-dependent Young's modulus of the substrate, deposition temperature, and substrate tensile stress on the maximum shear stress can be obtained. The input parameters required for the numerical analysis are listed in Table 9.1.

Effects of the substrate thickness and deposition temperature on the interfacial shear stress are shown in Fig. 9.17, where the initial stress on the substrate is kept at 0.5 kPa. As can be seen, under the condition of a constant substrate thickness,

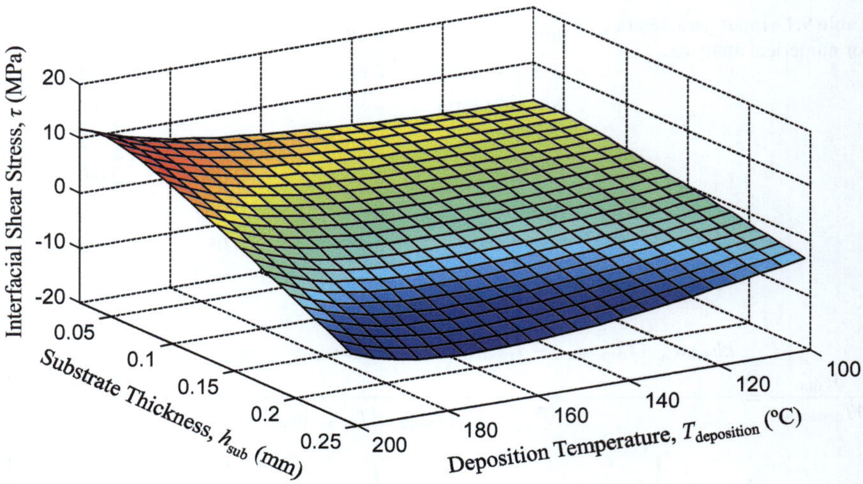

Fig. 9.17 Effects of the substrate thickness and the deposition temperature on the interfacial shear stress ($\sigma_{sub}^{initial} = 0.5\,\text{kPa}$). © 2011 IEEE. Reprinted, with permission, from Ref. [26]

the interfacial shear stress decreases monotonously with the increase of the deposition temperature when the substrate thickness is small, and becomes nonmonotonic with the increase of thickness. On the other hand, the interfacial shear stress is a monotonous decreasing function of the substrate thickness at a certain deposition temperature.

Effects of the substrate thickness and deposition temperature on the interfacial shear stress are shown in Fig. 9.18, where the initial stress on the substrate is kept at 0.25 kPa. The relationship of the shear stress with the substrate tension and substrate thickness is similar to that in Fig. 9.17, but the degree is different. By comparing Fig. 9.17 with Fig. 9.18, it can be found that the initial stress changes the distribution of shear stress related to the geometric dimension and deposition temperature, which is in some degree consistent with the results in Fig. 9.16.

Effects of the substrate thickness and substrate tension on the interfacial shear stress are shown in Fig. 9.19, where the deposition temperature is maintained at 150 °C. The interfacial shear stress is a monotonically decreasing function of the substrate tension when the substrate thickness is kept constant. Similarly, the interfacial shear stress is a monotonically decreasing function of the substrate thickness when the substrate tension remains constant.

Effects of the substrate thickness and substrate tension on the interfacial shear stress are shown in Fig. 9.20, where the deposition temperature is maintained at 200 °C. Variation trend of the interfacial shear stress with the substrate tension

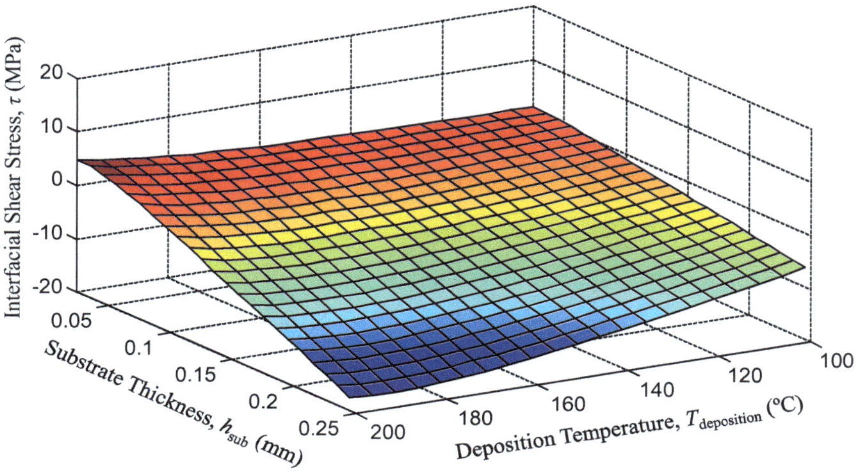

Fig. 9.18 Effects of the substrate thickness and the deposition temperature on the interfacial shear stress ($\sigma_{sub}^{initial} = 0.25$ kPa). © 2011 IEEE. Reprinted, with permission, from Ref. [26]

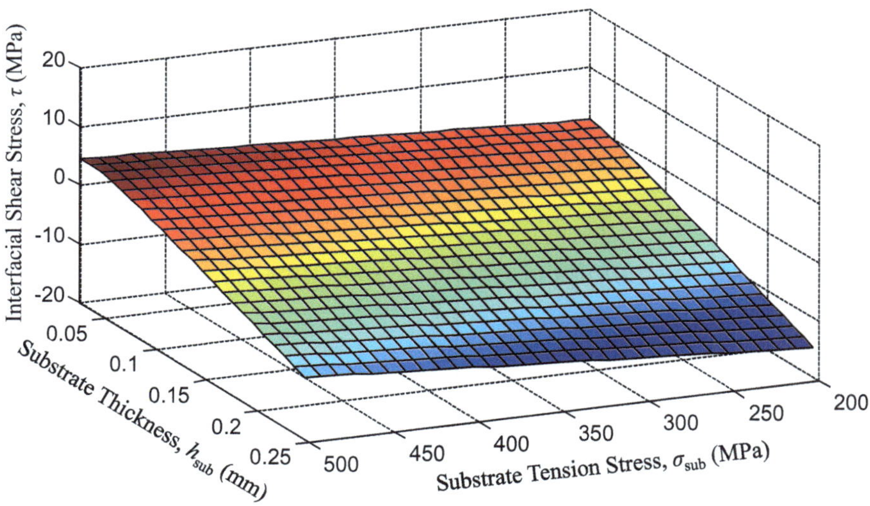

Fig. 9.19 Effects of the substrate thickness and the substrate tension on the interfacial shear stress ($T_{deposition} = 50$ °C). © 2011 IEEE. Reprinted, with permission, from Ref. [26]

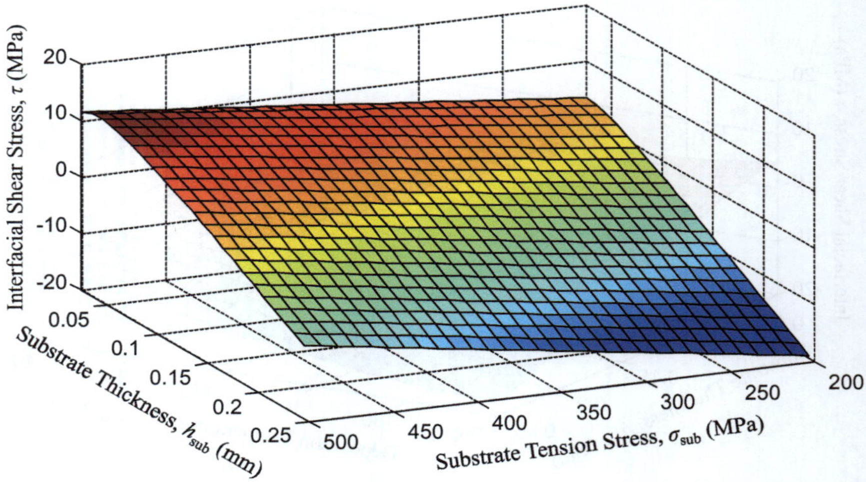

Fig. 9.20 Effects of the substrate thickness and the substrate tension on the interfacial shear stress ($T_{deposition} = 200$ °C). © 2011 IEEE. Reprinted, with permission, from Ref. [26]

and substrate thickness is similar to that in Fig. 9.19. By comparing Fig. 9.19 with Fig. 9.20. It can be found that the deposition temperature only affects the distribution of the interfacial shear stress related to the substrate thickness and substrate tension. Besides, the results in Figs. 9.19 and 9.20 are in similar variation trend.

9.3.4 Co-optimization on Material, Structure, and Process

In aspect of the material, the co-optimization refers to the selection of a suitable Young's modulus and CTE. As to the structure, it refers to the design of the thickness of the film and substrate. As to the process, it refers to the determination of the temperature and substrate tension. It can be noted from Figs. 9.17, 9.18 and 9.19 that there exist optimal combinations of material properties, geometric dimensions, and process parameters to attain a smallest shear stress. When thermal and mechanical loads simultaneously act on the substrate, a nonlinear phenomenon can be found. It gives an idea to reduce the interfacial shear stress by adjusting parameters of structures, materials, and process. By substituting $E_{sub} = E_{sub}^{initial} - k(T - T_{ref})$ into Eqs. (9.42)–(9.43), the maximum shear stress can be expressed as

$$\tau_{\max} = \left[(1+\nu_{\text{film}})\alpha_{\text{film}}\left(T_{\text{work}}+T_{\text{deposition}}\right)-(1+\nu_{\text{sub}})\alpha_{\text{sub}}(T_{\text{work}}+T_{\text{room}})-\frac{\sigma_{\text{sub}}^{\text{initial}}}{\overline{E}_{\text{sub}}h_{\text{sub}}}\right] \times \frac{\dfrac{1}{\overline{E}_{\text{film}}h_{\text{film}}}+\dfrac{1}{\overline{E}_{\text{sub}}h_{\text{sub}}}+3\dfrac{\overline{E}_{\text{sub}}h_{\text{sub}}^2-\overline{E}_{\text{film}}h_{\text{film}}^2}{\overline{E}_{\text{sub}}h_{\text{sub}}+\overline{E}_{\text{film}}h_{\text{film}}}\left[\dfrac{1}{\overline{E}_{\text{film}}h_{\text{film}}^2}+\dfrac{1}{\overline{E}_{\text{sub}}h_{\text{sub}}^2}\right]}{3\left(\dfrac{1}{\overline{E}_{\text{film}}h_{\text{film}}}+\dfrac{1}{\overline{E}_{\text{sub}}h_{\text{sub}}}\right)^2 l_{\text{film-on-sub-half}}} \tag{9.46}$$

The right side of Eq. (9.46) can be divided into two parts. One is closely related to the R2R processing, and mainly focused on the process temperature and substrate tension. The other is comparatively related to the device design, and explicitly includes the material and structural parameters such as the Young's modulus and thickness when temperature-dependent properties are not considered. It is a multivariable optimization issue involving material parameter k, film thickness, substrate thickness, tensile stress on substrate, and deposition temperature, as given by $f_{\min} = \min_{k,h_{\text{sub}},\sigma}\left\{\max\left[\left|\tau_{\max}(k, h_{\text{sub}}, \sigma_{\text{sub}}^{\text{initial}})\right|\right]\right\}$, where $0 < k < 0.01$, $0.01\ \text{mm} < h_{\text{sub}} < 1\ \text{mm}$, $0.1\ \mu\text{m} < h_{\text{film}} < 1\ \mu\text{m}$, $0 < \sigma_{\text{sub}}^{\text{initial}} < 1\ \text{MPa}$, $100\ °\text{C} < T_{\text{deposition}} < 200\ °\text{C}$. The thickness of flexible substrate is 50–80 μm, 50–200 μm, and 25–150 μm for the glass, polymer, and steel, respectively. The thickness of thin film is slightly less than 1 μm.

In order to show the significance of the co-optimization approach, the relationship among these parameters are discussed based on the analytical solution. The maximum shear stress in Eq. (9.46) may be positive or negative, which depends on the thickness and material properties of the film and substrate, deposition temperature, and substrate tension. In order to get $f_{\min}$, one can set $\tau_{\max} = 0$, namely, the first and second items of Eq. (9.46) are set to be zero. The first item results from the misfit related to processes and material, and the second term is related to the structure itself. The corresponding approach can be regarded as the zero-interface stress-based method.

First, the second item of Eq. (9.46) is adopted to be zero to make $\tau_{\max} = 0$, as follows:

$$\frac{1}{\overline{E}_{\text{film}}h_{\text{film}}}+\frac{1}{\overline{E}_{\text{sub}}h_{\text{sub}}}+3\frac{\overline{E}_{\text{sub}}h_{\text{sub}}^2-\overline{E}_{\text{film}}h_{\text{film}}^2}{\overline{E}_{\text{sub}}h_{\text{sub}}+\overline{E}_{\text{film}}h_{\text{film}}}\left[\frac{1}{\overline{E}_{\text{film}}h_{\text{film}}^2}+\frac{1}{\overline{E}_{\text{sub}}h_{\text{sub}}^2}\right]=0 \tag{9.47}$$

If the Young's modulus of the film and substrate are determined, the relationship between the thickness of the film and substrate can be obtained. A comparison of the results of neutral-plane-based method and zero-interface stress-based method are shown in Fig. 9.21. As can be seen, the ratio of the film thickness to substrate thickness by the zero-interface stress-based method is larger than that of the moment-based method.

Second, in order to attain the minimum interfacial shear stress, one can make the first item to be zero, and,

$$h_{\text{sub}} = \frac{\sigma_{\text{sub}}^{\text{initial}}}{\overline{E}_{\text{sub}}\left[(1+\nu_{\text{film}})\alpha_{\text{film}}(T_{\text{work}} - T_{\text{deposition}}) - (1+\nu_{\text{sub}})\alpha_{\text{sub}}(T_{\text{work}} - T_{\text{room}})\right]} \tag{9.48}$$

It can be noted from Eq. (9.48) that the interfacial stress can be reduced by adjusting the substrate thickness, deposition temperature, and substrate tension. The relationship between the substrate thickness, deposition temperature, and substrate tension under the condition of minimum interfacial shear stress is shown in Fig. 9.22. Above analytical results may be achieved for all parameters that are located in reasonable ranges. Nonideal disturbances are unavoidable in practical design and manufacturing process. In order to get an optimal result, it is recommended that Eq. (9.47) is first adopted to get the structure parameters set, then these parameters can be substituted into Eq. (9.48) to get the process parameters.

9.4 Thermal-Pressing Consolidation in Chip Placing-on Process

9.4.1 LWT-Based Mechanical Model

A chip-adhesive-substrate structure after the adhesive curing is shown in Fig. 9.23a, where the thermal-compression produces a small quantity of overflowed ACA around the chip. Due to the very small metal bumps on chip and particles in Fig. 9.23b, the adhesive area in the middle layer is dominated as illustrated in Fig. 9.23c. In this case, the warpage of the ultra-thin assembly mainly depends on thermal stresses. The effect

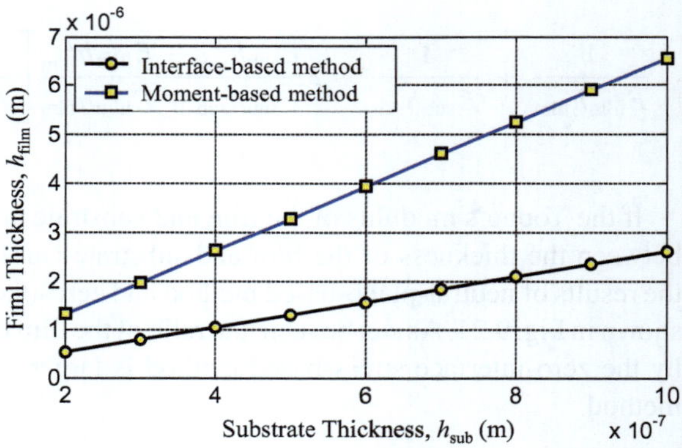

Fig. 9.21 Comparison results of neutral-plane-based method and zero-interface stress-based method. © 2011 IEEE. Reprinted, with permission, from Ref. [26]

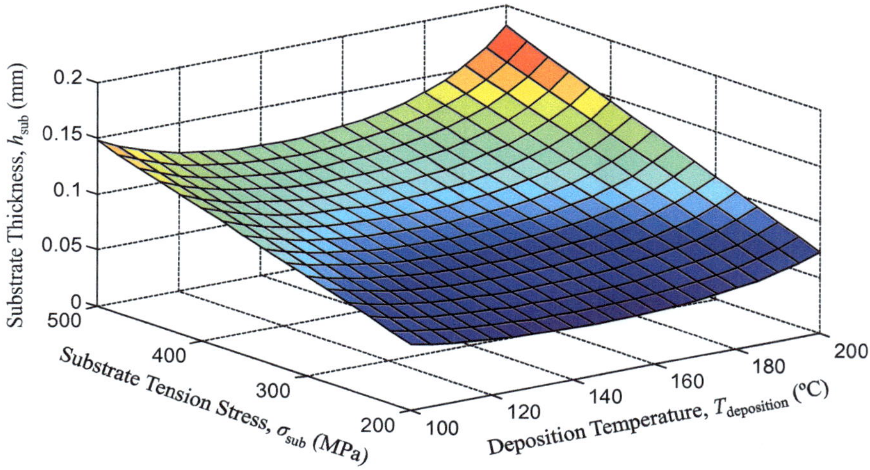

Fig. 9.22 Co-optimization design based on material, structure, and process. © 2011 IEEE. Reprinted, with permission, from Ref. [26]

of the compressing can thus be ignored [31]. A trilayer structure is constructed, as shown in Fig. 9.23d. Verifications of these assumptions and two-dimensional/three-dimensional models have been conducted [32, 33]. The length of each layer at the stage of thermal stresses calculation is assumed to be equal to the chip length. Further, in consideration of the symmetry, only the right half of the structure is modeled, as illustrated in Fig. 9.23e, in which the temperature changes are, respectively, represented by ΔT_{chip}, ΔT_{adh}, and ΔT_{sub} for the chip, adhesive, and substrate layers.

Within the framework of the LWT, the physical trilayer structure is divided into a finite number of mathematical plies through the thickness direction, denoted by n as illustrated in Fig. 9.23e, resulting in $n + 1$ interfaces. In detail, the number of mathematical plies for the substrate, adhesive, and chip layers is denoted by n_{sub}, n_{adh}, and n_{chip}, respectively. The corresponding thickness of kth ply h_k is denoted as h_{sub}/n_{sub} at $1 \le k \le n_{sub}$, h_{adh}/n_{adh} at $n_{sub} + 1 \le k \le n_{sub} + n_{adh}$, h_{chip}/n_{chip} at $n_{sub} + n_{adh} + 1 \le k \le n$, where $n = n_{sub} + n_{adh} + n_{chip}$. Likewise, the material properties and temperature change of kth ply denoted as E_k, v_k, α_k, and ΔT_k are in accordance with the corresponding physical layer. Furthermore, the trilayer laminate has a rectangular coordinate with the origin at the midpoint of the left end as shown in Fig. 9.23e, given total thickness of the trilayer $h_{total} = h_{chip} + h_{adh} + h_{sub}$. In plain strain model, the layerwise displacement components $u(x, z)$ and $w(x, z)$ are functions of x coordinate and z coordinate:

$$u(x, z) = U_k(x)\Phi_k(z) \tag{9.49}$$

$$w(x, z) = W_k(x)\Phi_k(z) \tag{9.50}$$

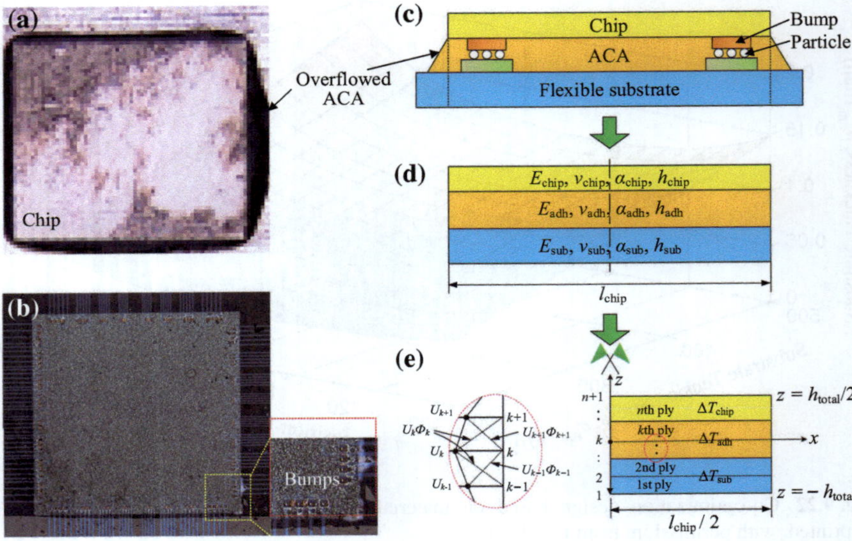

Fig. 9.23 Mechanical model of the chip-adhesive-substrate structure: **a** actual adhesive structure with overflowed ACA around the chip, **b** small metal bumps on the chip, **c** schematic illustration of the chip assembly, **d** a simplified model with a uniform trilayer structure, **e** right-half portion of the symmetric structure. Here, the trilayer is divided into a finite number of mathematical plies based on the LWT, and the enlarged schematic illustration represents the layerwise displacement and linear Lagrangian interpolation functions. Reprinted by permission from Springer Nature, Ref. [12], Copyright 2016

where $k = 1, 2, \ldots, n + 1$, $z \in [-h_{total}/2, h_{total}/2]$, $U_k(x)$, and $W_k(x)$ represent the x-displacement and z-displacement components of any point on the kth surface, respectively, $\Phi_k(z)$ is the local Lagrangian interpolation functions that are continuous in the z-direction [34], and can be represented by

$$\Phi_k(z) = \begin{cases} 0, & z \le z_{k-1} \\ (z - z_{k-1})/h_{k-1}, & z_{k-1} \le z \le z_k \\ (z_{k+1} - z)/h_k, & z_k \le z \le z_{k+1} \\ 0, & z \ge z_{k+1} \end{cases} \tag{9.51}$$

Substituting Eq. (9.49) into the strain–displacement relationship of linear elasticity, there exist

$$\begin{bmatrix} \varepsilon_x \\ \varepsilon_z \\ \gamma_{xz} \end{bmatrix}^k = \begin{bmatrix} U_k' \Phi_k \\ W_k \Phi_k' \\ U_k \Phi_k' + W_k' \Phi_k \end{bmatrix} - \begin{bmatrix} \alpha_k \Delta T_k \\ \alpha_k \Delta T_k \\ 0 \end{bmatrix}, \tag{9.52}$$

and the stresses are

$$\begin{bmatrix} \sigma_x \\ \sigma_z \\ \tau_{xz} \end{bmatrix}^k = \begin{bmatrix} C_{11}^k & C_{13}^k & 0 \\ C_{31}^k & C_{33}^k & 0 \\ 0 & 0 & C_{55}^k \end{bmatrix} \begin{bmatrix} \varepsilon_x \\ \varepsilon_z \\ \gamma_{xz} \end{bmatrix}^k \tag{9.53}$$

where $C_{11}^k = C_{33}^k = E_k/(1 - \nu_k^2)$, $C_{13}^k = C_{31}^k = \nu_k E_k/(1 - \nu_k^2)$, and $C_{55}^k = E_k/2(1 - \nu_k)$.

9.4.2 Equilibrium Equations and Solution Strategy

If the potential energy of the system is defined as U_{system}, and the potential energy of external loads is denoted by V_{external}, based on the principle of minimum total potential energy, $\delta U_{\text{system}} + \delta V_{\text{external}} = 0$, equilibrium equations of the trilayer laminate can be derived as

$$\delta U_k : \frac{\mathrm{d}M_x^k}{\mathrm{d}x} - Q_x^k = 0 \tag{9.54}$$

$$\delta W_k : \frac{\mathrm{d}R_x^k}{\mathrm{d}x} - N_x^k = 0 \tag{9.55}$$

where the results of stress are defined as

$$M_x^k = \int_{-h_{\text{total}}/2}^{h_{\text{total}}/2} \sigma_x \Phi_k \mathrm{d}z \tag{9.56}$$

$$N_z^k = \int_{-h_{\text{total}}/2}^{h_{\text{total}}/2} \sigma_z \Phi_k' \mathrm{d}z \tag{9.57}$$

$$Q_x^k = \int_{-h_{\text{total}}/2}^{h_{\text{total}}/2} \tau_{xz} \Phi_k' \mathrm{d}z \tag{9.58}$$

$$R_x^k = \int_{-h_{\text{total}}/2}^{h_{\text{total}}/2} \tau_{xz} \Phi_k \mathrm{d}z \tag{9.59}$$

Substituting Eq. (9.53) into Eqs. (9.56)–(9.59), expressions of the forces can be obtained as

$$M_x^k = U_j' \int_{-h_{\text{total}}/2}^{h_{\text{total}}/2} C_{11}^j \Phi_j \Phi_k \mathrm{d}z + W_j \int_{-h_{\text{total}}/2}^{h_{\text{total}}/2} C_{13}^j \Phi_j' \Phi_k \mathrm{d}z + M_x^{kT} \tag{9.60}$$

$$N_z^k = U_j' \int_{-h_{\text{total}}/2}^{h_{\text{total}}/2} C_{31}^j \Phi_j \Phi_k' \mathrm{d}z + W_j \int_{-h_{\text{total}}/2}^{h_{\text{total}}/2} C_{33}^j \Phi_j' \Phi_k' \mathrm{d}z + N_z^{kT} \tag{9.61}$$

$$Q_x^k = U_j \int_{-h_{\text{total}}/2}^{h_{\text{total}}/2} C_{55}^k \Phi_j' \Phi_k' \mathrm{d}z + W_j' \int_{-h_{\text{total}}/2}^{h_{\text{total}}/2} C_{55}^k \Phi_j \Phi_k' \mathrm{d}z \tag{9.62}$$

$$R_x^k = U_j \int_{-h_{\text{total}}/2}^{h_{\text{total}}/2} C_{55}^k \Phi_j' \Phi_k \mathrm{d}z + W_j' \int_{-h_{\text{total}}/2}^{h_{\text{total}}/2} C_{55}^k \Phi_j \Phi_k \mathrm{d}z \tag{9.63}$$

where $M_x^{kT} = \int_{-h_{\text{total}}/2}^{h_{\text{total}}/2} \left[\left(C_{11}^j + C_{13}^j\right)\alpha_j \Delta T_j \Phi_k\right] \mathrm{d}z$, and $N_z^{kT} = \int_{-h_{\text{total}}/2}^{h_{\text{total}}/2} \left[\left(C_{11}^j + C_{13}^j\right)\alpha_j \Delta T_j \Phi_k'\right] \mathrm{d}z$.

Taking Eqs. (9.60)–(9.63) into Eq. (9.54), the system equilibrium equations can be obtained as

$$D_{11}^{kj} U_j'' + \left(B_{13}^{kj} - B_{55}^{jk}\right) W_j' - A_{55}^{kj} U_j = 0 \tag{9.64}$$

$$D_{55}^{kj} W_j'' + \left(B_{55}^{kj} - B_{31}^{jk}\right) U_j' - A_{33}^{kj} W_j = N_z^{kT} \tag{9.65}$$

where N_z^{kT} indicates the result of thermal stress, A_{33}^{kj}, A_{55}^{kj}, B_{13}^{kj}, B_{31}^{jk}, B_{55}^{kj}, B_{55}^{jk}, D_{11}^{kj}, and D_{55}^{kj} as the rigidity terms are the constant coefficients of the equilibrium equations, and can be expressed as follows:

$$A_{33}^{kj} = \int_{-h_{\text{total}}/2}^{h_{\text{total}}/2} C_{33}^j \Phi_j' \Phi_k' \mathrm{d}z \tag{9.66}$$

$$A_{55}^{kj} = \int_{-h_{\text{total}}/2}^{h_{\text{total}}/2} C_{55}^j \Phi_j' \Phi_k' \mathrm{d}z \tag{9.67}$$

$$B_{13}^{kj} = \int_{-h_{\text{total}}/2}^{h_{\text{total}}/2} C_{13}^j \Phi_j' \Phi_k \mathrm{d}z \tag{9.68}$$

$$B_{31}^{jk} = \int_{-h_{\text{total}}/2}^{h_{\text{total}}/2} C_{31}^j \Phi_j \Phi_k' \mathrm{d}z \tag{9.69}$$

$$B_{55}^{kj} = \int_{-h_{\text{total}}/2}^{h_{\text{total}}/2} C_{55}^j \Phi_j' \Phi_k \mathrm{d}z \tag{9.70}$$

$$B_{55}^{jk} = \int_{-h_{\text{total}}/2}^{h_{\text{total}}/2} C_{55}^j \Phi_j \Phi_k' \mathrm{d}z \tag{9.71}$$

$$D_{11}^{kj} = \int_{-h_{\text{total}}/2}^{h_{\text{total}}/2} C_{11}^j \Phi_j \Phi_k \mathrm{d}z \tag{9.72}$$

$$D_{55}^{kj} = \int_{-h_{\text{total}}/2}^{h_{\text{total}}/2} C_{55}^j \Phi_j \Phi_k \mathrm{d}z \tag{9.73}$$

Carrying out the integrations into Eqs. (9.66)–(9.73), the rigidity terms can be calculated by

$$A_{pq}^{kj}(pq=33,55)=\begin{cases}-\dfrac{C_{pq}^{(k-1)}}{h_{k-1}}, & (j=k-1)\\ \dfrac{C_{pq}^{(k-1)}}{h_{k-1}}+\dfrac{C_{pq}^{(k)}}{h_k}, & (j=k)\\ -\dfrac{C_{pq}^{(k)}}{h_k}, & (j=k+1)\\ 0, & (j<k-1 \quad \text{or} \quad j>k+1)\end{cases} \tag{9.74}$$

$$B_{pq}^{kj}(pq=13,55)=\begin{cases}-\dfrac{C_{pq}^{(k-1)}}{2}, & (j=k-1)\\ \dfrac{C_{pq}^{(k-1)}}{2}-\dfrac{C_{pq}^{(k)}}{2}, & (j=k)\\ \dfrac{C_{pq}^{(k)}}{2}, & (j=k+1)\\ 0, & (j<k-1 \quad \text{or} \quad j>k+1)\end{cases} \tag{9.75}$$

$$D_{pq}^{kj}(pq=11,55)=\begin{cases}\dfrac{C_{pq}^{(k-1)}}{6}h_{k-1}, & (j=k-1)\\ \dfrac{C_{pq}^{(k-1)}}{3}h_{k-1}+\dfrac{C_{pq}^{(k)}}{3}h_k, & (j=k)\\ \dfrac{C_{pq}^{(k)}}{6}h_k, & (j=k+1)\\ 0, & (j<k-1 \quad \text{or} \quad j>k+1)\end{cases} \tag{9.76}$$

The above ordinary differential equations can be further expressed in a concise form

$$[A]\{\Phi''\}+[B]\{\Phi'\}+[C]\{\Phi\}=\{F\} \tag{9.77}$$

where $[A]=\begin{bmatrix}D_{11}^{kj} & 0\\ 0 & D_{55}^{kj}\end{bmatrix}_{2(N+1)\times 2(N+1)}$, $[B]=\begin{bmatrix}0 & \left(B_{13}^{kj}-B_{55}^{jk}\right)\\ \left(B_{55}^{kj}-B_{31}^{jk}\right) & 0\end{bmatrix}_{2(N+1)\times 2(N+1)}$, $[C]=\begin{bmatrix}-A_{55}^{kj} & 0\\ 0 & -A_{33}^{kj}\end{bmatrix}_{2(N+1)\times 2(N+1)}$, $\Phi=\{U_k, W_k\}^T$, $F=\{0, N_z^{kT}\}^T$.

The solution to Eq. (9.77) can be obtained by superimposing general solution $\{\Psi\}$ of the corresponding set of homogeneous ordinary differential equations to a particular solution $\{\Phi_s\}$, that is $\{\Phi\}=\{\Psi\}+\{\Phi_s\}$. Thus,

$$\{\Phi_s\}=[C]^{-1}\{F\} \tag{9.78}$$

$$[A]\{\Psi''\}+[B]\{\Psi'\}+[C]\{\Psi\}=\{0\} \tag{9.79}$$

To solve Eq. (9.79), assuming that the general solution $\{\Psi\} = \{\Psi_0\}\exp(\lambda x)$, where λ and $\{\Psi_0\}$ are, respectively, the eigenvalue and eigenvector of the characteristic equation corresponding to Eq. (9.79). There exists

$$\lambda^2[A]\{\Psi_0\} + \lambda[B]\{\Psi_0\} + [C]\{\Psi_0\} = \{0\} \tag{9.80}$$

Assuming $\{\Psi_1\} = \lambda\{\Psi_0\}$, a generalized eigenvalue problem can be converted from Eq. (9.80)

$$\lambda\begin{bmatrix} I & 0 \\ 0 & A \end{bmatrix}_{4(N+1)\times 4(N+1)} \begin{Bmatrix} \Psi_0 \\ \Psi_1 \end{Bmatrix} = \begin{bmatrix} 0 & I \\ -C & -B \end{bmatrix}_{4(N+1)\times 4(N+1)} \begin{Bmatrix} \Psi_0 \\ \Psi_1 \end{Bmatrix} \tag{9.81}$$

Final expression of the general solution to Eq. (9.77) is

$$\begin{Bmatrix} \Phi_0 \\ \Phi_1 \end{Bmatrix} = [V]_{4(N+1)\times 4(N+1)}[Q(x)]\{K\}_{4(N+1)\times 1} + [C]^{-1}\{F\}_{4(N+1)\times 1} \tag{9.82}$$

where $\{\Phi_0, \Phi_1\}^T = \left[U_k, W_k, U_k', W_k'\right]^T$, $Q(x) = \mathrm{diag}(e^{\lambda_1 x}, e^{\lambda_2 x}, \ldots, e^{\lambda_{4(N+1)} x})$. $[V]_{4(N+1)\times 4(N+1)}$ are eigenvectors corresponding to eigenvalues λ, $\{K\}_{4(N+1)\times 1}$ are the real or complex constants to be determined by the boundary conditions in Fig. 9.23e.

9.4.3 Boundary Conditions

(a) Symmetric conditions at the left end

$$U_k = 0, \quad Q_x^k = \int_{-h_k/2}^{h_k/2} \tau_{xz}^k \mathrm{d}z = 0 \tag{9.83}$$

(b) Free boundary conditions at the right end

$$M_x^k = 0, \quad R_x^k = 0 \tag{9.84}$$

(c) Temperature distribution and evolution of the trilayer system, ΔT_{chip}, ΔT_{adh}, and ΔT_{sub}, are obtained by the transient thermal finite element analysis.

9.4.4 Numerical Approach

In order to verify the proposed theoretical approach, a finite element model of the UTCOF assembly is defined using Abaqus software. The numerical approach associated with the actual ultra-thin chip bonding process involves one step of thermal and two steps of thermal-mechanical analysis. In the first step, the transient thermal analysis is performed to characterize the thermal profile of the UTCOF assembly induced in the curing process, where effects of the air convection and platform on heat transferring are also considered. The characterized thermal profile is then taken as the temperature loading of the subsequent thermal-mechanical analysis. A subroutine is compiled to set up Young's modulus of the adhesive layer at a very small value before the adhesive curing. At last, the thermal-mechanical model continues to be used in the cooling process.

A common case is provided here based on the numerical simulation to verify analytical results. Material properties of the chip-ACA-substrate structure are listed in Table 9.2. Geometric dimensions are set as: $h_{chip} = h_{adh} = 50\ \mu m$, $h_{sub} = 100\ \mu m$, $l_{chip} = 5$ mm, and $l_{sub} = 15$ mm. The thickness and length of the platform in thermal analysis are set as 10 mm and 15 mm, respectively. All layers are assumed to be linear elastic and isotropic. The element DC2D4 is employed in the heat transfer analysis. In the thermal-mechanical analysis, the whole structure is considered to deform under the condition of plane strain using the element CPE4R. Only the right-half portion of the structure is constructed considering both the symmetry and computational cost.

In the step of thermal finite element modeling, the required heating time and temperature for the curing process are determined by chemical properties of the adhesive layer. Here, considering the demand of actual ACA curing, temperature profile of the transient load is illustrated in Fig. 9.24a, which starts with a temperature arose from 120 to 190 °C at an increasement of 3.5 °C/s, and followed by a 5 s dwell time at high temperature. Meanwhile, the temperature of the platform maintains constant at 80 °C throughout the curing process. Temperature distribution of the assembly at heating time $t = 25$ s and the temperature history of marked points A, B, C, and D during the ACA curing process are shown in Fig. 9.24b. As can be seen, there exists a significant temperature gradient across the trilayer assembly. Here, it is assumed that the adhesive curing completes once the average temperature of the adhesive layer exceeds the glass transition temperature $T_g = 140$ °C. The critical heating time for the average temperature of the adhesive layer reaching the glass transition temperature is termed as t_g, which is also the time to yield the initial adhesive stresses. Therefore,

Table 9.2 Material properties in the thermal and thermal-mechanical finite element modeling

Material (–)	Thermal conductivity (W/m °C)	Specific heat (J/kg °C)	Density (kg/m^3)	Young's modulus (GPa)	Poisson ratio (–)	CTE (ppm/°C)
Chip	120	708	2330	129	0.28	2.7
Adhesive	0.43 at 21 °C	1100 at 21 °C	1550	3.30 at 21 °C	0.42 at 21 °C	61 at 21 °C
	0.47 at 100 °C	1250 at 100 °C		1.65 at 100 °C	0.42 at 100 °C	61 at 100 °C
	0.51 at 140 °C	1500 at 140 °C		0.8 at 140 °C	0.45 at 140 °C	180 at 140 °C
	0.56 at 190 °C	1500 at 190 °C		0.8 at 190 °C	0.45 at 190 °C	180 at 190 °C
Substrate	0.5	1700	1470	9.6	0.3	14
Platform	120	900	3100	–	–	–

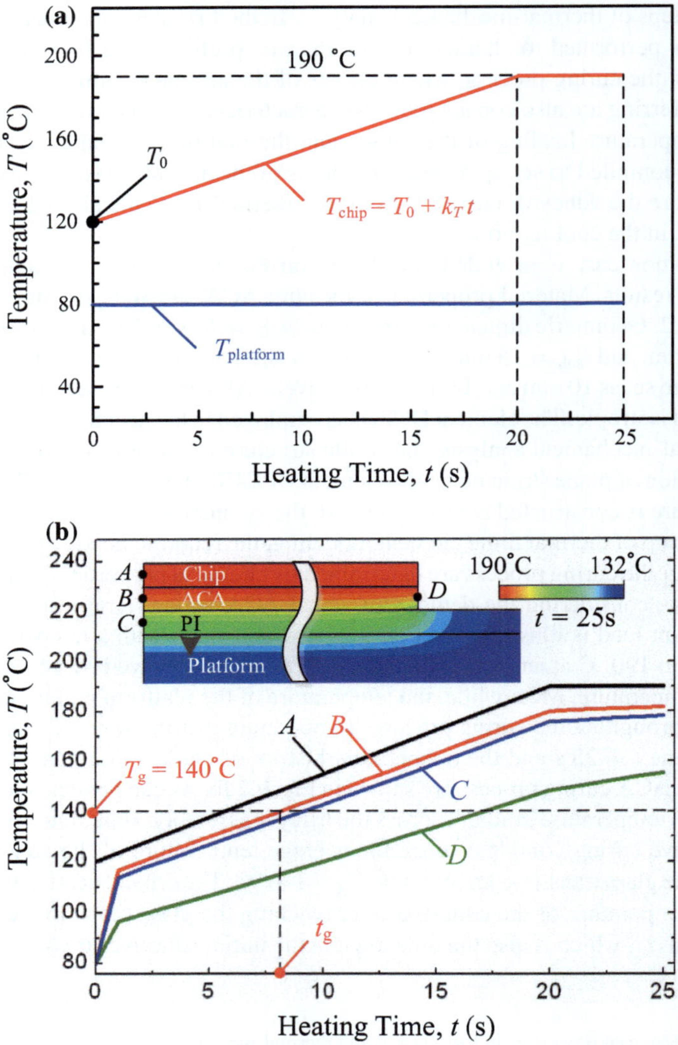

Fig. 9.24 Numerical results of transient thermal analysis: **a** variation of the bonding temperature profile in the adhesive curing process, **b** variation of the curing-induced temperature with the heating time. Reprinted by permission from Springer Nature, Ref. [12], Copyright 2016

the average temperature at the critical heating time across the chip/substrate thickness is, respectively, 148/113 °C, while the average temperature at $t = 25$ s across the chip/adhesive/substrate thickness is, respectively, 190/182/155 °C. The thermal results are further adopted as inputs of the subsequent thermal-mechanical analysis and thermal boundary conditions of the theoretical model in Fig. 9.23e.

9.4.5 Mechanism of Ultra-thin Chip Warpage

Based on results of the thermal finite element analysis, the thermal boundary condition of the heating process is temperature changes of each layer from t_g to $t = 25$ s, that is, $\Delta T_{chip} = \Delta T_{adh} = \Delta T_{sub} = 42$ °C. The final chip warpage induced during the subsequent cooling process should superimpose the result induced during the heating process, whose thermal boundary is equivalent to a cooling process from the temperature of assembly at t_g to the ambient temperature 21 °C. Thus, the thermal boundary condition of the cooling process can be given as $\Delta T_{chip} = -127$ °C, $\Delta T_{adh} = -119$ °C, and $\Delta T_{sub} = -92$ °C. Taking these thermal conditions into the analytical model and setting the structural dimensions and material properties the same as above analysis, the ultra-thin chip warpage induced during the UTCOF packaging can be theoretically predicted.

Here, it should be noted that some of the adhesive material properties listed in Table 9.2 are temperature-dependent. In order to obtain a relatively precise analytical result, the temperature range is divided into several small temperature intervals (for example, $\Delta T = 5$–10 °C), and the material properties are adopted as average values in corresponding small temperature interval. The total warpage value can be obtained approximately by summarizing all warpage values at each divided temperature interval. Additionally, based on the LWT, the larger number of subdivided mathematical plies is beneficial to a more precise prediction of the theoretical model, despiting a larger amount of calculation [35–37]. Here, taking into account the efficiency and precision of calculation, n_{sub}, n_{adh}, and n_{chip} are selected as 4, 2, and 2, respectively, in the theoretical model.

The mechanism of the ultra-thin chip warpage under conditions of high and normal temperature is depicted in Fig. 9.25, where the former is under thermal loading at $t = 25$ s. The theoretical and numerical results represented, respectively, by solid and dash lines agree well with each other. It is obvious that a concave warpage generates at the end of the thermo-curing process under the condition of high temperature, while a convex warpage generates at the end of the cooling process under the condition of normal temperature. Besides, the extent of the chip warpage under the condition of normal temperature is larger than that at high temperature. It can also be indicated that the temperature difference between the temperature state at $t = t_g$ and the high/normal temperature determines the extent of the ultra-thin chip warpage. Thus, the application of the ACA with a lower glass transition temperature and heating temperature is recommended to reduce the extent of the chip warpage.

9.4.6 Effects of Structure Dimensions and Materials

Effects of the chip length and the chip thickness on the maximum warpage under conditions of high and normal temperature are shown in Fig. 9.26. Here, unless specified otherwise, other geometric dimensions and material properties are set the

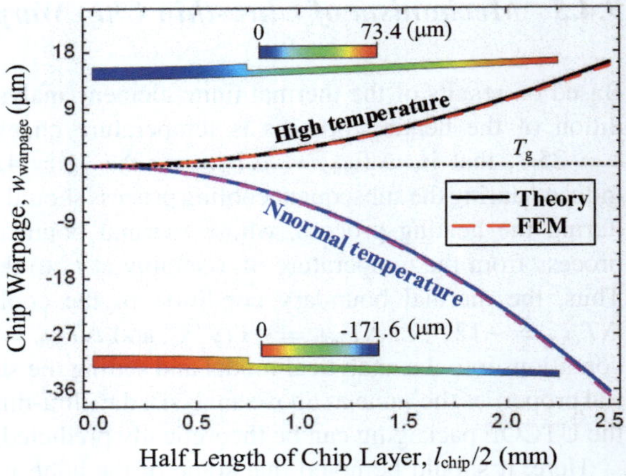

Fig. 9.25 Warpage at high and normal temperature versus the chip length, where the theoretical predictions and numerical results are represented by solid and dotted lines, respectively. The corresponding displacement contours are also presented. Reprinted by permission from Springer Nature, Ref. [12], Copyright 2016

same as above discussion. The analytical and numerical results are represented by solid lines and scatter plots, respectively, and they match each other very well. It is shown that the larger chip length and thinner chip thickness promote a more serious extent of the chip warpage, which is due to the fact that the ultra-thin chip with a larger ratio of length to thickness is more likely to be warped under the condition of mismatched stress.

Effects of the thickness, the CTE, and Young's modulus of the adhesive layer on the maximum value of the ultra-thin chip warpage is shown in Fig. 9.27. The analytical results are verified by finite element simulation. It is shown that the effect of the adhesive thickness on the extent of warpage is very limited, while the larger CTE and Young's modulus of the adhesive layer lead to a more serious extent of the warpage of the ultra-thin chip. It can be concluded that the ACA with small CTE and Young's modulus of the adhesive layer should be adopted to reduce the warpage of the ultra-thin chip.

The maximum value of the ultra-thin chip warpage versus the thickness and Young's modulus of the substrate is shown in Fig. 9.28, where the analytical and numerical results are represented by solid lines and scatter plots, respectively. As can be seen, the theoretical results match well with the finite element analysis. Obviously, it can be found that the extent of the chip warpage decreases with the increase of the substrate thickness and Young's modulus of the substrate layer. Actually, both the chip and substrate layers are demanded to be thinner to achieve a higher flexibility and lower height of the chip-ACA-substrate assembly. As deduced from the results in Fig. 9.28b, the extent of warpage can be reduced by packaging the ultra-thin chip on the rigid but ultra-thin substrate.

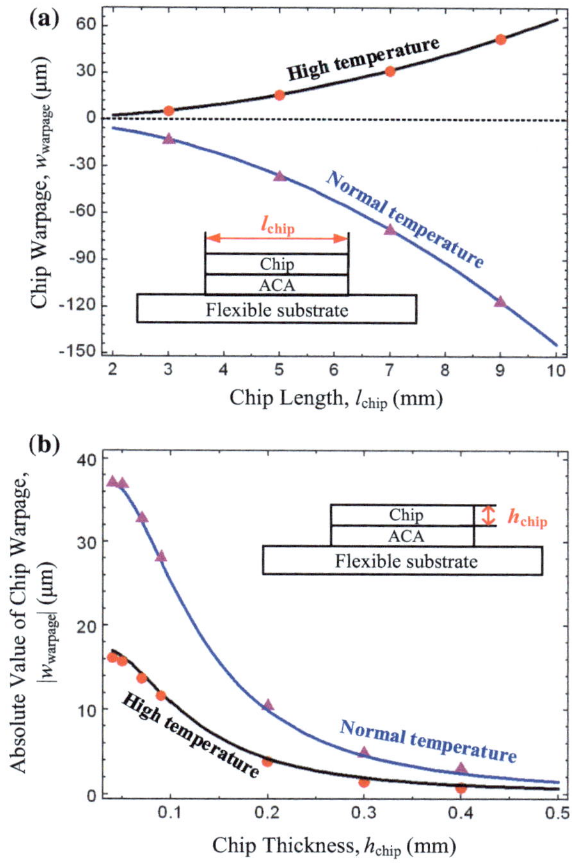

Fig. 9.26 Results of the maximum warpage under conditions of high and normal temperature considering the effects of: **a** the chip length, **b** the chip thickness. The analytical results and finite element simulations are represented by solid lines and scatter plots, respectively. Reprinted by permission from Springer Nature, Ref. [12], Copyright 2016

9.4.7 Experimental Verification

In order to verify the proposed theoretical approach, corresponding experiments for the packaging processes of quadrate chips with the thickness of 50 μm and length of 5 mm on polyimide (PI) substrates with the thickness of 100 μm are conducted, where the thickness of the ACA decided approximately by the total height of the bump on the chip and the circuit on the substrate is about 45–50 μm. In the curing process, the uniform pressure on the top surface of the chip is set at 20 N. The heating conditions on the chip and substrate are set the same as that in Fig. 9.24a. After the curing and cooling processes, the top surface profile of the chip assembly is observed

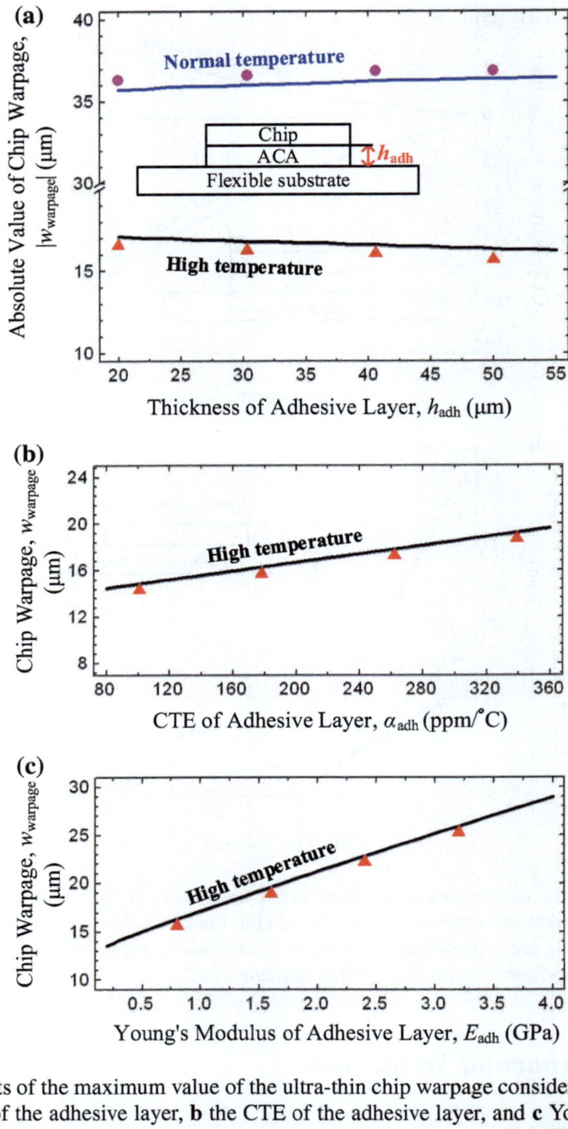

Fig. 9.27 Results of the maximum value of the ultra-thin chip warpage considering the effects of: **a** the thickness of the adhesive layer, **b** the CTE of the adhesive layer, and **c** Young's modulus of the adhesive layer. The analytical and numerical results are represented by solid lines and scatter plots, respectively. Reprinted by permission from Springer Nature, Ref. [12], Copyright 2016

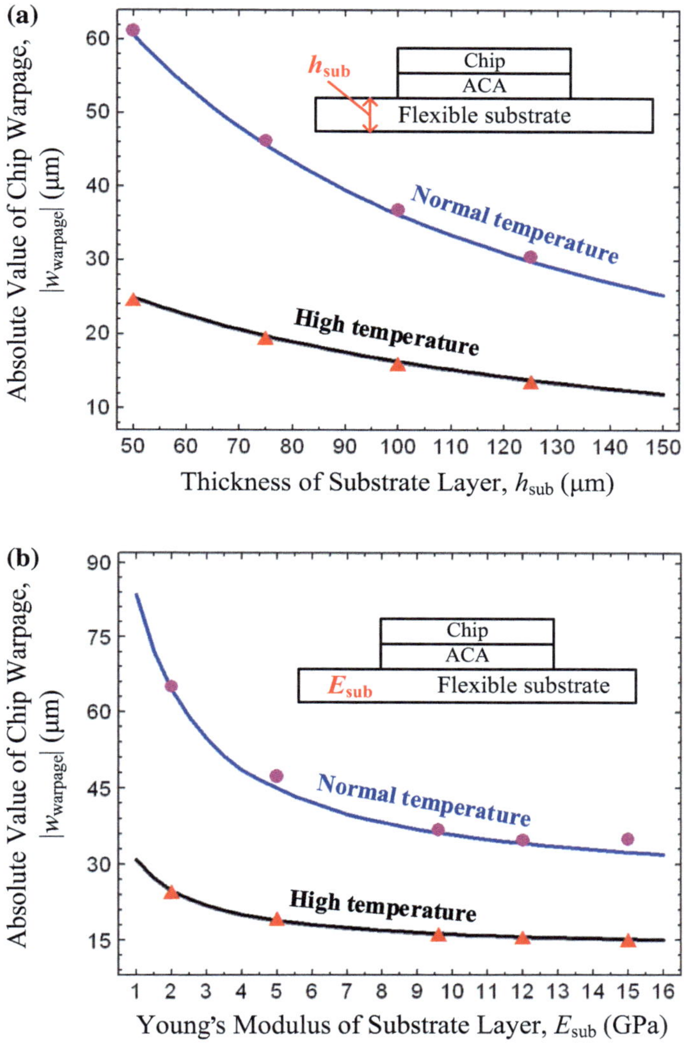

Fig. 9.28 Results of the maximum value of the ultra-thin chip warpage considering the effects of: **a** the thickness of the substrate layer, **b** Young's modulus of the substrate layer, where the analytical and numerical results are represented by solid lines and scatter plots, respectively. Reprinted by permission from Springer Nature, Ref. [12], Copyright 2016

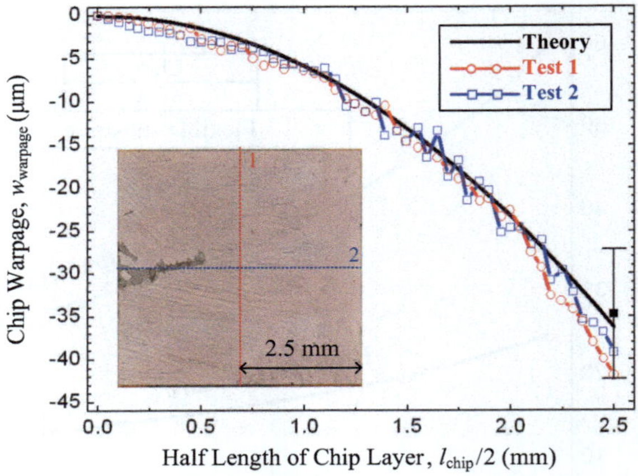

Fig. 9.29 Measured warpage of a 50-μm-thick and 5-mm-length quadrate chip assembly after the curing and cooling processes. Here, the solid line represents the theoretical prediction, and the dotted lines are testing results with the laser scanning confocal microscope. Due to the adoption of the quadrate chip, the warpage value along the half of longitudinal and transverse midline of the top surface of the chip (denoted by 1 and 2 in the scanned profile separately) is extracted. Reprinted by permission from Springer Nature, Ref. [12], Copyright 2016

by the laser scanning confocal microscope (LSCM, Keyence, VK-X200K), followed by measuring the warpage value of the profile through the image analysis software (VK Analyser, 2.1.0, Keyence).

A typical experimental result of the ultra-thin chip warpage after the thermal-compressing and cooling processes is shown in Fig. 9.29. Considering the top surface of the chip, the effective testing results of the ultra-thin chip warpage are along the half of longitudinal and transverse midline of the top surface profile. It is shown that the testing results match the above theoretical predictions very well, even though there are small difference of the maximum warpage between the testing (39.2 μm and 41.1 μm on the testing line 1 and 2, respectively) and theoretical (36.3 μm) results. Additionally, it should be noted that the testing results of the chip warpage are not smooth due to the surface toughness of the chip and the contaminant existence like small amount of cured ACA and particulates. The average value of maximum warpage from nine testing samples is about 34.7 μm, which approximates to the theoretical prediction. Therefore, it can be concluded that the proposed LWT-based theory is available for predicting the warpage of the ultra-thin chip effectively.

9.5 Summary

Theoretical models were first presented to investigate the thin chip picking-up process under hypotheses of constant and variable length of the substrate. Then, an optimization method for the R2R processing was proposed to reduce the interfacial stress of multilayer structure. A LWT-based theoretical model for the UTCOF assembly was last developed to analyze the behavior of thermally induced warpage.

It was found that the higher VAP below the substrate accelerated the chip detachment from the substrate. The chip detachment was mainly driven by the picking-up displacement, and the condition of the VAP also had a great effect on the delamination process. Typically, a reliable picking-up process could be achieved by decreasing the VAP below the substrate and increasing the VAP above the chip. Moreover, the type of the VAP was an important factor affecting the chip picking-up performance. The chip should be kept as flat as possible during the process considering that the chip warpage seriously affected the chip delamination from the substrate. The picking-up head with distributed hole-matrix was thus recommended. The thinner and more compliant substrate could be used to more effectively reduce the picking-up force.

In the analysis of the R2R processing, the interfacial shear stress and peel stress related to parameters of material, structure, and process were given in analytical form. Material selection and structure optimization were studied based on the interfacial stresses. The results highlighted the dramatic effect of the film thickness, substrate thickness, deposition temperature, and substrate tension on the interfacial shear stress. The co-optimization model was established to obtain the minimum interfacial stress. Low interfacial stress could be achieved even when high temperature, 100–200 °C, was adopted for the film deposition. It offers an opportunity to resolve the problem of residual stress in the R2R processing of flexible electronics.

Effects of geometric dimensions and material properties on the chip warpage of the UTCOF assembly were last investigated. The proposed theoretical model could provide an effective way to accurately predict the ultra-thin chip warpage. To reduce the extent of the ultra-thin chip warpage, the ACA with a lower glass transition temperature, smaller CTE, and smaller Young's modulus was recommended. The use of the rigid and ultra-thin flexible substrate like the metal foil promoted not only the flexibility but also the reduction of the ultra-thin assembly warpage.

References

1. Peng B, Huang Y, Yin Z, Xiong Y (2011) Analysis of interfacial peeling in IC chip pick-up process. J Appl Phys 110(7):073508
2. Peng B, Huang Y, Yin Z, Xiong Y (2011) On the analysis of dynamic effect in the die pick-up process. In: 12th international conference on electronic packaging technology and high density packaging, 8–11 Aug 2011. IEEE, pp 1–4
3. Liu Z, Huang Y, Chen J, Yin Z (2014) Tunable peeling technique and mechanism of thin chip from compliant adhesive tapes. IEEE Trans Compon Packag Manuf Technol 4(4):560–568
4. Feil M, Adler C, Hemmetzberger D, Konig M, Bock K (2004) The challenge of ultra thin chip assembly. In: 54th Electronic Components and Technology Conference (IEEE Cat. No. 04CH37546), 4–4 June 2004. IEEE, pp 35–40
5. Peng B, Huang Y, Yin Z, Xiong Y (2012) Competing fracture modeling of thin chip pick-up process. IEEE Trans Compon Packag Manuf Technol 2(7):1217–1225

6. Liu Z, Wan X, Huang Y, Chen J, Yin Z (2018) Theoretical and experimental studies of competing fracture for flexible chip-adhesive-substrate composite structure. IEEE Trans Compon Packag Manuf Technol 8(1):57–64
7. Liu Z, Huang Y, Xiao L, Tang P, Yin Z (2015) Nonlinear characteristics in fracture strength test of ultrathin silicon die. Semicond Sci Technol 30(4):045005
8. Lin YJ, Hwang SJ (2005) Static analysis of the die picking process. IEEE Trans Electron Packag Manuf 28(2):142–149
9. Cheng TH, Du CC, Tseng CH (2006) Study in IC chip failure during pick-up process by using experimental and finite element methods. J Mater Process Technol 172(172):407–416
10. Xu Z, Liu Z, Liu H, Yin Z, Huang Y, Chen J (2015) Analytical evaluation of interfacial crack propagation in vacuum-based picking-up process. IEEE Trans Compon Packag Manuf Technol 5(11):1700–1708
11. Hassan M-U, Angelopoulos EA, Rempp H, Endler S, Burghartz JN (2010) Packaging challenges associated with warpage of ultra-thin chips. In: 3rd electronics system integration technology conference ESTC, 13–16 Sept 2010. IEEE, pp 1–5
12. Chen J, Xu Z, Huang Y, Duan Y, Yin Z (2016) Analytical investigation on thermal-induced warpage behavior of ultrathin chip-on-flex (UTCOF) assembly. Sci China Technol Sci 59(11):1646–1655
13. Lu S-T, Chen W-H (2010) Reliability and flexibility of ultra-thin chip-on-flex (UTCOF) interconnects with anisotropic conductive adhesive (ACA) joints. IEEE Trans Adv Packag 33(3):702–712
14. Yang SY, Kwon W-S, Lee S-B (2012) Chip warpage model for reliability prediction of delamination failures. Microelectron Reliab 52(4):718–724
15. Liu Y (2010) Trends of power semiconductor wafer level packaging. Microelectron Reliab 50(4):514–521
16. Tu K (2011) Reliability challenges in 3D IC packaging technology. Microelectron Reliab 51(3):517–523
17. Liu Z, Valvo PS, Huang Y, Yin Z (2013) Cohesive failure analysis of an array of IC chips bonded to a stretched substrate. Int J Solids Struct 50(22–23):3528–3538
18. Liu Z, Huang Y, Yin Z, Bennati S, Valvo PS (2014) A general solution for the two-dimensional stress analysis of balanced and unbalanced adhesively bonded joints. Int J Adhes Adhes 54:112–123
19. He M, Evans A, Hutchinson J (1997) Convergent debonding of films and fibers. Acta Mater 45(8):3481–3489
20. Xu Z, Liu Z, Huang Y, Chen J, Liu H, Yin Z (2015) Vacuum-based picking-up of thin chip from adhesive tape. J Adhes Sci Technol 29(13):1315–1329
21. Medding J, Stalder R, Niederhauser M, Stoessel P (2004) Thin die bonding techniques. In: IEEE/CPMT/SEMI 29th international electronics manufacturing technology symposium (IEEE Cat. No. 04CH37585), 14–16 July 2004. IEEE, pp 68–73
22. Liu Z, Huang YA, Liu H, Chen J, Yin Z (2014) Reliable peeling of ultrathin die with multineedle ejector. IEEE Trans Compon Packag Manuf Technol 4(9):1545–1554
23. Cheung YM, Chong ACM, Huang B (2006) Determination of the interfacial fracture toughness of laminated silicon die on adhesive dicing tape from stud pull measurement. In: International conference on electronic materials and packaging, 11–14 Dec 2006. IEEE, pp 1–10
24. Chen J, Liu H, Huang Y, Yin Z (2016) High-rate roll-to-roll stack and lamination of multilayer structured membrane electrode assembly. J Manuf Process 23:175–182
25. Huang Y, Yin Z, Xiong Y (2010) Thermomechanical analysis of thin films on temperature-dependent elastomeric substrates in flexible heterogeneous electronics. Thin Solid Films 518(6):1698–1702
26. Huang Y, Chen J, Yin Z, Xiong Y (2011) Roll-to-roll processing of flexible heterogeneous electronics with low interfacial residual stress. IEEE Trans Compon Packag Manuf Technol 1(9):1368–1377
27. Gleskova H, Cheng IC, Wagner S, Sturm JC, Suo Z (2006) Mechanics of thin-film transistors and solar cells on flexible substrates. Sol Energy 80(6):687–693

28. Huang Y, Yin Z, Xiong Y (2010) Thermomechanical analysis of film-on-substrate system with temperature-dependent properties. J Appl Mech Trans ASME 77(4):041016
29. Wang KP, Huang YY, Chandra A, Hu KX (2000) Interfacial shear stress, peeling stress, and die cracking stress in trilayer electronic assemblies. IEEE Trans Compon Packag Technol 23(2):309–316
30. Jiang ZQ, Huang Y, Chandra A (1997) Thermal stresses in layered electronic assemblies. J Electr Packag Trans ASME 119(2):127–132
31. Lu S-T, Chen W-H (2010) Experimental/numerical analysis of thermally induced warpage of ultrathin chip-on-flex (UTCOF) interconnects. IEEE Trans Compon Packag Technol 33(4):819–829
32. Yao Q, Qu J (1999) Three-dimensional versus two-dimensional finite element modeling of flip-chip packages. J Electron Packag 121(3):196–201
33. Zhao J-H, Dai X, Ho PS (1998) Analysis and modeling verification for thermal-mechanical deformation in flip-chip packages. In: 48th electronic components and technology conference (Cat. No. 98CH36206), 25–28 May 1998. IEEE, pp 336–344
34. Nosier A, Kapania RK, Reddy J (1993) Free vibration analysis of laminated plates using a layerwise theory. AIAA J 31(12):2335–2346
35. Yousefsani SA, Tahani M (2013) Accurate determination of stress distributions in adhesively bonded homogeneous and heterogeneous double-lap joints. Eur J Mech A/Solids 39:197–208
36. Yousefsani SA, Tahani M (2014) An analytical investigation on thermomechanical stress analysis of adhesively bonded joints undergoing heat conduction. Arch Appl Mech 84(1):67–79
37. Yousefsani SA, Tahani M (2013) Analytical solutions for adhesively bonded composite single-lap joints under mechanical loadings using full layerwise theory. Int J Adhes Adhes 43:32–41

Appendix

Meaning	Symbol
Absolute value of maximum chip warpage	$\lvert w_{\text{warpage}} \rvert$
Actual transverse displacement of TPB test specimen	w_{absolute}
Adhesion energy	U_{adhesion}
Adhesive fracture energy of i (i = adh, exten, inexten) layer	G_i^c
Adhesive fracture energy of i (i = exten, inexten) layer at reference peel speed	G_i^0
Adhesive fracture energy at peel speed of 0 m/s	G_0
Ambient temperature	T_{ambient}
Arc length of curved TPB test specimen	s_{TPB}
Axial displacement at central axis in film layer	u_{film}
Axial force at crack-tip	$N_{\text{crack-tip}}$
Axial force at film layer	N_{film}
Axial force at left boundary of i (i = chip, sub) layer	N_i^{left}
Band gap	E_{band}
Bending moment applied on adhesively bonded joint	M_{bend}
Bending moment at crack-tip	$M_{\text{crack-tip}}$
Bending moment at left boundary of i (i = chip, sub) layer	M_i^{left}
Bending moment at right boundary of substrate and chip layers	$M_{\text{sub-chip}}^{\text{right}}$
Bending moment in TPB test	M_{TPB}
Bending moment of i (i = adh, film) layer	M_i
Blade radius	r_{blade}
Bubble height	h_{bubble}
Built-in strain	$\varepsilon_{\text{built-in}}$
Built-in stress	$\sigma_{\text{built-in}}$
Chip cracking index	Γ_{crack}
Classical linear solution of deflection angle of TPB test specimen	θ_{TPB}^0

(continued)

Y. Huang et al., *Modeling and Application of Flexible Electronics Packaging*,
https://doi.org/10.1007/978-981-13-3627-0

(continued)

Meaning	Symbol
Classical linear solution of maximum specimen deflection in TPB test	$\delta_{max-TPB}$
Competition index	$\Gamma_{fracture}$
Concentrated force applied at reference point	$P_{concentrated}$
Conformal angle	$\theta_{conformal}$
Constant length of region i (i = 2, 3)	$l_{reg-i}^{constant}$
Crack length	a
Crack length at the end of peeling-off process	$a_{peel\text{-}off}$
Crack length during chip picking-up process	$a_{pick\text{-}up}$
Crack propagation increment	Δa
Critical competition index	Γ_c
Critical chip length separating chip peeling from chip cracking	l_{chip}^c
Critical chip thickness for chip cracking	h_{crack}
Critical chip thickness for chip peeling	h_{peel}
Critical chip thickness for transformation of peeling mode (mode I or II cracking) under actions of ejector needle and prestress	$\bar{h}_{chip}^c$
Critical chip thickness for transformation of peeling mode (mode I or II cracking) under pure action of ejector needle	h_{chip}^c
Critical peel radius for device peeling under condition of substrate tension of i N (i = 6, 10, 15)	$r_{blade}^{T_{sub}=i\,N}\vert G_{adh}^c$
Critical peel radius to avoid device cracking under condition of substrate tension of i N (i = 6, 10, 15)	$r_{blade}^{T_{sub}=i\,N}\vert \sigma_{dev}^c$
Critical peel radius to avoid device cracking under condition of peel radius of i mm (i = 6, 10)	$T_{sub}^{r_{blade}=i\,mm}\vert \sigma_{dev}^c$
Critical picking-up displacement for accomplishing chip detachment	$h_{pick-up}^{complete}$
Critical picking-up height for generating initial crack	$h_{pick-up}^{initial-crack}$
Critical substrate tension for device peeling under condition of peel radius of i mm (i = 4, 6, 10)	$T_{sub}^{r_{blade}=i\,mm}\vert G_{adh}^c$
Critical Young's modulus of adhesive layer for transformation of peeling mode (mode I or II cracking) under actions of ejector needle and prestress	$\bar{E}_{adh}^c$
Critical Young's modulus of adhesive layer for transformation of peeling mode (mode I or II cracking) under pure action of ejector needle	E_{adh}^c
Cross-section rotation of i (i = chip, sub) layer	ϕ_i
Cross-section rotation of i (i = chip, sub) layer under j (j = B, U) condition	ϕ_i^j
CTE of i (i = chip, adh, sub, film) layer	α_i
Deflection angle of TPB test specimen	θ_{TPB}
Deflection angle of TPB test specimen with vertical load	$\theta_{0\text{-}TPB}$
Deflection parameter of TPB test specimen	λ_{TPB}
Deposition temperature	$T_{deposition}$
Diameter of ejector needle	Φ_{needle}
Displacement of any point in x direction on ith (i = 1, 2, …, n) surface	U_i
Dissipated energy	$U_{dissipated}$

(continued)

(continued)

Meaning	Symbol
Distance between adjacent needles	d_{needle}
Distance between two adjacent chips	l_{adja}
Distance from neutral axis to film-substrate interface	h_{bend}
Ejector needle displacement	w_{needle}
Ejector needle force per unit width	P_{needle}
Equivalent axial force	N_{eq}
Equivalent double transverse shear load	Q_{double}
Equivalent single transverse shear load	Q_{single}
ERR	G
ERR corresponds to equivalent axial force	G_N
ERR corresponds to equivalent bending moment	G_M
ERR corresponds to equivalent double transverse shear load	G_Q^{double}
ERR corresponds to equivalent single transverse shear load	f_Q^{single}
ERR of interfacial peeling at crack tip	$G_{\mathrm{crack\text{-}tip}}$
ERR of interfacial peeling in chip-adhesive-substrate model	$G_{\mathrm{adh}}^{\mathrm{crack-tip}}$
External work of applied load	U_{external}
Externally applied bending moment	M_0
Glass transition temperature	T_{g}
Glass transition time	t_{g}
Half distance between two adjacent chips	$l_{\mathrm{adja\text{-}half}}$
Half length of i (i = chip, adh, sub, film-on-sub) layer	$l_{i\text{-half}}$
Half thickness of i (i = chip, adh, sub, dev) layer	$h_{i\text{-half}}$
Horizontal distance from origin in TPB test	x_{TPB}
Increase of crack length from the end of peeling-off process during chip picking-up process	$\Delta a_{\mathrm{peel\text{-}off}}$
Increase of upward height from the end of peeling-off process during chip picking-up process	$\Delta h_{\mathrm{pick\text{-}up}}$
Increment of upward height by vacuum adsorption pressure above chip	Δh_{lift}
Initial half span length of TPB test specimen	$l_{\mathrm{TPB\text{-}half}}$
Initial length of substrate layer	$l_{\mathrm{sub}}^{\mathrm{initial}}$
Initial span length of TPB test specimen	l_{TPB}
Initial temperature	T_0
Initial tension strain of substrate layer	$\varepsilon_{\mathrm{sub}}^{\mathrm{initial}}$
Initial tension stress of substrate layer	$\sigma_{\mathrm{sub}}^{\mathrm{initial}}$
Integration constant	F_i
Integration constant under balanced condition	F_i^{B}
Integration constant under unbalanced condition	F_i^{U}
Interfacial peeling index	Γ_{peel}
Internal axial force per unit width at i (i = chip, sub) layer under j (j = B, U) condition	N_i^j

(continued)

(continued)

Meaning	Symbol
Internal axial force per unit width at i (i = chip, sub, dev) layer	N_i
Internal axial force per unit width at region $i\,(i = 1, 2, 3, 4)$ of j (j = chip, sub) layer under k condition (k = pick-up, stretch, single, multiple)	$N^k_{\text{j-reg-i}}$
Internal bending moment per unit width at i (i = chip, sub) layer under j (j = B, U) condition	M^j_i
Internal bending moment per unit width at i (i = chip, sub, dev) layer	M_i
Internal bending moment per unit width at region $i\,(i = 1, 2, 3, 4)$ of j (j = chip, sub) layer under k condition (k = pick-up, stretch, single, multiple)	$M^k_{\text{j-reg-i}}$
Internal shear force per unit width at i (i = chip, sub) layer under j (j = B, U) condition	Q^j_i
Internal shear force per unit width at i (i = chip, sub, dev) layer	Q_i
Internal shear force per unit width at region $i\,(i = 1, 2, 3, 4)$ of j (j = chip, sub) layer under k condition (k = pick-up, stretch, single, multiple)	$Q^k_{\text{j-reg-i}}$
Kinetic energy of system	U_{kinetic}
Laser beam width	w_{beam}
Laser fluence	I_{laser}
Laser pulse frequency	f_{pulse}
Laser scanning speed	v_{scan}
Length of film-on-substrate structure	$l_{\text{film-on-sub}}$
Length of i (i = chip, adh, sub, strip, film) layer	l_i
Length of region i (i = 1, 2, 3)	$l_{\text{reg-}i}$
Lower limit of competing index	Γ_{low}
Maximum bending stress of TPB test specimen	$\sigma_{\text{max-TPB}}$
Maximum competing index	Γ_{max}
Maximum length of substrate layer	$l^{\text{max}}_{\text{sub}}$
Maximum normal stress of i (i = chip, adh, dev) layer	σ^{max}_i
Maximum shear stress	τ_{max}
Maximum shear stress of adhesive layer	$\tau^{\text{max}}_{\text{adh}}$
Maximum third principle stress	σ^{max}_3
Maximum transverse displacement of TPB test specimen	$w_{\text{max-TPB}}$
Maximum value of chip warpage	w_{warpage}
Mid-plane displacement along axial direction at region $i\,(i = 1, 2, 3, 4)$ of j (j = chip, sub) layer under k condition (k = pick-up, stretch, single, multiple)	$u^k_{\text{j-reg-i}}$
Mid-plane displacement along transverse direction at region region $i\,(i = 1, 2, 3, 4)$ of j (j = chip, sub) layer under k condition (k = pick-up, stretch, single, multiple)	$w^k_{\text{j-reg-i}}$
Mid-plane displacement of i (i = chip, sub) layer along axial direction under j (j = B, U) condition	u^j_i
Mid-plane displacement of i (i = chip, sub) layer along transverse direction under j (j = B, U) condition	w^j_i
Mid-plane displacement of i (i = chip, sub, dev) layer along axial direction	u_i

(continued)

(continued)

Meaning	Symbol
Mid-plane displacement of i (i = chip, sub, dev, film) layer along transverse direction	w_i
Minimum length of substrate tape layer	$l_{\text{sub}}^{\min}$
Mismatch strain	ε_{mis}
Mixed-mode angle corresponds to equivalent axial force	ψ_N
Mixed-mode angle corresponds to equivalent bending moment	ψ_M
Mixed-mode angle corresponds to equivalent double transverse shear load	ψ_Q^{double}
Mixed-mode angle corresponds to equivalent single transverse shear load	ψ_Q^{single}
Mode I ERR	G_{I}
Mode II ERR	G_{II}
Mode-mixity angle at crack tip	$\psi_{\text{crack-tip}}$
Mode-mixity angle corresponds to ERR of interfacial peeling in chip-adhesive-substrate model	$\psi_{\text{adh}}^{\text{crack-tip}}$
Half span length after specimen deflection in TPB test	$l'_{\text{TPB-half}}$
Span length after specimen deflection in TPB test	l'_{TPB}
Nondimensional specimen deflection in TPB test	δ_{TPB}
Nondimensional function corresponds to equivalent axial force	f_N
Nondimensional function corresponds to equivalent bending moment	f_M
Nondimensional function corresponds to equivalent double transverse shear load	f_Q^{double}
Nondimensional function corresponds to equivalent single transverse shear load	f_Q^{single}
Normal strain along z direction at adhesive layer	$\varepsilon_{zz}^{\text{adh}}$
Normal strain at i (i = sub, film) layer	ε_i
Normal stress at i (i = adh, sub) layer	σ_i
Normal stress at adhesive layer under i (i = B, U) condition	σ_{adh}^i
Normal stress at crack tip	$\sigma_{\text{crack-tip}}$
Normal stress at top surface of chip layer	$\sigma_{\text{chip}}^{\text{top-surf}}$
Number of mathematical plies at i (i = chip, adh, sub) layer	n_i
Peel angle	θ_{peel}
Peel angle for i (i = exten, inexten) extensible tape	θ_i
Peel force	F_{peel}
Peel force in experimental test on adhesive fracture energy	P_{peel}
Peel speed	v_{peel}
Peel speed for in peeling an i (i = exten, inexten) tape	v_i^{peel}
Peel stiffness	k_σ
Photon energy	E_{laser}
PI film thickness	$h_{\text{PI-film}}$
Picking-up force	$P_{\text{pick-up}}$
Placing-on force	$P_{\text{place-on}}$

(continued)

(continued)

Meaning	Symbol
Point mass of ejector needle	m_{needle}
Poission's ratio of *i* (*i* = chip, adh, sub, dev, film) layer	ν_i
Potential energy of external loads	V_{external}
Potential energy of system	U_{system}
Predefined stress	P_{pre}
Process margin of chip peeling-off	$\Gamma_{\mathrm{process}}$
Radius of curvature of TPB test specimen	ρ_{TPB}
Reaction force in TPB test	F_{TPB}
Reference peel speed	v_0
Reference peel speed in peeling an *i* (*i* = exten, inexten) tape	v_i^0
Reference temperature	T_{ref}
Resultant axial force at right boundary of substrate and chip layers	$N_{\mathrm{sub-chip}}^{\mathrm{right}}$
Resultant shear force at right boundary of substrate and chip layers	$Q_{\mathrm{sub-chip}}^{\mathrm{right}}$
Room temperature	T_{room}
Root of characteristic equation under *k* (*k* = B, U) condition	λ_i^k
Rotation angle of *i* (*i* = sub, dev) layer	θ_i
Rotation of cross section at region *i* ($i = 1, 2, 3, 4$) of *j* (*j* = chip, sub) layer under *k* condition (*k* = pick-up, stretch, single, multiple)	$\phi_{\mathrm{j-reg-i}}^k$
Shear angle of *i* (*i* = sub, dev) layer	φ_i
Shear correction factor	k_{shear}
Shear force at crack-tip	$Q_{\mathrm{crack\text{-}tip}}$
Shear force at film layer	Q_{film}
Shear force at left boundary of at *i* (*i* = chip, sub) layer	Q_i^{left}
Shear modulus of *i* (*i* = chip, adh, sub, dev) layer	$\bar{G}_i$
Shear stiffness	k_τ
Shear strain at adhesive layer	$\gamma_{xz}^{\mathrm{adh}}$
Shear stress at adhesive layer	τ_{adh}
Shear stress at adhesive layer under *i* (*i* = B, U) condition	τ_{adh}^i
Shear stress at crack tip	$\tau_{\mathrm{crack\text{-}tip}}$
Moment of inertia of TPB test specimen	I_{TPB}
Speed of needle impacting	v_{needle}
Speed of substrate layer	v_{sub}
Speed of substrate layer at left part of R2R manufacturing system	$v_{\mathrm{sub}}^{\mathrm{left}}$
Speed of substrate layer at right part of R2R manufacturing system	$v_{\mathrm{sub}}^{\mathrm{right}}$
Stored elastic energy	U_{elastic}
Support radius	r_{sup}
Supporting force at peel blade	N_{blade}
Switch level of competing index	Γ_{switch}
Temperature change of *i* (*i* = chip, adh, sub) layer	ΔT_i

(continued)

(continued)

Meaning	Symbol
Temperature of chip layer	T_{chip}
Temperature of platform	$T_{platform}$
Tensile force applied on adhesively bonded joint	$P_{tension}$
Tensile fracture strength of *i* (*i* = chip, dev, silicon) layer	σ_i^c
Tension load	P_0
Tension of substrate layer	T_{sub}
The third principle stress	σ_3
Thickness of *i* (*i* = chip, adh, sub, film, strip, ACA, TPB)	h_i
Threshold laser fluence	I_{laser}^c
Total system energy	U_{total}
Total thickness	h_{total}
Transverse displacement of TPB test specimen	w_{TPB}
Upper limit of competing index	Γ_{high}
Upward height at the end of chip peeling-off process	$h_{peel\text{-}off}$
Upward height during chip picking-up process	$h_{pick\text{-}up}$
Vacuum adsorption pressure above chip layer	p_{chip}
Vacuum adsorption pressure below substrate layer	p_{sub}
Variation of half span length of TPB test specimen	Δl_{TPB}
Variation of transverse displacement of TPB test specimen	Δw_{TPB}
Vertical load in TPB test	P_{TPB}
Width of *i* (*i* = sub, film, strip, TPB)	b_i
Window of substrate thickness under substrate tension of *i* N (*i* = 6, 10, 15)	$h_{sub}^{win}\|T_{sub=i\,N}$
Work temperature	T_{work}
Young's modulus of *i* (*i* = chip, adh, sub, film) layer under plane strain condition	$\bar{E}_i$
Young's modulus of *i* (*i* = chip, adh, sub, film, strip, dev, TPB)	E_i

Note Unless specified otherwise, the full-spelling form corresponding to the abbreviated subscript in Appendix is as follows: ACA → anisotropic conductive adhesive, adh → adhesive, B → balanced, chip → chip, dev → device, exten → extensible tape, film → film, film-on-substrate → film-on-substrate structure, inexten → inextensible tape, multiple → multiple-needle ejecting, pick-up → pick-up process, single → single-needle ejecting, stretch → stretched substrate, strip → strip, sub → substrate, TPB → TPB test specimen, U → unbalanced

MIX
Papier aus verantwortungsvollen Quellen
Paper from responsible sources
FSC® C105338
FSC
www.fsc.org

If you have any concerns about our products,
you can contact us on
ProductSafety@springernature.com

In case Publisher is established outside the EU,
the EU authorized representative is:
Springer Nature Customer Service Center GmbH
Europaplatz 3, 69115 Heidelberg, Germany

Printed by Libri Plureos GmbH
in Hamburg, Germany